D1824751

Frontier
The Definitive Guide

Frontier
The Definitive Guide

Matt Neuburg

O'REILLY™

Cambridge · Köln · Paris · Sebastopol · Tokyo

Frontier: The Definitive Guide
by Matt Neuburg

Copyright © 1998 O'Reilly & Associates, Inc. All rights reserved.
Printed in the United States of America.

Published by O'Reilly & Associates, Inc., 101 Morris Street, Sebastopol, CA 95472.

Editor: Tim O'Reilly

Production Editor: Nancy Wolfe Kotary

Printing History:

> February 1998:   First Edition.

Nutshell Handbook and the Nutshell Handbook logo are registered trademarks and The Java™ Series is a trademark of O'Reilly & Associates, Inc. The association between the image of a buffalo and the topic of Frontier is a trademark of O'Reilly & Associates, Inc.

Many of the designations used by manufacturers and sellers to distinguish their products are claimed as trademarks. Where those designations appear in this book, and O'Reilly & Associates, Inc. was aware of a trademark claim, the designations have been printed in caps or initial caps.

While every precaution has been taken in the preparation of this book, the publisher assumes no responsibility for errors or omissions, or for damages resulting from the use of the information contained herein.

This book is printed on acid-free paper with 85% recycled content, 15% post-consumer waste. O'Reilly & Associates is committed to using paper with the highest recycled content available consistent with high quality.

ISBN: 1-56592-383-9

To my father, Ned Neuburg, who gave me my love of programming.

To my mother, Greenie Neuburg, who gave me my love of teaching.

To both my parents, who together gave me my love of learning.

Table of Contents

Preface ....*xv*

I.   *Getting Acquainted* *1*

1.   *What Is Frontier?* *3*
    A Day in the Life *3*
    Days in Other Lives *8*
    Tool of All Trades *10*

2.   *Edit Windows* *13*
    Window Types and Objects *14*
    Outlines *14*
    Wptexts *19*
    Tables *19*

3.   *The Database* *25*
    Main Window and Root Table *27*
    Direct Database Navigation *27*
    Searching the Database *29*
    Guide to the Database *29*

II.   *UserTalk Basics* *33*

4.   *What a UserTalk Script Is Like* *35*
    First Principles *35*
    Scripts as Database Entries *37*

Scripts as Outlines *39*

Comments in Scripts *44*

5. *Handlers and Parameters* *47*

What Is an Eponymous Handler? *47*

The Handler Rule *48*

Stubs *51*

Call Results *52*

Line Evaluation *53*

Parameters *54*

6. *Referring to Database Entries* *58*

How Variables and Database Entries Differ *58*

Being Careful with the Database *59*

Calculated Names *60*

Partial References *62*

with *66*

7. *The Scope of Variables and Handlers* *69*

local *70*

Variable Scope *73*

Handler Scope *74*

with, again *76*

Implicit Locals *77*

Dynamic Scoping *78*

Recursion *81*

8. *Addresses* *82*

Address Syntax *82*

Name Construction *85*

Address Parameters *85*

Further Uses of Address Parameters *88*

Addresses of Verbs *88*

Being Careful with Address Parameters *89*

9. *Special Evaluation* *91*

Non-Evaluation *91*

Pre-Evaluation *92*

Constants *93*

evaluate( ) *93*

10. *Datatypes* **95**
    Summary of the Datatypes 97
    Coercion 100
    String and Char Literals 109

11. *Arrays* **112**
    Array Notation 112
    Array Verbs 114
    Lists and Records; Strings and Binaries 115

12. *Control Structures* **118**
    Looping Constructs 118
    Visit Constructs 125
    Conditional Constructs 128
    Errors 131

13. *Running and Debugging Scripts* **135**
    Compiling 135
    Running a Script 138
    Debug Mode 140
    Runtime Errors 143
    Desperation 144
    Getting Help from DocServer 144
    Getting Help from the Database 145

III. *Data Manipulation* **147**

14. *Strings and Chars* **149**
    Substrings 149
    Case and Character Type 151
    Find and Replace 152
    Pad and Trim 155
    Parsing 155
    Formatted Numbers 158
    HTML-Related Conversions 158

15. *Math* **161**

16. *Dates* **164**

17.   *Objects* *169*
         Creation and Destruction *170*
         Information *171*
         Moving and Copying *172*
         Age and Dirtiness *173*

18.   *Non-Scalars* *174*
         The Target *175*
         Verb Types, Window Types, and Modes *177*
         Verbs Operating in Edit Windows *178*

19.   *Datafiles* *188*
         Utilities *188*
         Basic Verbs *190*
         Finding in Textfiles *191*

20.   *Resources* *193*
         Inspecting Resources *193*
         Resource Data *194*
         The Resource Chain *196*
         Resource Attributes *197*

21.   *Yielding, Pausing, Threads, and Semaphores* *198*
         Yielding and Pausing *198*
         Threads *198*
         Semaphores *202*

22.   *Stacks* *205*

23.   *Extending the Language* *207*
         Why XCMDs and UCMDs Exist *208*
         Importing and Glue *209*

IV. *Interface* *211*

24.   *Windows* *213*
         Manual Window Manipulation *213*
         Programmatic Window Manipulation *216*

25. Dialogs *223*

    Dialog Types *223*

    Status Messages *223*

    Preconfigured Modal Dialogs *224*

    Resource-Based Dialogs *226*

    Card-Based Dialogs *232*

    Dialogs in Other Applications *238*

26. Menus and Suites *240*

    Menu Categories *241*

    Modifying Menus *241*

    Where the Menubars Are *242*

    Menubar Edit Windows *242*

    Shared Menus *246*

    Modal Menus *250*

    Suites *250*

    Programmatic Menu Manipulation *252*

27. Agents and Hooks *255*

    Agents *255*

    Hooks *258*

    User-Based Pseudo-Hooks *260*

28. Pictures *261*

    PICTs and Frontier Pictures *261*

    Text in Pictures *262*

    Other Graphic Formats *263*

29. Import/Export *264*

    Types of Exported Object *264*

    Packed Objects *265*

    Desktop Scripts *268*

    Droplets *270*

    Implementation of Exported Objects *272*

    Backing Up the Database *273*

    Clean Root *277*

30. Multiple Databases *279*

    Manual Interface *280*

    Programmatic Interface *280*

V. Border Crossings *283*

31. Driving the System *285*
     Frontier *285*
     System *286*
     File System *290*

32. Driving Other Applications *300*
     How to Use This Chapter *300*
     Apple Events *302*
     Apple Event Suites *311*
     Object Model *315*
     Glue *319*

33. AppleScript *332*
     AppleScript Scripts *332*
     OSAXen *334*
     Import/Export *335*

34. Driving Frontier from Outside *336*
     OSA Scripting Applications *337*
     Standard Apple Events *340*
     Custom Apple Events *344*

35. External Editors *347*
     The odbEditor Suite *348*
     Script Debugger *349*
     BBEdit and PageSpinner *349*
     Scalars *350*
     Other External Editors *352*

VI. Applied Frontier *355*

36. ToDo List *357*

37. Scheduler *358*
     Missed Tasks *360*
     Logging *360*
     Making a Task *360*

38. *TCP/IP* *362*
　　　Server *362*
　　　Client *368*

39. *NetFrontier* *371*
　　　Programming Example *371*
　　　Behind the Scenes *373*
　　　Further Exploration *374*

40. *CGIs* *375*
　　　Configuring *376*
　　　Making a CGI Script *377*
　　　Testing *378*
　　　Macros *379*
　　　Form Example *380*
　　　Other Servables *382*
　　　Search Example *383*
　　　Security *386*
　　　Keep It Up *386*
　　　Further Explorations *388*

41. *Web Site Management* *389*
　　　Architecture *390*
　　　Web Site Tables *391*
　　　Directives *393*
　　　Outline Renderers *401*
　　　Templates *404*
　　　Web Page Object Datatypes *406*
　　　Macros *407*
　　　Glossaries *412*
　　　Escape Character *413*
　　　Filters *414*
　　　Relative Links *416*
　　　Previewing and Releasing *418*
　　　Loading an Existing Site *420*
　　　Important Routines *421*
　　　Some Utility Scripts *424*
　　　BBEdit Front-End *428*

42. *Dynamic Web Sites* *431*
       News Page *431*
       Classified Ads *433*
       Further Explorations *437*

VII. *Reference* *439*

43. *XCMDs and UCMDs* *441*
       XCMDs *442*
       UCMDs *446*

44. *Operators* *449*
       Operators Used in Boolean Expressions *450*

45. *Punctuation* *453*

46. *Verbs* *455*

47. *Apple Event Suites* *554*

Index *563*

Preface

A Little History

Before there was the Open Scripting Architecture, before there was AppleScript, before there were Apple events, before there was System 7, way back in 1988, UserLand began development of Frontier. Its earliest incarnation, UserLand IAC Toolkit, appeared in 1989. IAC stands for *inter-application communication*; the program was dedicated to the proposition that if it was going to be too hard for the ordinary user to write an application, it should still be possible to tie existing applications together, acting in concert and communicating with one another.

After System 7 and Apple events had a created a system-level standard for inter-application communication, Frontier 1.0 appeared, one of the earliest programs to exploit them. With its quietly revolutionary scripting language, milieu for editing and storing information in various formats, and database that tied it all together, it was not only the first system-level scripting environment for the Macintosh, but in its own right a tool of manifold usage and great versatility. Frontier 2.0 appeared in October 1992, and Version 3.0 in November 1993.

By this time, AppleScript had been released. At first, AppleScript wasn't commonly available; it could be obtained cheaply, but few people were aware of this. However, Apple Computer soon bundled it with their system software. UserLand responded by doing something revolutionary: in 1995, they began giving Frontier away for free (the so-called "Aretha" release).*

Frontier 4.0 shipped in May 1996, 4.1 shipped in October 1996, and 4.2 shipped in January 1997. These versions firmly established Frontier in the realms of Web

---

* See *http://www.scripting.com/davenet/95/05/beingfree.html.*

site management and CGI scripting. It is a remarkable testament to Frontier's power and flexibility that it had all along been applicable to these areas; the main obstacle was that someone had to realize this.[*]

With Frontier 5.0, UserLand will embark upon a mission to bring Frontier's scripting power to the Windows platform. At first, this will consist of a subset of Mac OS capabilities, but growth and expansion in new directions are to be expected in the future.

To obtain Frontier, see *http://www.scripting.com/frontier/*. Here you will learn how to join the Frontier mailing lists, how to access the huge library of download-able third-party Frontier tools, and what the latest developments are in the ongoing evolution of Frontier.

Flattening the Learning Curve

Frontier is big; even people extremely familiar with it comment routinely on the power and surprises tucked away in the crevices of its database. The new surge of interest in Frontier generated by the Aretha release brought a new problem: Frontier was hard to learn. The printed documentation was outdated and unavail-able. Online documentation of new features, though excellent, was sporadic and scattered. The primary paths to learning Frontier were study of the database itself and an oral culture that emerged in the Frontier mailing lists.

It was a frustrating situation. The same questions were being asked repeatedly on the lists as new users came on board, because the answers were documented obscurely or not at all. (To give but one example: the syntax of handler declara-tions with on, the fundamental linchpin of parameter-passing, was essentially undocumented.) Frontier was getting a reputation as a "geek tool": the only way to learn it was by a sink-or-swim immersion course of intensive study, for which few people had the leisure.

Meanwhile, I had become acquainted with Frontier starting with the free Aretha release, and then, while serving as editor of *MacTech Magazine*, had developed a Frontier-driven system for automating the transformation of a year's worth of published articles from QuarkXPress to a customized RTF format for republication on the 1996 *MacTech* CD. The lack of documentation had made the task much harder than it should have been, and soon after, I resolved to systematize my own understanding of Frontier.

---

[*] This point is eloquently witnessed by Mason Hale, *MacTech Magazine* 11:12 (December 1995), p. 58: "[I came across] an article Dave [Winer] had written for *MacTech Magazine* titled 'A Nerdy Guide to Fron-tier'… I couldn't believe what I was reading; it was like a laundry list of all the features I had been looking for in a CGI environment…" The article that Mason discovered in 1995 had been published in August 1993. On the vision behind Frontier 4.0, see *http://www.scripting.com/davenet/96/05/watchthis.html*.

Thus, after some weeks of reading all the existing documentation and then studying the entire Frontier database, I developed an outline that brought together, under topic headings for easy reference, the sum total of all that I had been able to discover about Frontier. Armed with this reference, I was quickly able to answer quite a large proportion of the questions that beginners were asking on the mailing lists, even though I was little more than a beginner myself.

At that point, one thing became manifestly obvious to me: Frontier was *not* difficult. It was not, for example, anywhere near as hard as Ancient Greek, a subject I had for many years prided myself on being able to render crystal clear to anyone possessed of a willing determination to learn it. But learners of Greek did have a resource that learners of Frontier did not: reference books. There were Greek grammars, and Greek lexicons. For Frontier, there was almost nothing. What this program needed was documentation. It needed a teacher. The existence of such resources would prove, I insisted, that the "Frontier learning curve" was largely illusory.

Before I knew what was happening, I had written two online tutorials directed at complete beginners. My original outline was then distributed to users as an online reference; together with the tutorials, it is also the basis of this book.

I have had the joy and privilege of writing the book that I myself wished to read. This is the book I wanted by my side as I was desperately trying to cobble together that automated system for *MacTech*. This is the book that should have lived beside my computer as a handy reference. This is the book that ought to have been my *vade mecum*, in cafés, at the beach, in the wee hours. I love reading books like this; I would have loved to read one about Frontier, and now there is one. On the off chance that there exist others who have wanted the same thing, this book is for them, too.

How to Use This Book

This book is structured in seven parts, as follows:

Part I: Getting Acquainted

> For those who have never used Frontier before, a gentle introduction to what it does, and how to navigate and work with its most important windows and its database.

Part II: UserTalk Basics

> For those who want to start learning Frontier programming: what a script is like, where scripts live and how they interact, the fundamentals of the User-Talk language, and how to run and debug a script.

Part III: Data Manipulation

Deeper explorations in Frontier programming: how UserTalk manipulates and edits various types of data, such as strings, numbers, objects, non-scalars, and files. Also, how UserTalk handles threads, and how UserTalk can be extended with compiled code fragments.

Part IV: Interface

A guided tour of the Frontier interface: windows, dialogs, menus, agents, and pictures, plus how to import and export database objects as separate files. There is discussion of how these interface features are driven and managed with UserTalk (which is why the explanation of the interface comes after User-Talk itself has been introduced), but this part of the book will be of interest also to the general Frontier user.

Part V: Border Crossings

How UserTalk reaches outside Frontier to give commands to the system and to other applications, and how other applications can give commands to Frontier.

Part VI: Applied Frontier

About the application-like features packaged inside Frontier, especially using Frontier across a network, Frontier as a CGI application, and how to use Frontier to construct and manage Web sites of any size.

Part VII: Reference

Technical details about UserTalk, gathered for convenient access.

The book expounds Frontier fully and in logical order, so someone wishing to learn the program completely could read the whole book sequentially, Parts I through VI, consulting Part VII, *Reference*, as necessary.

But this is not the only way to use this book. Someone wishing to begin immediately with the practical application of Frontier as a Web tool might read Part I and then leap immediately to Part VI, perhaps occasionally turning back to topics in Part IV in order to understand the interface more fully, and only later, when the need and desire arises for more sophisticated customization, advancing to the study of UserTalk through Parts II, III, and V.

Then again, an already experienced Frontier user might see the whole book as a kind of reference, to be consulted piecemeal on individual topics.

Now a word about how UserTalk verbs are explained in the course of the book. Technical details of the sort that would impede the flow of thought and exposition are located in Part VII. In the body of the book, UserTalk verbs are described in connection with *what they do*: "to achieve this end, use this verb." The parameters are listed, denoted by names that should make their usage easy to divine,* and there is some supplementary discussion, with advice and examples of usage; but for the details, such as what parameter values will cause a verb to raise an error, see Chapter 46, *Verbs*, where the verbs appear in alphabetical order. Having discovered in the main part of the book that a particular verb fits your need, you will probably wish to look it up in the Reference section before using it in a script.

Versions

A practical and philosophical problem was raised by the fact that during the writing of this book, UserLand was creating version 5.0 of Frontier and releasing public alphas for user comment.

The practical aspect was that since 5.0 promised to represent in many ways a strong break with the past, there was a chance that it would falsify some of the book's information. The situation was made still more complicated by the fact that the 5.0 release was ultimately to include a Windows version.

The philosophical aspect was that I am averse to writing about beta software, let alone alpha: to do so amounts to writing about the future, which is unknown, whereas the fully existing past version can be described with certainty.

The latest fully finished version of Frontier, version 4.2.3, was not outdated or unreliable, and it had the advantage of a stable feature set that could be studied at leisure, whereas 5.0 was a moving target, experimental and constantly being revised. At the same time, it did not seem kind to wait for 5.0 to be completely finished and then write about it; the timetable for 5.0 was unknown, and production of a book after 5.0 was finished would involve still further delay in publication, whereas it seemed to me that users and potential users of Frontier needed a book *now*.

Accordingly, what this book describes is *Frontier 4.2.3 for Mac OS.*

Conventions

The following typographic conventions are used in this book:

Italic
> is used for emphasis, citations (including chapter names and URLs), file and folder names, first occurrences of technical terms, and placeholders for values and variable names in text.

`Constant Width`
> is used for syntax definitions, code, keywords, literals, and object names.

---

* If you want to call a verb using named-parameter syntax (see "Handlers and Parameters"), you need to know the *real* parameter names; for these, consult the verb's implementation within the database.

Constant Width Italic
> is used within syntax definitions for placeholders for values and variable names.

Bold
> is used for text of GUI items such as menus, buttons, and dialogs.

We'd Like to Hear from You

We have tested and verified all of the information in this book to the best of our ability, but you may find that features have changed (or even that we have made mistakes!). Please let O'Reilly know about any errors you find by writing:

> O'Reilly & Associates, Inc.
> 101 Morris Street
> Sebastopol, CA 95472
> 800-998-9938 (in U.S. or Canada)
> 707-829-0515 (international/local)
> 707-829-0104 (fax)

You can also send us messages electronically. To be put on the mailing list or request a catalog, send email to:

> *nuts@oreilly.com*

To ask technical questions or comment on the book, send email to:

> *bookquestions@oreilly.com*

Acknowledgments

The first acknowledgment must be to Frontier itself. This is a really cool program. The more familiar I have become with it, the more I have been astonished at its cleverness, its power, its elegance, its architectural beauty; every chapter in this book has been an eye-opening experience to write. The makers of the Frontier application, of the UserTalk language, and of the scripts that constitute the database are brilliant, clear-thinking programmers. To study Frontier is to do no less than to look into their minds; and it has been an honor and a privilege to be permitted to do so.

Without Dave Winer and Doug Baron, there wouldn't be anything for this book to be about. Doug could have written this book, and he would have written it much better than I have. Luckily for me, he's too busy to write it. But he has shared his knowledge, with clarity, humor, and rich insight.

The book owes a tremendous debt to the Frontier mailing lists. I have learned from the questions people asked, for these were the yardstick against which the value of the book had always to be measured; I have learned from the answers people gave, for they taught me what I did not know; I have learned from trying to answer questions myself, for in this way I was compelled to codify and clarify my own understanding; I have learned from being mystified by conversations that were utterly over my head, for they reminded me how limited that understanding was.

In particular, I wish to thank the following people. Some of them helped me explicitly. Others probably had no idea that I was looking over their metaphorical shoulders, busily taking notes. They are: Brian Andresen, Henri Asseily, John Baxter, David Bayly, Jay Bourland, Jim Correia, John Delacour, Seth Dillingham, Mason Hale, Preston Holmes, Nobumi Iyanaga, Scott Lawton, Brent Simmons, and Cameron Smith. John Baxter also generously gave of his time to study a final draft of the book and, with his keen eyes and encyclopedic knowledge, saved me from numerous errors.

Sandra Schneible of Bare Bones Software helped at a crucial moment by giving me access to BBEdit 4.5.

Adam Engst and Tonya Engst not only endured my absence from regular editorial contribution to *TidBITS* without so much as a word of reproof; they actually contributed materially to my stock of tools to get the job done.

At O'Reilly & Associates, special thanks go to Tim O'Reilly, for his insight, kindness, and courage; and to Steve Clark, who tirelessly championed Frontier and this book, and bound up my wounds when, as often happened, I was ready to collapse.

O'Reilly & Associates also lived up to its reputation by bringing to bear the attention and intelligence of its production experts; it has been a pleasure to watch them at work. Mike Sierra converted the book from Microsoft Word to Adobe FrameMaker and provided extensive tools support. Nancy Wolfe Kotary was the production editor and project manager, and did the copyediting. Madeleine Newell was the proofreader. Mary Anne Mayo, Nicole Gipson Arigo, and Sheryl Avruch performed quality checks. Seth Maislin wrote the index.

More than anything else, I was sustained during the writing of this book by people who took the time to write in appreciation of my earlier efforts. I am more grateful than I can possibly express.

Finally, the classic disclaimer: the mistakes are mine. This is not an "insider" book; I have not been privy to special information, but am just an ordinary user, deducing how Frontier works through experience, experiment, discussion, the existing documentation, and study of the database. Sometimes I have been helped, like other users, by the inside experts who haunt the Frontier mailing lists; but sometimes I have had to guess, and it is perfectly possible now and then that I have guessed wrong.

I

Getting Acquainted

This section introduces Frontier to someone who has never seen it. First, Frontier is described in general terms. What sort of a program is it? What does it do? What sort of task might one use it for? Next, its user interface is introduced, including the different sorts of windows the user will encounter, how to work with them, and the keyboard shortcuts that are not listed in the menus. Also discussed is the database that is the basis of so much of Frontier's power.

On the whole, the purpose of this section is, as its title says, to acquaint the reader with Frontier. After reading this section and playing with Frontier a little, even a complete beginner will be comfortable with the program and ready to learn to put it to work.

Because the reader of this section is assumed to be a total beginner, much of the material here is deliberately elementary and even sketchy. Technical details are almost completely omitted; they are provided in later parts of the book. The intent is to tell you enough to get you started and no more.

1

What Is Frontier?

A Day in the Life

Up about 7:30. Coffee. Turn on the computer, which also causes Frontier to start up; I like to have Frontier always running.

Check my mail with Eudora Pro. Several mailing list digests have arrived during the night. I subscribe to a lot of mailing lists; I don't like to receive all the individual messages from these over the course of the day, so each subscription is to a single digest that arrives at night, containing all the messages from the previous day (see Figure 1-1).

Who	Date	K	Subject
BBEdit Scripting	11:03 PM	5	Digest for 9/15/97
Frontier Central	11:04 PM	30	Digest for 9/15/97
Info-Mac	12:35 PM	23	Info-Mac Digest V15 #199

Figure 1-1. Some digest messages

Now I want those digests saved to my hard disk as text files, each mailing list in its own particular folder. To do this by hand, I'd have to select a message, choose **Save As**, use the Save dialog to find the right folder, press the **Save** button, and delete the message—over and over, for each message. I'd rather do it automatically. But there's no way to script that kind of action with Eudora's built-in filters.

However, I've got Frontier. Frontier gives Eudora an extra menu; I can modify that menu, and in it I've put an item **DigestsGo**. I choose that item, and Frontier quickly runs through all the messages in my Eudora In-box, checking each one to see if it's a digest; if it is, Frontier saves the digest as a text file into the right folder and deletes the message.

Frontier can drive any scriptable application, and can communicate with the computer's filing system, make textfiles, and so on. As for the specific functionality of **DigestsGo**, I wrote it myself in Frontier's scripting language, UserTalk. How hard was it to write the script? Not very. Frontier's scripting language includes an online command reference and integrated debugger.

Last night I was using my other computer, a portable, in the living room, writing some articles. I work on my articles sometimes on my main computer, sometimes on my portable. I work on several articles at once, and I keep them all in one folder—except it's two folders, because there's a copy on my portable and a copy on my main computer. Now I need to bring both folders up to date. I hook the two computers together with a serial cable, with File Sharing turned on; now my main computer can "see" my portable.

To know which files to copy where, I don't need to compare the modification times on the two computers for every single file. I've got Frontier. Frontier gives the Finder an extra menu, and in that menu is an item **Reconcile Folders**. I just select the two folders and choose that item; Frontier does the rest, copying files to make both folders identical, with the most recent versions of everything.

> Again, I'm using Frontier's ability to drive the filing system. The **Reconcile Folders** functionality is a UserTalk script, but I didn't have to write it—Frontier comes with this and lots of other utility scripts ready to run.

So much for that. Now . . . whoops, Frontier has just put up a dialog. Ah, it's reminding me I've got a bicycle club meeting tonight. I've set up Frontier to notify me on the first Wednesday of each month, except in June, July, and August, when the club doesn't meet. I also own a dedicated calendar program, but it can't handle concepts this complicated; Frontier can, because it's programmable. Good thing I've got Frontier.

> Frontier comes with several "mini-applications" built from UserTalk scripts. One of these is the Scheduler, which lets any scriptable action be triggered at a specific date or interval; all I had to do was write a script that puts up the reminder dialog and then adjusts its own next trigger date.

Every so often I like to make sure I don't have any strange invisible files on my hard drive. I don't need to search the Internet for a special application to do this; I've got Frontier. I run a script which looks for invisible files and arranges the results as an outline, showing the hierarchy of folders that each invisible file lives in (Figure 1-2). The program is configured to exclude invisible files called *Icon*, because there are so many of them and they're harmless. There are certainly some invisible files here that I don't understand, though; they might bear watching. Time needed to examine over 3000 files: 12 seconds. Did I mention that Frontier is fast?

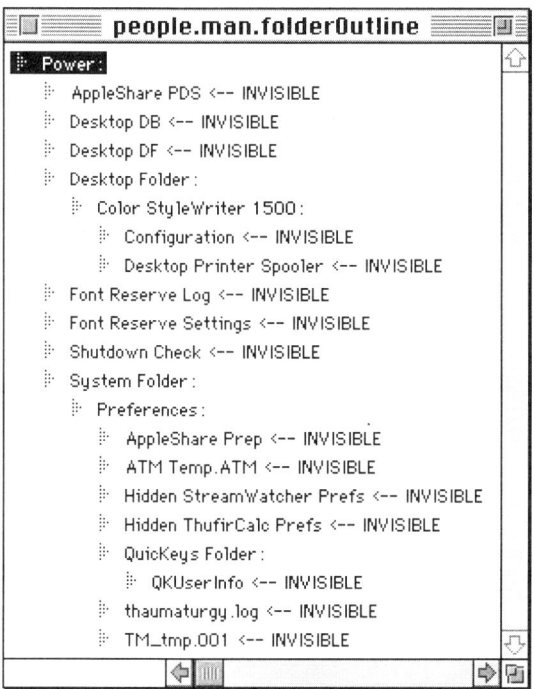

Figure 1-2. Frontier lists my invisible files

I wrote the script that does this myself, based on an example included with Frontier. Again, it takes advantage of Frontier's ability to communicate with the filing system. In addition to dialogs, Frontier can put up various sorts of windows; this one is an outline, one of Frontier's most important ways of structuring information. Frontier itself is completely scriptable, so it can be programmed to generate the contents of this outline.

Time to get some real work done. Today I want to add a new feature to my Web site. I've got three Web pages describing my past career, in case some prospective employer comes looking. I want these to be really snappy and impressive, so I'm going to add a row of "smart" navigation links across the top and bottom of each page—the pages will all be linked to each other, but no page will contain a link to itself (Figure 1-3). There's nothing I hate so much as a Web page containing a link to itself, and yet I see that sort of thing all the time. I guess those Webmasters don't have Frontier.

The three pages initially live inside the database that comes with Frontier, and they're configured to share a "template" (Figure 1-4) containing the material to be placed at the start and end of each page. So, to add this feature to all three pages, I have to add it to the template only once.

Figure 1-3. Smart links—each page has links to the three pages other than itself

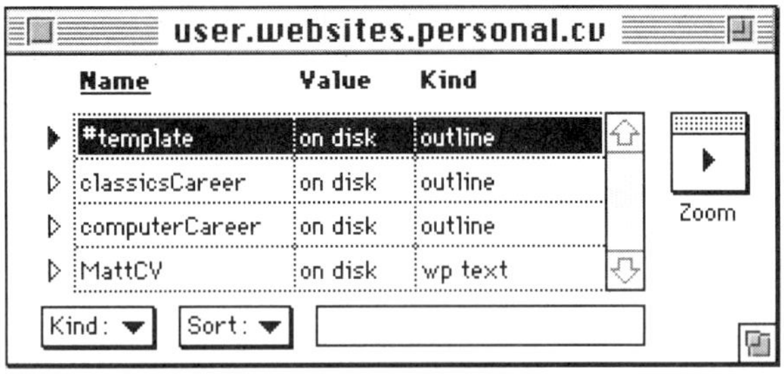

Figure 1-4. Inside Frontier's database, where the template and three Web pages live

Frontier lets you easily generate, without any programming at all, Web sites that include such advanced features as relative mutual links, "next" links, and site outlines, but this particular job calls for a little custom program in the template which calculates the HTML text for the smart links. The first line showing in Figure 1-5 is partly literal HTML, centering the smart links and making horizontal rules above and below them, and partly a call to the new program to generate the HTML of the links appropriate to the particular page that's being created.

Now I simply choose a menu item, **Release Table**, in Frontier's **Web** menu. In a few seconds, Frontier has made three HTML files on my hard disk; viewing them in my Web browser shows the links working perfectly (as shown in Figure 1-3).

> Frontier's database is good for storage and retrieval of all sorts of data; in this case, the data consist of outlines that act as the source for Web pages. Frontier comes with a Web site generator, one of its most powerful "mini-applications," Frontier is perfectly suited to this task, since, as we've already seen, it can make

Figure 1-5. Adding the smart links

text files, and a Web page's text can be "calculated" with a script. This particular example involves adding some customized calculation; you can add any further features you like to Frontier's already full-featured Web site generation, because Frontier's scripting system is completely open.

The HTML files are ready to upload to my Web site. Since I don't feel like bothering with a separate FTP client program, I have Frontier perform the upload for me. Frontier can act as an Internet client for mail, HTTP, FTP, and so on.

Frontier is a self-contained network citizen. It can act as a client or a server over a TCP/IP network, such as the Internet, and it comes with scripts implementing the most common Internet protocols, such as FTP.

Another feature I want to add to my Web page is one of those "counter" graphics (Figure 1-6) that states the number of visitors to my site. Personally I think these are a silly waste of cycles and bandwidth, but they've somehow become *de rigueur*, so my Web page may as well conform.

You are visitor number:

Figure 1-6. Counter graphic

Such graphics are GIFs generated by a CGI. Many Web servers come with CGI plug-ins, but I don't need those: I've got Frontier. Frontier comes as a CGI application, out of the box; it simply has to be running on the server machine, with the server application configured to route to Frontier any request for a document whose name ends with *.fcgi*. So, to add a counter to my Web page, I just add this HTML:

<img src=counter.fcgi>

Now I'll write the script that makes the GIF. Frontier can't make a GIF, but it can drive any scriptable application, and Clip2GIF, a scriptable application, can create GIFs on demand. So the Frontier script (shown in Figure 1-7) increments a counter in the database and drives Clip2GIF to draw the counter's value as white-on-black text.

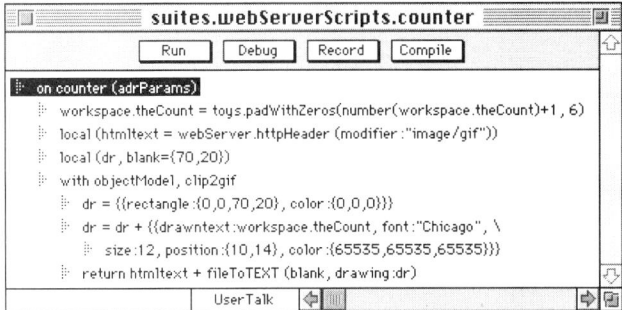

Figure 1-7. CGI script returning a counter graphic

First, I test the script on my machine. Once it's working, I need to make it available to the copy of Frontier running on my Web server machine. One copy of Frontier can speak to another across the Internet, so it takes only a moment to transfer the script from my local Frontier database to the remote one.

> Frontier can be driven by other applications; that's why it can function as a CGI application, responding to messages from the Web server. And although omitted from the example for simplicity, Frontier is multithreaded, so it responds asynchronously to requests from other applications—that is, it deals with them simultaneously, without having to wait for one to finish processing before starting on the next. This feature, combined with UserTalk's speed, makes Frontier a remarkably rapid CGI application. The example also brings together many of the other Frontier features we've already seen: database storage, driving other applications, and communicating across a network.

Time for a break, perhaps a little early lunch. Wonder what I'll do with the rest of the day? One thing's for sure: I'll be productive. I've got Frontier.

Days in Other Lives

The "Day in the Life" narrative illustrates Frontier's main features, variously combined and applied to a wide range of tasks. You might combine them in some completely different way—which is exactly the point. Frontier is a tool, not an end in itself; each person's needs being different, each user of Frontier sees the program differently. It's hard for me to imagine all the ways in which someone *might* use Frontier, so I solicited input from other actual users. Perhaps the best insight on the wide-ranging capabilities of Frontier comes from real-life stories like these:

I run a weekly newsletter. I have Frontier generate folders in a calendar structure (weeks within months) so I can drop in articles intended for each issue. Periodically, I have Frontier add up the number of characters in all the articles in each folder, so that I know how long each issue is getting.

• • •

I supervise a public bank of computers. Each night, the copy of Frontier on my computer automatically wakes up, and drives copies of Frontier on the other computers to clean out files that don't belong there. It also reports on any anomalies on the other computers, such as important files that appear to be missing or moved.

• • •

We went to a computer expo where we were demonstrating our Web site. Our booth had a computer where users could navigate our site over the Internet using a Web browser. When a user got done playing and walked away, Frontier would notice that the computer was idle, and would tell the browser to go back to our site's main page, making it the starting place for the next user.

• • •

I manage a monthly journal. I have a FileMaker database of authors who are supposedly working on articles. Each month, Frontier uses the FileMaker database to construct and send personalized email messages to each author, asking how the article is going.

• • •

Using Frontier's Web and CGI capabilities, I construct and administer college tests on the Web; each student uses a Web browser to take the test, whereupon the Web server gives that student's responses to Frontier, which puts all responses into a FileMaker database, where we grade them. Then Frontier constructs Web pages showing the results and statistics, and places them on our Web site.

• • •

I'm a molecular biologist. My colleagues often get DNA and protein sequences off the Net and then need them parsed and formatted in special ways so they can use them in certain computer programs. I've written little Frontier routines that take care of this task for them.

• • •

Here's how Frontier helped me get a job. A grocery store was periodically putting out a price list with bar-codes, making the price list by hand, using old lino equipment. I used Frontier to process their source data, including calculations to make the bar-code in a custom font, then drive QuarkXPress to generate the pages. A task that was taking them days on the lino was done in fifteen minutes. When I demonstrated this to the supervisor, he was blown away.

• • •

To me, Frontier is CGIs. When you come to my Web page, Frontier uses the current date and time to perform trigonometric calculations that work out the current phase of the moon, and puts a picture in the Web page showing the moon at that phase. Also, I listen to CDs on my computer. My CD player application is scriptable. So I wrote a CGI, and now one of my Web pages tells you what music I'm listening to at that moment. I do also use Frontier CGIs for serious purposes, like SSL transaction processing, error redirection, and IP screening to control access to intranet documents.

• • •

I work at an elementary school. At the start of each year, I want students to be able to use their own email accounts immediately. A Frontier script runs through a FileMaker database of all the students, generating usernames and passwords for all of them. Then Frontier creates a folder for each student, with a folder called *email* in each one containing a *Eudora Settings* file customized for that student; and it sets up file sharing and access privileges for each folder.

• • •

We sell online photo albums; a customer mails us photos, and we scan them to build a password-protected Web site where the customer can display and sell the photos. Frontier automates the entire workflow process. Dialogs prompt for the customer's name and address; Frontier stores these in our customer database, then generates a unique URL for the new site, plus passwords for it, and creates and prints customized instructions telling that customer how to maintain the site. Next, Frontier drives graphics software to equalize, resize, crop, and convert the images, uploading the results via FTP and compressing and archiving the original scans. Finally, it builds and uploads the Web pages themselves.

Tool of All Trades

Just what is Frontier, and what sorts of thing can you do with it?

Keeping and Organizing Information

You can make an *outline*, where chunks of text are arranged hierarchically; subordinate material can be hidden, and if you move material, whatever is subordinate to it travels automatically with it. This makes for easy navigation and rearrangement.

You can make a word-processing *text*, where paragraphs can have widths and characters can have styling.

You can make a *table*, holding any number of pieces of information, each piece being any type of data: a number, a string, an entire file from disk, even an outline or a word-processing text. Each piece of information in a table has a unique name, so you can designate it meaningfully and retrieve it easily. And an entry in a table can be a table, providing another dimension for the hierarchical organization of information.

Frontier comes with one particular table, full of other tables and pieces of information, called the *database*. You can store information here. No matter how big or deep the database gets, every piece of information in it is uniquely named and instantly accessible.

Writing Programs

You write a program by making a *script*. Like everything else, the script lives in a table in the database. The script is written in a special language, UserTalk. Frontier understands UserTalk, and runs it quickly. There is also a complete debugging environment. So you can develop scripts, test them, debug them, and run them.

Scripts can work together. A script can see all the other scripts in the database, and can run any of them as a subroutine of itself. So complicated programs can be constructed from many small simple scripts.

You might not feel like writing any scripts at first. You don't have to: Frontier comes with plenty of scripts already. They range from basic functions that make up the UserTalk language to sophisticated tools for accomplishing large tasks with little or no programming on the part of the user.

All the same, you'll soon realize that writing scripts is the way to access and customize the full power of Frontier. So you'll want to learn UserTalk. Part II of this book teaches it to you.

What Can a Script Do?

One of the things a Frontier script can control is Frontier itself. All parts of Frontier are scriptable. A script can see everything in the database. It can retrieve or change the value of a table entry, delete or create a table entry, and move or rename a table entry. It can construct an outline, enter information in an outline, extract data from an outline, and reorganize an outline. It can make a word-processing text, change any part of it, and use the text as input. It can open a Frontier window, and change its size and position. It can put up dialogs.

Frontier scripts can perform math calculations. They can also manipulate strings: change case, find-and-replace, concatenate, and extract. A Frontier script can make a file, write to it, read from a file, store a file in the database, rename a file, move a file, and delete a file. That's why Frontier is a natural tool for managing large Web sites: an HTML file is merely text that Frontier "calculates" for you.

Like most applications, Frontier has menus. A Frontier script can do whatever would happen if you chose a menu item from one of those menus: change the font or size of selected text, copy and paste, save the database, and even quit Frontier. Also, the opposite is also possible: a Frontier script can live inside a menu item. You can make new menu items and store scripts inside them. That makes it easy to run any commonly needed script: just choose the menu item you've stored it in.

A Frontier script can be made to run all the time, quietly, in the background. Then, when the right moment comes, it acts. A script might wait until two o'clock in the morning and then run. Or it could run every minute, checking something, and reacting under certain conditions.

Frontier scripts can reach outside Frontier, into the world of your computer. A Frontier script can check the time, beep the speaker, store data directly onto the clipboard, and watch for a keypress.

Frontier can communicate over a TCP/IP network. It can function as a client: it could talk to an FTP server to upload a file for you, or to an HTTP server to check whether some links are still good. It can function as a server, listening for a client to contact it. The client can be another copy of Frontier: anything a script can make Frontier do, it can make another copy of Frontier do, remotely, across the Internet.

Frontier can talk to other applications, on Mac OS, using Apple events. Many applications are "scriptable," and respond to commands and queries. A single Frontier script might operate upon a Frontier outline, get information from a File-Maker database, and write and send an email with Eudora.

Other applications can also talk to Frontier and give it commands. A FileMaker script might run a Frontier script as a subroutine. Frontier is multithreaded, so it can perform many tasks simultaneously. So Frontier is a natural CGI application, responding to a Web server by constructing Web pages on demand.

What Is Frontier?

It's the command-line to your computer. It's a tool to make tools. It's for people who believe that a computer is for you to control, not the other way round.

2

Edit Windows

To work comfortably with Frontier, you need to be familiar with the various kinds of edit window in which you'll be doing most of your work. An *edit window* is a window in which you can perform editing actions, such as typing, cutting and pasting, and so forth. In this chapter, three main kinds of edit window are discussed.

First, we talk about *outline* edit windows. These allow you to construct and arrange text data in outline form; you can keep notes as outlines, and outlines can be used as input for automated processes (for example, an outline might tell Frontier what Web pages you want to generate). Also, two other kinds of Frontier edit window not discussed in this chapter—*script* edit windows and *menubar* edit windows—are basically just varieties of outline edit windows. (They are further described in Chapter 4, *What a UserTalk Script Is Like*, and Chapter 26, *Menus and Suites*, respectively.) In script edit windows, you write UserTalk programs that Frontier can execute, and in menubar windows, you customize Frontier's menus. So in order to work with these, you first need to know your way around an outline edit window.

Then, we discuss *word-processing text* (hereinafter referred to as *wptext*) edit windows. These are simple text document windows, often used as data for automated processes; for example, Web pages are often wptexts.

Finally, we talk about *table* edit windows. These are very important, because the database that accompanies Frontier (see Chapter 3, *The Database*) consists of tables within tables within tables, and navigation within the database and operations upon it are performed primarily through table edit windows.

Window Types and Objects

When you create an edit window, you must specify its type. An edit window of one type cannot be transformed into an edit window of another type. So, for example, if you create a new window as an outline edit window, you can create and edit an outline in that window; you cannot create a wptext in that window, and you cannot change the window to a wptext edit window.

The reason is as follows. When you create an edit window, you are really creating two things. One is the edit window; the other is the *object*, the thing to which the edit window gives access. Each type of object has its own appropriate variety of edit window, and object types that have edit windows cannot be transformed into one another. For instance, in the case of an outline edit window, there is also an outline object; you can't change an outline object into a wptext object, so you can't change its edit window into a wptext edit window.

Objects are primary; edit windows are secondary—they are just the means of accessing the object. It helps to think of the object as possessing an edit window, which it continues to possess whether or not that window is open. Assuming that you don't throw away an object, it persists; so you can dismiss its window, and then open it at a later time. Changes in an object's edit window are remembered with the object, so when an edit window is opened, it will appear in the same state as when it was last closed.

Objects are created in Frontier's database. The object may be created anywhere in the database, but until you are familiar with the database and the conventions about where to store objects in it, you should limit yourself to creating objects in the workspace table. To do so, choose **Workspace** from the **Open** menu, and then, from the **Table** menu that appears, choose **New Outline**, or **New Text** (for a wptext), or **New Sub-Table** (for a table). In the dialog that appears, assign the new object a name.* The object will be created in the database, and its edit window will open.†

Outlines

Figure 2-1 shows an outline edit window. When an outline edit window is frontmost, the **Outline** menu also appears at the top of the screen.

---

* Frontier object names can be anything at all, provided the name is unique within its table; object naming conventions are discussed further in Chapter 6, *Referring to Database Entries*.

† The workspace window is a table edit window; table edit windows are explained later in this chapter. The database itself is discussed in Chapter 3.

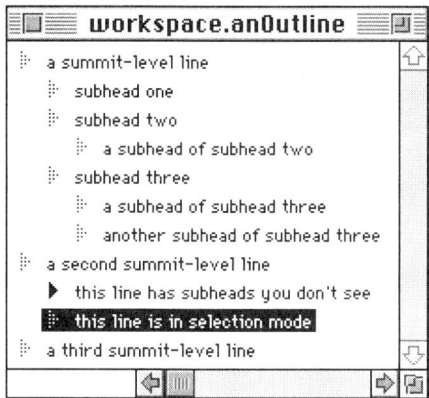

Figure 2-1. An outline edit window

Outlines are hierarchical arrangements of lines of text. Each line's text may consist of at most 255 characters. If the edit window is too narrow to show the whole line, the line is not "soft-wrapped," but simply extends past the right edge of the window; to see more of such a line, widen the window, or scroll it horizontally.

You may change the font and size of the text in an outline edit window; this changes the font and size of *all* the text in the window. To do so, choose from the **Font** and **Size** submenus of the **Edit** menu; or, in the **Main** menu, choose any of the items in the **Common Styles** submenu.

Each line of an outline is preceded by a triangle. The triangle is a sort of "handle" for manipulating the line, as well as an indicator as to its status; the details are explained throughout these sections.

Modes

Distinguish between a line and its text. When you click on a triangle, it darkens and the text of the line is reversed; this means that the line itself is selected, and that the outline is in *selection mode* (illustrated in Figure 2-1, "this line is in selection mode"). In selection mode, you edit the outline by working with its lines—rearranging lines, making hidden lines visible, and so on.

If, on the other hand, you click *within* the text of a line, a blinking insertion bar appears; or you can click-drag to select some text, double-click to select a word, or triple-click to select all the text in a line. The outline is now in *content mode*. In content mode, you edit the text, which you do as you would in any word processor, by typing, cutting and pasting, navigating with arrow keys, and so on. (For arrow key modifiers, see "Wptexts," later in this chapter.)

To toggle between content mode and selection mode with the keyboard, hit the Enter key. Or, in selection mode, simply start typing: you will enter content mode automatically, and what you type will replace the contents of the selected line.

In selection mode, the physical line may be thought of as extending invisibly to both edges of the window. This means that when I say to click or drag a line's triangle, you can actually click or drag (in selection mode) *any* region to the left or right of the line. This is useful to know; many beginners think that the triangle is small, and hard to "grab" or "hit" with the mouse, whereas in fact there is no need to be so precise.

Hierarchy

The indentation of a line's triangle indicates the hierarchical relationship between lines. A line whose triangle is as far to the left as possible is not hierarchically subordinate to any other line, and is called a *summit-level* line.

A line not at summit level is hierarchically subordinate to the line that is most immediately above it and to its left. Such a line is called a *subhead* of the line to which it is subordinate. So, for example, in Figure 2-1, both the line "subhead one" and the line "subhead two" are subordinate to "a summit-level line." From the other point of view, the line to which a line is subordinate is called its *parent*. So "a summit-level line" is the parent of "subhead three."

The sum total of lines subordinate to a given line—its subheads, and any subheads of those subheads, and so on—is referred to as the *bundle* subordinate to that line. Lines whose triangles line up one above the other are at the same *level*. Lines at the same level and part of the same bundle are called *siblings*. So, "subhead one," "subhead two," and "subhead three" are siblings.

Subheads may be *expanded* or *collapsed*, meaning that they are shown or hidden. This ability of subheads to be shown or hidden is what makes an outline a useful mechanism for viewing hierarchically organized text. No matter how much material the outline contains, its collapsible hierarchies help you see as much or as little of it as is convenient at any particular time.

In Figure 2-1, there are collapsed subheads subordinate to the line "this line has subheads you don't see." The line's triangle is differently shaded, to signify this.

Table 2-1 lists the methods of collapsing and expanding lines.

Table 2-1. Collapsing and Expanding in Outlines

In Order to . . .	You Should . . .
Collapse a line's subheads	Double-click the line's triangle; or select the line and hit the keypad minus key; or select the line and choose **Toggle Expand**
Expand a line's immediate subheads	Double-click the line's triangle; or select the line and hit the keypad plus key; or select the line and choose **Toggle Expand**
Expand all of a line's subheads	Select the line and hit the keypad asterisk key
Collapse a subhead and its siblings	Select the subhead and choose **Collapse To Parent**
Collapse all subheads, leaving just summit-level lines visible	Choose **Full Collapse**
Expand all subheads	Choose **Full Expand**

Navigation and Reorganization

Navigation and reorganization in content mode work as in a wptext or any basic word processor—cut and paste, typing and deletion of text, and so forth. Our concern here is with selection mode.

In selection mode, the arrow keys move as described in Table 2-2.

Table 2-2. Arrow-Key Navigation in Outlines

In Order to Move . . .	Hit This Key . . .
Up or down among siblings[a]	Up-arrow, down-arrow
To first or last sibling	Command-up-arrow, Command-down-arrow
Up or down absolutely ("flatup" and "flat-down") among visible lines	Left-arrow, right-arrow
To first or last line	Command-left-arrow, Command-right-arrow
To parent	Option-left-arrow
To parent summit	Command-Option-left-arrow

[a] It is possible to change the behavior of the up-arrow and down-arrow keys so that they cause movement up or down absolutely, just like left-arrow and right-arrow. To do so, choose **Quick Script** from the **Open** menu, type `op.flatCursorKeys(true)`, and hit Enter. In this mode, additionally, Command-left-arrow moves to the current line's parent and collapses the parent's subordinate bundle; Command-right-arrow expands the current line's subordinate bundle (if there is one) and moves to the first line of that bundle. To move to the first or last line, use Command-up-arrow and Command-down-arrow. The change in behavior will persist until Frontier is quit or until you say `op.flatCursorKeys(false)`.

When a line is moved, copied, cut, or deleted, its subordinate bundle is automatically included, regardless of collapsed or expanded state. This is what makes an outline a useful mechanism for hierarchical storage and organization of textual data.

Multiple lines can be selected by Shift-clicking triangles or by adding the Shift key to arrow-key navigation. (To select all siblings of a given line, select the line and choose **Select All** from the **Edit** menu, or Shift-double-click its triangle.) The multiple selection may then be moved, copied, cut, or deleted. In what follows, it is presumed for simplicity's sake that the selection consists of a single line.

Copying, cutting, and pasting are performed with the usual commands in the **Edit** menu. Material pasted in selection mode is placed downwards from, and as a sibling of, the currently selected line. Pasting a line never causes the deletion of a line.

A line may be moved by dragging its triangle. An arrow appears, showing at each moment where the line would go if you were to release the mouse button. If you move a line so that it becomes subordinate to a line with collapsed subheads, the subheads expand and the line becomes the last of them. (You can change this behavior by holding the Option key: in that case, the line you are moving vanishes, as it becomes the last subhead and is collapsed all in one motion.)

To move the currently selected line with the keyboard, use Command-u, Command-d, Command-l, and Command-r to move the line up, down, left, or right respectively.* Movement up and down is among siblings, and is thus restricted to the current bundle. Movement right means that the line becomes the last subhead of the sibling presently above it. Movement left means that the line becomes the next sibling of its current parent. Suppose, for example, in Figure 2-1, you select "subhead two" and hit Command-l. The line becomes summit-level, and occupies the first available "slot" after what was its parent, namely, just above "a second summit-level line."

Other reorganizational commands appear in the **Outline** menu. **Promote** causes all subheads of the currently selected line to become its siblings. **Demote** causes all siblings below the currently selected line to become its subheads. **Sort** sorts siblings.

To create a line, hit Return, either in selection mode or in content mode. The new line is created as a sibling down from the current line, unless the current line possesses expanded subheads, in which case the new line becomes the current line's first subhead.

To delete a selected line, hit Delete (or choose **Clear** from the **Edit** menu). This also deletes the line's subordinate bundle, of course, so be careful.

---

* It is also possible to use the Tab key to move a line to the right, and Shift-Tab to move it to the left. You can disable this feature, perhaps in order to type a tab character in the text of an outline, which is not normally possible. To do so, choose **Quick Script** from the **Open** menu, type `op.tabKeyRe-org(false)`, and hit Enter. The change in behavior will persist until Frontier is quit, or until you type `op.tabKeyReorg(true)`.

Wptexts

Figure 2-2 shows a wptext edit window. When a wptext edit window is front-most, the **WP** menu also appears at the top of the screen.

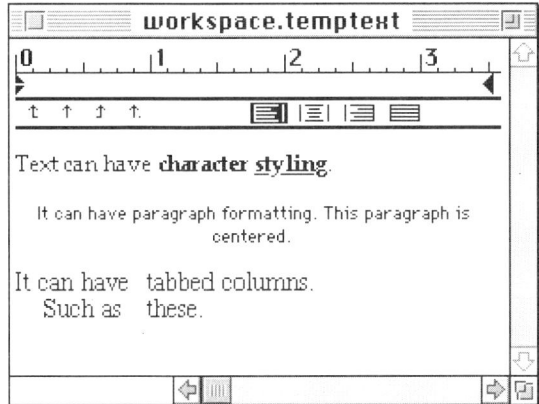

Figure 2-2. A wptext edit window

The window works like that of any simple word processor. Text can have different fonts, sizes, and styles; use the **Edit** menu to apply them. Leading (the space between lines) and justification can also be set from the **Edit** menu. A ruler can be shown at the top of the window by choosing **Ruler On/Off** from the **WP** menu; with the ruler, paragraph justification, margins, and tabs can be set. The most common use of a wptext edit window, though, is simply as a way to work with long texts; the formatting is generally secondary.

Copying, cutting, and pasting all work in the usual way. When navigating with arrow keys, adding the Option key causes navigation by word; adding the Command key causes navigation to the start or end of the text. You can add the Shift key to arrow key navigation to select text.

Tables

Figure 2-3 shows two table edit windows. When a table edit window is frontmost, the **Table** menu also appears at the top of the screen.

A table is a collection of *entries*. An entry in a table has a name which is unique within the table; it also has a value, which is of some single type. In formal terms, then, a table entry is a *name-value pair*. We may say of a table that it *contains* its entries; from the opposite point of view, the table containing a given entry is called that entry's *parent*, or its *parent table*.

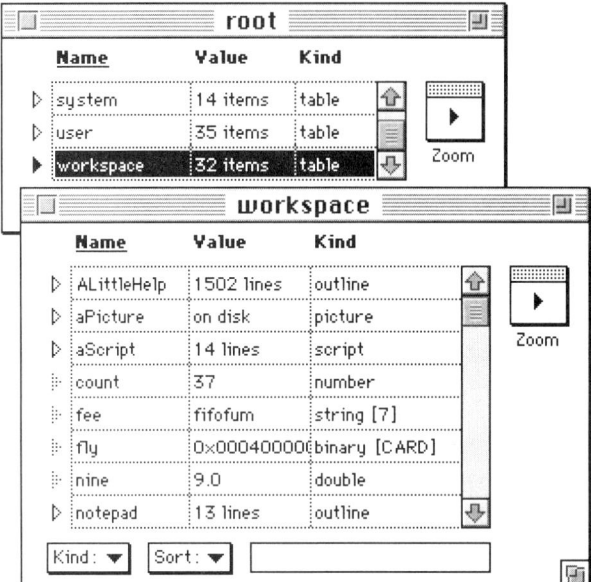

Figure 2-3. Table edit windows

One possible type of table entry is a table; thus, one can have a table within a table (sometimes called a *subtable*). This nested containment may be carried out to any desired depth, providing a containment hierarchy somewhat analogous to that of parents and subheads in an outline. Figure 2-3 illustrates this: the rear window contains three tables, one of which has been opened as the front window.

The job of a table edit window is to image the information about a table and its entries in a format that may be readily understood and conveniently manipulated. To this end, a table edit window has rows and columns. Each row corresponds to an entry; the columns correspond to the three pieces of information about each entry: its name, its value, and its value's type (referred to, in the window, as its "kind").

Kinds of Entry

Referring again to Figure 2-3, consider the visible entries in **workspace**. Their values are of various types, as listed in the Kind column: an outline, a picture, a script, a number, a string, a binary, and a double. These are some of the many *datatypes* that an object's value can have in Frontier; they are discussed in detail in Chapter 10, *Datatypes*, and the details need not concern us for the present.

Now observe that, in some cases, the Value column actually shows us the entry's value; in other cases, though, it shows us only some information about the entry. For example, the entry `count` has a numeric value 37; the entry `nine` has a numeric value `9.0`; the entry `fee` has a string value `"fifofum"` (and is of length 7, as the Kind column informs us). But the entry `notepad` is an outline, and an outline cannot be shown in a row of a table. To see the value of `notepad`, you would double-click its triangle; this causes `notepad`'s edit window to open, showing the actual outline.

Not only is this a difference in how you *see* the values of entries; it is also a difference in how you *edit* the values of entries. An outline is edited in an outline edit window. A number, a string, or a double, on the other hand, is edited directly in its parent table; you simply alter what's in the Value column. A table entry like a string or a number, which is edited directly in its parent table, is called a *scalar*; a table entry like an outline or a wptext, which is edited in its own edit window, is called a *non-scalar*. Scalar entries in a table edit window have a dotted triangle; non-scalar entries have a white triangle (which becomes black if the entry is selected).

Some types of entry can be transformed into other types of entry. For example, a string whose value is `"31"` can be transformed into the number 31. This is called *coercion*, and is studied in great detail in Chapter 10. You can perform any legal coercions on a table entry by selecting it and then choosing from the **Kind** popup at the bottom of the window; those datatypes into which the currently selected entry may be legally coerced are enabled, and choosing one performs the coercion.

Modes

Like an outline, a table edit window may be in selection mode or content mode.

In selection mode, an entry is selected as a whole. An entry selected as a whole can be cut or copied, or simply deleted by hitting Delete or choosing **Clear** from the **Edit** menu. Pasting takes place into the same table as the currently selected entry. If an entry is copied and then pasted right back into the same table, the pasted copy is renamed, to ensure uniqueness of naming; for example, `myEntry` would be pasted as `myEntry copy`. Pasting an entry will never cause automatic deletion of an existing entry (if Frontier thinks this might be what you want to do, it puts up a dialog and asks).*

---

* It is impossible to create a table entry with the same name as an existing entry, so in general Frontier responds helpfully when you try to do so. An interesting situation arises if you copy text, then enter selection mode in a table edit window and paste. This ought to mean that you are trying to create a new table entry, but you do not have a full table entry on the clipboard. Frontier responds by creating a new wptext entry containing what was on the clipboard.

In content mode, the information about an entry may be edited: you can change its name; if it is a scalar, you can change its value. Editing the name or (if it is a scalar) value of a table entry works as in any word processor. You can type, you can select text, you can copy, cut and paste; navigate with the arrow keys, and adding the Shift key selects. After editing an entry in content mode, it is important to return to selection mode; this causes the changes to "take."

As in an outline, you can toggle between selection mode and content mode by hitting the Enter key. Or you can click on an entry's triangle to select it in selection mode; click in its Name column text or (if it is a scalar) in its Value column to select it in content mode.

When you change a scalar value, Frontier has to guess what datatype you intend; if, after returning to selection mode, you find (by looking in the Kind column) that Frontier has guessed wrong, you can change the datatype with the **Kind** popup.

If you change a scalar value so that the new value is a valid UserTalk Expression, then when you return to selection mode Frontier evaluates the expression and replaces the value with the result. This is very convenient, and perfectly understandable once you know UserTalk (see Part II, *UserTalk Basics*); but it can be surprising at first. For example, if you type 3+4 into a number's Value column, the value becomes 7. If this is not what you want, you may have to write a more explicit UserTalk expression; for example, to assign a string the value "3+4", you need to include the double quotes.*

Creation of Entries

There are two ways of creating a new table entry. One is to choose from the **Table** menu. Here, you must decide upon the entry's datatype beforehand. The various datatypes are arranged in the **Table** menu, in three groups: those visible as items within the **Table** menu itself, beginning with the word **New**; those dependent on the **New Scalar** submenu; and those dependent on the **New Special** submenu. Choosing any of these items presents a dialog in which to name the new table entry; the entry is then created in the same table as the currently selected entry.

A second method of creating a table entry is to select a table entry and hit Command-Return. This creates a new entry just below the currently selected entry; the new entry has no name and no datatype, so you need to assign them

---

* It is not possible to give a table entry a "live" value; the value is evaluated and assigned when you return to selection mode. For example, if you give a table entry the value clock.now(), its value will become the date and time when you made this assignment—it will not become a live, "ticking" clock.

now. The entry is created in content mode, waiting for you to type a name. After typing the name, you can hit Enter to cause the name to "take." This will also cause the entry to be sorted into position, so it may jump as the name moves into its proper alphabetical place. Then you can edit the entry's value.

Navigation

Table 2-3 summarizes keyboard shortcuts for table navigation.

Table 2-3. Keyboard Table Navigation

In Order to . . .	You Should . . .
In selection mode, select any entry in the same table	Type the first few letters of its name
Select the previous or next entry in same table	Up-arrow, down-arrow
In selection mode, select the first or last entry in the same table	Command-up-arrow, Command-down-arrow
In content mode, select the next or previous editable cell	Tab, Shift-Tab
Open a non-scalar entry's edit window	Select the entry and Command-Enter; select the entry and hit the **Zoom** button; or double-click the entry's triangle (hold Option key to dismiss the current window at the same time)
Open an edit window's parent table	Shift-Command-Enter; or double-click the window's titlebar[a] (hold Option key to dismiss the current window at the same time)
Open an edit window's parent table, or its parent table, or its parent table (and so on)	Command-click on the title in the titlebar, and choose from the popup menu that appears (hold Option key to dismiss the current window at the same time)

[a] Unfortunately, the Mac OS system's "window shade" mechanism may, depending on your settings, interpret Command-double-clicking a window's titlebar as a signal to show or hide that window's content region, thus overriding this feature of Frontier.

With a little practice, the user will soon become adept at the various techniques for navigating and viewing database entries, so as to move comfortably and swiftly through the database without leaving too many windows cluttering up the screen.

Nevertheless, there is a certain tendency, in the heat of editing, for open windows to accumulate. A desired open window may be chosen from the **Window** menu, to bring to it to the front. A window may be closed by hitting Command-w; all open windows may be closed by hitting Option-Command-w when the Main

Window is not frontmost. Windows are discussed in detail in Chapter 24, *Windows*.

Sort Order and Aesthetics

The order in which a table's entries are displayed within its parent table edit window or beneath its parent row is called its *sort order*. There are three such orders, corresponding to the three columns. To sort entries, choose the desired order from the **Sort** popup at the bottom of the window, or simply click on the column name at the top of the window. The name of the column corresponding to the current sort order is underlined.

Columns may be widened and narrowed by dragging the column divider. The font and size of the text in the window may be changed; as with an outline, this affects *all* the text.

3

The Database

Accompanying the Frontier application is a database file called *Frontier.root.* Whenever Frontier is started up, it automatically opens the copy of *Frontier.root* that is in the same folder with itself. It is possible to close a database; it is possible to open a second database (see Chapter 30, *Multiple Databases*). But in general, any running copy of Frontier will have exactly one database open: the *Frontier.root* that is shipped by UserLand.

Most of the features that give Frontier its actual behavior are located here: its menus, its scripts, its ability to interpret those scripts, and all the tools and features that make Frontier useful. Without a database open—without *this* database open—Frontier will be virtually incapable of doing anything. It is for this reason that we speak in the singular of "the database." (It is referred to sometimes as the *database*, sometimes as the *root*; these terms are interchangeable. It is also sometimes referred to by Frontier mavens as the *odb*, for "online database.")

In this chapter, the database is introduced. There is some discussion of its contents, though the details are left to the rest of the book (by the book's end, there will be scarcely a cranny of the database left unexplored); and there is considerable discussion of how to navigate the database. After studying this chapter, the reader will be comfortable with the nature and presence of the database, and will be ready to explore and modify it.

WARNING   Be very, *very* careful with the database! If you read nothing else in this book, read the next few paragraphs. Remember, and obey!

This is your database. You are free to explore it, and you are expected to do so; Frontier is an "open system," so the database is deliberately exposed to your gaze, and studying it and tracing its workings is one of the chief ways of learning about Frontier. You are free to store things in the database, and you are expected to do so; you will *need* to do so if you wish to liberate Frontier's full potential.

Nevertheless, it is also Frontier's database, and you can bring the program completely to its knees by making wanton changes. A single false move, such as deleting the `system` entry from the `root` table, will cause Frontier to choke completely, and you may have to download a fresh copy of the database to get it working again. More subtle changes could cause specific tools to break, or make Frontier behave unpredictably, and since it has great power over things you value, such as what's on your hard disk, that could be disastrous.*

To minimize risks, I suggest you follow two simple rules:

- Make frequent backups of the database. You do this by choosing **Backup** from the **Main** menu; Frontier maintains a *Backups* folder for this purpose. For safety, Frontier does not overwrite earlier backups when you make a new one, so after a while you will probably want to delete or compress some older backups manually in the Finder, as they are rather large.

- As my father said to me when he first taught me to use an electric saw: "Never Stop Being Careful."† This chapter will describe what parts of the database are yours to play with (within reason) and which parts are crucial to Frontier's basic functionality. Know what you are doing at all times! If you don't know that what you are doing is safe, don't do it.

If, despite your best efforts, you think you've screwed something up, choose **Revert** from the **File** menu, thus fetching the most recently saved changes from disk and reversing recent changes that have not been saved. (It is also possible to quit Frontier without saving the database, by clicking **No** when you are offered a chance to save it.) To return to a previously backed up version of the database, quit Frontier, rename *Frontier.root* to something else (or even throw it away), move or copy a backup of the root into same folder as Frontier, and rename it *Frontier.root*. If the worst comes to the worst, download a new clean root from UserLand's site.

I'm sorry if this little lecture has made you edgy. But Frontier, unlike most other programs, exposes most of its inner workings to you. That gives you a lot of power, which is the point; but that power is accompanied by increased responsibility.

---

* Remember, in the movie *2001: A Space Odyssey*, the astronaut Dave wandering among HAL 9000's memory banks with a screwdriver? That's you in the database.

† He somehow said this so that I was able to hear the capital letters, so I have reproduced them here.

The care and feeding of the database is further discussed in Chapter 29, *Import/ Export.*

Main Window and Root Table

The presence of the database is indicated by the presence of the Main Window. Figure 3-1 shows the Main Window.

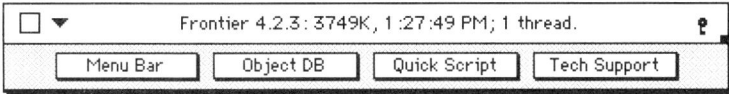

Figure 3-1. The Main Window

If you don't see the Main Window at all, choose the first item in the **Window** menu. If you don't see the row of buttons, click the "mailbox flag" icon at the right end of the window. If you don't see a status message like the one here, click the triangle at the right end of the window to bring up the popup menu and choose **StatusMessage**. You can change the font and size of the text in the window. You can widen or narrow the window by dragging the little dark square in the middle of the right edge. You can drag the whole window by its top half.

You can dismiss the main window by clicking in the go-away box at its top left, but if you do, Frontier thinks you wish to close the database—so don't.

The status message tells some useful things. One is how much heap memory Frontier has free (here, 3749K). Experience has shown that it is wise to keep this number large (3000K is a good minimum); to achieve this, you may need to save the database, or quit Frontier and increase its Finder memory allocation using **Get Info**. This matter is only partly related to the size of the database, which can grow very large without having much effect on Frontier's use of RAM or on the speed with which Frontier can access database entries. But Frontier does need RAM memory as soon as it is called upon to *do* anything, and it is wise to give it plenty.

From the Main Window, you can access the contents of the database by pressing the **Object DB** button or hitting Command-Enter. This brings up the `root` table edit window, showing the top level of the database; you can then "drill down" to any entry in the database, table by subtable by subtable, using the table edit window navigation techniques described in Chapter 2, *Edit Windows*.

Direct Database Navigation

To reach a desired database entry, it is not actually necessary to "drill down" to it through successive subtables; it is possible to navigate to it directly, because every database entry is uniquely named.

To see that this is so, recall that every table entry is uniquely named within its parent table. By the same token, the parent table is also uniquely named within its parent table, and so on, all the way up to the root table (which has no parent). Therefore, every database entry is uniquely named by the names of all the tables containing it plus its own name. The names are concatenated by means of "dot-notation." For example, in the root table is a table called system, which contains a table called verbs, which contains a table called colors, which contains an entry called blue. The full unique name of this entry is therefore system.verbs.colors.blue. A full name like this is called a *path*, and the parts between the dots, such as verbs and colors, are the *elements* of that path. Notice that the element root may be omitted from the start of a path.

In practice, a number of paths are so commonly referred to that, for the sake of convenience, Frontier makes it possible to refer to their contents without stating the whole path explicitly. For example, it is possible to refer to system.verbs.colors.blue by saying simply blue. After a time, the user tends to internalize a knowledge of which paths these are. The matter is discussed in great detail in Chapter 6, *Referring to Database Entries*.

To reach any database entry directly by name is called *jumping*. To jump to a database entry, choose **Jump** from the **Open** menu to bring up the Jump dialog, and type the path into the edit field (or choose it from the popup menu, if it is already present), and hit **OK**.

The popup menu in the Jump dialog remembers the last several jump destinations. You can change the maximum number of remembered destinations by changing the value at user.jump.maxItemsInJumpList. You can introduce permanently preset destinations by inserting them in the list at user.jump.permanentJumpList; they must be literal (double-quoted) strings, separated by commas, between the curly-brace list delimiters, like this:

```
{"system.verbs.apps", "people.man", "suites.html"}
```

To jump again to the destination most recently jumped to, without bringing up the Jump dialog, hold the Shift key while choosing **Jump**, or hit Shift-Command-J.

Another place to store frequent jump destinations is in the **Bookmarks** submenu of the **Open** menu. In a table edit window, select the desired jump destination, and choose **Add Bookmark** from the **Bookmarks** submenu. From then on, that jump destination will appear as a menu item in the **Bookmarks** submenu; choosing the item jumps to the destination that it names.

Still another means of jumping is by Command-double-clicking the name of a database entry in the text of any edit window. Basically, if you can see an entry's name, you can jump to it by this method. This is particularly useful when debugging scripts.

In the **UserLand Suites** submenu of the **Suites** menu is an item **Object Database Map**. When you choose it, an outline appears listing all tables in the database; the outline is not "live," but can be rebuilt by choosing **Rebuild** from the **Map** menu. Command-double-clicking the name of any table in the outline jumps to that table.

Searching the Database

The database can be searched by choosing the **Find & Replace Dialog** item of the **Find & Replace** submenu of the **Main** menu. This brings up a dialog which works as Find dialogs generally do. The process looks for text within the name or value of any database entry, so searching the whole database can be slow.

The Find dialog gets its context from what is frontmost at the time the dialog is first summoned. So, for example, to search the whole database, have the `root` table edit window frontmost beforehand; to search the `suites` table, have the `suites` table edit window frontmost beforehand. What entry in the table is selected may make a difference, too.

Once a search finds a match, you can continue the same search by hitting Command-g.

If text you wish to search for is already visible within an edit window, you can conveniently search for further instances of that text, without having to copy it into the Find dialog, by selecting the text and hitting Shift-Command-G.

Guide to the Database

Later parts of this book discuss just about every area of the database, so this section provides a mere sketch of the database's contents. If a part of the database is said to "belong to you," it is available for you to store material in. If it "belongs to UserLand," you generally should not store material in it or modify what is there, unless you are very sure of what you are doing. One reason for this is that what you modify might be some crucial link in the chain of a Frontier behavior, so that altering it might break that behavior. Another reason has to do with updates. From time to time, UserLand may update a part of the database that belongs to it, and you will wish to download and install the update into your copy of the database. When you do this, if you have made any changes in that part of the database, they may be overwritten and lost. However, UserLand will never distribute an updater that will overwrite a part of the database that belongs to you. This is discussed further in Chapter 29.

workspace

> This table belongs to you, absolutely. It is readily available by choosing **Workspace** from the **Open** menu. Here you may keep anything, of any description.

For instance, this is where I develop scripts, keep notes lying around, and store all kinds of material both temporary and permanent.

`workspace.notepad`

This is an outline that you can use for any purpose you like. It is readily available by choosing **Notepad** from the **Open** menu.

`people.[user.initials]`

In the `people` table there is a table with your initials. This will have been created automatically when you first started up Frontier with a clean root.[*] Mine, for instance, is called `people.man`. This table belongs to you, absolutely. Its most important use is for scripts and other objects to which you wish to refer conveniently in scripts.

This is because (as discussed in Chapter 6, *Referring to Database Entries*) anything you place here can be referred to without a complete path. For example, on my machine, a script at `people.man.myAmazingScript` can be called by saying `myAmazingScript()`. Or, if I store my favorite color as `people.man.myFaveColor`, I can refer to it in a script by saying `myFaveColor`.

`scratchpad`

This table belongs to everyone, but only for temporary storage. In general, it is meant to be accessed by scripts rather than humans. The idea is that scripts can place anything they like here, secure in the knowledge that they won't be tromping on anything, because whatever is already here was put here on the understanding that it *might* be tromped on. At any time (unless a script is running) you could delete everything in the `scratchpad` table and nothing would break.

`suites`

This table contains the self-contained tools and mini-applications that come with Frontier, as well as the implementation for a number of menu-driven functionalities. On the whole, any existing entry in `suites` belongs to Frontier; but you are free to write your own suites and place them here, and other users are free to write suites and distribute them. Suites are further discussed in Chapter 26, *Menus and Suites*.

`user`

This table is shared by you and UserLand. Various suites and other functionalities place entries here. But you are allowed and expected to modify certain of those entries, because they will be checked to learn what your wishes are.

---

[*] Your initials are used for a range of identifying purposes. If you decide you want to use a different set of initials, you need to change the value of `user.initials`, then select `login` in the `user` table and choose **Run Selection** from the **Main** menu, save the database, and quit and restart Frontier.

So you cannot store material freely in the user table, but you can store particular material that Frontier expects to find here.

Thus, you shouldn't modify anything in the user table without knowing what you're doing, but once you do know, there will be lots of things you will in fact modify. Some of these will be small things, such as changing a preference like the value of user.jump.maxItemsInJumpList. Others will be very large, like keeping an entire Web site in user.websites.

system

This table belongs to UserLand. Here beats the heart of Frontier; the system table holds Frontier's menus, the definitions of UserTalk verbs and constants, "glue" tables for driving other applications, and many other basic functions. On the whole, you should tread warily here; but there are places where you might make modifications. Here are some of the highlights of what's in the system table:

system.compiler

This is the *sanctum sanctorum*, where UserTalk itself is implemented at the lowest level. On the whole, you should keep out.

system.agents, system.startup, system.suspend, system.resume, and system.shutdown

Scripts here are automatically called periodically in the background or in response to certain external events; you might write a script and put it in one of these places. See Chapter 27, *Agents and Hooks.*

system.extensions

Scripts here provide access to "language extensions"; you might write or install one of these. See Chapter 23, *Extending the Language.*

system.misc.menubar and system.menubars

You might customize a menu here. See Chapter 26.

system.verbs.apps

These are "glue" verbs that Frontier uses to communicate with other applications. It is possible that you will "tweak" existing glue, or generate new glue of your own. See Chapter 32, *Driving Other Applications.*

II

UserTalk Basics

This section describes the basics of UserTalk, Frontier's native programming language, in which most Frontier scripts are written for performing calculations, for operating on the Frontier database, for driving the Frontier program, and for driving and communicating with other programs.

The full power of Frontier is realized primarily through UserTalk scripts located in the Frontier database. UserTalk is simple but powerful; this section will make you comfortable with its use, regardless of whether you have previously done any programming or scripting.[*]

Part II is about how to write, debug, and run a UserTalk script, and about the rules of the language: its syntax, its keywords, and the core verbs required in order to drive the most fundamental aspects of the language. It is essential reading for those wishing to write or understand UserTalk.

This section is the only place in the book where UserTalk keywords and datatype coercion verbs are explained. UserTalk operators and punctuation are also further summarized in chapters of their own in Part VII, *Reference*. The rest of the User-Talk verbs appear, categorized by subject, in subsequent parts of the book, with technical details reserved for Part VII.

---

[*] The UserTalk language may be characterized on the surface as C-like, and its simplicity and orthogonality is said to put it squarely in the ALGOL family. On the other hand, Frontier has some distinctly Lisp-like features, such as its use of local handlers and its scoping and typing rules; ultimately, there are enough significant differences of detail between UserTalk and C that the comparison is unhelpful. Rather than present a cumbersome conspectus of the similarities and differences between C and UserTalk here, we shall simply comment occasionally on them as they arise in the course of the discussion.

4

What a UserTalk Script Is Like

The physical milieu in which scripts are most often written is the edit window of a script object. This chapter describes this physical milieu and how it is intimately bound up with UserTalk syntax. Since scripts are essentially outlines, you should familiarize yourself first with editing and navigating outlines; see Chapter 2, *Edit Windows.*

First Principles

UserTalk is an "imperative" language, meaning that a UserTalk program is basically a sequence of commands. These commands are executed consecutively from the start of a script to its end, or until a command to stop execution is reached.

Some UserTalk commands just introduce syntactic constructs which help define the path of execution. For instance, such a command might declare certain other commands to be a loop. But these commands, which are called *keywords*, don't really take significant action. Every UserTalk command that takes action does so by one of two methods: it performs an assignment, or it calls a verb.

A *variable* is a kind of container, a named object whose job is to hold a value. An *assignment* is a command to set the value of a variable; the equal sign (=) is the main operator denoting this command.

A *verb* is a script, or part of a script: it is a named sequence of commands. A *verb call* is an instruction causing a particular verb to be executed. It consists of the name of the verb followed by parentheses, which must contain any *parameters* expected by that verb; it is the use of parentheses which signifies that this is a verb call. (Parameters are discussed in detail in Chapter 5, *Handlers and Parameters.*)

Here, for instance, is a two-line UserTalk script. The first line performs an assignment, giving a variable s the string value `"Hello, World!"`; the second line then calls a Frontier verb whose job is to put up a dialog displaying whatever text was passed to it as its parameter. In this case, that text is the value of the variable s, which was assigned in the first line. Therefore, executing this script causes a dialog to appear that reads **Hello, World!**

Example 4-1. Our first script

```
s = "Hello, World!"
dialog.notify(s)
```

TIP   You may wish to try typing and executing this script (and others). You will have to create a script object in the database; I generally keep "scratch" script objects in the **workspace** table. To create such a script, choose **Workspace** from the **Open** menu, then choose **New Script** from the **Table** menu, and name the script. To execute a script whose edit window is frontmost, click the window's **Run** button.

It is not an error for a UserTalk command to consist of neither an assignment nor a verb call; but such a command will perform no action. Here, for instance, is the same two-line script, bereft of assignment in the first line (the variable and equal sign are gone) and of a verb call in the second line (the parentheses have been removed).

Example 4-2. A script without actions

```
"Hello, World!"
dialog.notify
```

This script runs without error, but nothing happens.* The reason there is no error is because what Frontier is really looking for whenever you speak UserTalk to it is a legal UserTalk expression. Frontier's reaction is to evaluate the expression; it chokes only if it cannot do so. In Example 4-2, each expression can be evaluated, so there's no problem. If an expression involves an assignment or a verb call, Frontier also performs the command, as part of the evaluation process; that's how a script actually takes action.

To illustrate, open the Quick Script window (choose **Quick Script** from the **Open** menu). This is a place to speak UserTalk directly to Frontier. Type 3+4 into the main part of the window, and hit its **Run** button (or the Enter key). In the lower

---

* Indeed, a common source of trouble for beginning (and not-so-beginning) UserTalk programmers is accidentally writing lines like the second one, expecting that this will cause an action because it is supposedly a verb call, when in fact it is nothing of the sort.

part of the window, Frontier responds: 7. You typed a legal UserTalk expression; Frontier evaluated it and told you its value. Now select and delete the contents of the main part of the window, and type instead: `workspace.x=8`. Frontier responds: 8. The assignment command itself has a value—the value of its right-hand side—so Frontier evaluated it and told you that value. But the evaluation of an assignment *also* causes the assignment to be performed. Examine the `workspace` table: you will see that it now has an entry called `x`, and that `x`'s value, sure enough, is 8.

There is one more thing about assignment that is useful to understand right from the start. At every moment, a variable's value has a particular type, and the set of legal types, UserTalk's *datatypes*, is well defined. But a variable is not glued to its type; it doesn't mind changing it. The following is a legal UserTalk script:

```
s = "Hello, World!"
s = 7
```

Frontier is not at all upset that the variable `s` starts life with a string value and is then assigned a number value; it happily turns `s` into a number. We'll talk much more about the datatypes in Chapter 10, *Datatypes*.

Scripts as Database Entries

Did you, observant reader, notice in Example 4-1 a certain similarity between the name of the verb `dialog.notify` and the way one refers to an entry in the database? Both `workspace.x` (for instance) and `dialog.notify` use a dot-notation, the names being composed of elements with a period between them.

The vast majority of UserTalk's built-in verbs are named in this way (though some just have single-element names). This is a great convenience, as it makes the verbs easier to name uniquely and to remember. There are about three dozen pre-defined categories of verb, making up the first part of the name; within each category, the verbs are named appropriately to their function. For instance, `dialog.notify()` and `dialog.alert()` both put up dialogs; they differ in the look of their respective dialogs.

But the use of dot-notation in naming verbs is not merely a matter of convenience, nor is the use of dot-notation both in naming verbs and in referring to database entries a coincidence: all UserTalk verbs are in fact database entries. To see this, take a look at them: they're located mostly at `system.verbs.builtins`. That table, you will see, has a `dialog` entry, itself a table which does indeed contain a `notify` script and an `alert` script.* Other verbs, those with "simple" names, are entries in `system.verbs.globals` or `system.macintosh.globals`;

---

* The reason why a script can refer to the database entry `system.verbs.builtins.dialog.notify` simply as `dialog.notify()` is explained in Chapter 6, *Referring to Database Entries*.

and all verbs constituting the core of UserTalk's functionality are ultimately entries in `system.compiler.language.builtins` and `system.compiler.kernel`, where they are translated into tokens at compile time.

This helps to explain why it is not an error to leave the parentheses off a verb call. The expression:

```
dialog.notify
```

instead of:

```
dialog.notify()
```

refers, quite legally, to an entry in the database: namely **system.verbs.** **builtins.dialog.notify**. But since the parentheses are omitted, it fails to inform Frontier that what you want to do is not merely to talk about that database entry, but to execute the script that is located there. (Conversely, it *is* a runtime error to make a verb call to a database entry that is not a script object.)

Scripts that you create can be script objects in the database, too, just like built-in UserTalk verbs; then you can call them from other scripts just as you call built-in UserTalk verbs. From Frontier's point of view, there is no distinction between built-in verbs and verbs that you create by writing scripts. So when you write a script in UserTalk, you are effectively extending the UserTalk language.

Therefore, it is not entirely clear, with respect to the repertory of UserTalk verbs, just where the boundaries of the UserTalk language are! One tends to speak of "built-in verbs" on the one hand and "utility scripts" on the other, but the distinction is contingent and artificial. The only verbs that can be said to be *truly* "built-in" are those whose ultimate implementation is hard-coded into Frontier: you can spot these because all their scripts do is pass the call directly along to the "kernel." For instance, `dialog.hideItem()` works this way.

But `dialog.notify()` and `dialog.alert()` are mere scripts; their chief action is performed by calling a different verb, `card.run()`, which calls the kernel. Yet `dialog.notify()` and `dialog.alert()` are considered "built-in verbs," in the sense that they are a standard part of the UserTalk language. On the other hand, `suites.toys.readWholeFile()`, which reads a textfile from disk into a variable, is generally considered a "utility script," not a basic verb of the language, even though its implementation doesn't differ significantly in nature from that of `dialog.alert()`—and though it performs a valuable, even fundamental, function.

The truth is that the repertory and implementation of UserTalk verbs are moving targets. Historically, `dialog.alert()` was a "kernel" verb at one time, but it was rewritten as a script when a new method of constructing dialogs (MacBird) was integrated into Frontier. In just the opposite way, **string.iso8859encode()** was created as a script, and now it is a "kernel" verb (for speed). Thus, the official database, the "clean root" distributed each time a new version of Frontier comes out, constantly evolves.* Meanwhile, each user is evolving her or his "unofficial" parts of the database, adding utility scripts and other objects.

All UserTalk verbs, then, are script objects in the database. Any useful UserTalk program will probably contain some verb calls, so a UserTalk program can be envisioned as database script objects calling one another. The user is free, and encouraged, to extend the UserTalk verb repertory by writing scripts. The user is also able to explore the database, and so to examine the implementation of "built-in" verbs and utility scripts; this can be a way of learning more about the language, and a help in debugging scripts.

Because the "clean root" is constantly evolving, and because the distinction between built-in verbs and utility scripts is not clearcut, this book sets itself some arbitrary limits on what parts of the database to document. All verbs in `system` that are common currency are documented, except for `system.verbs.apps` which are specialized and mostly self-documenting (I do discuss this area in some detail, but without documenting every verb); many utility verbs from `suites` are also listed. Generally speaking, whenever valuable functionality is squirrelled away in the database in places where you would have had to explore systematically to discover it, I have tried to save you the trouble.

Scripts as Outlines

The window in which one edits a script object is nearly identical to that in which one edits an outline object; in essence, a script is a kind of outline. Frontier makes the outline integral to one's experience of the UserTalk language. This turns out to be hugely sensible, powerful, and convenient—indeed, the integration of outlining with script editing is one of Frontier's most characteristic and brilliant innovations.†

How do scripts relate to their outline milieu? First, outlines come in distinct lines. Therefore, each line of a script is considered a separate command.‡

---

* In fact, in Frontier 5.0, further evolution will probably falsify these very examples! `dialog.notify()` and `dialog.alert()` will probably become kernel verbs again, and `suites.toys.readWholeFile()` will probably be promoted to become a file verb at last.

† One wonders why outline editing has not been incorporated into other coding environments. There is no reason why a C program could not be edited as an outline in just the same way that a UserTalk program can be; and it would be much easier to edit a C program as an outline than as mere text, which is how most C editors treat it.

‡ In C and a number of other text-based languages, there has to be a command delimiter of some sort, usually a semicolon. A UserTalk script outline works more like BASIC, in which a new line indicates a new command.

Since a line of an outline is limited to 255 characters, very long lines of code, which are sometimes unavoidable (for instance, in a call to a verb that takes very many parameters) are impossible. To get around this, the backslash (\) is used as a line continuation indicator: a line of UserTalk ending with a backslash is considered to continue on the next line. This device is useful also just to improve readability of scripts, to get around the fact that lines do not "soft-wrap" in a script edit window.

Second, UserTalk takes advantage of the hierarchical nature of outlines by using subordination in its control-flow and definitional keyword constructs. Control-flow and definitional commands apply to the bundle that is subordinate to them.

This UserTalk script illustrates all the points just made.

Example 4-3. A greeting customized to the time of day

```
local (theDay, theMonth, theYear, theHour, theMin, theSec)
date.get (clock.now(), @theDay, @theMonth, \
    @theYear, @theHour, @theMin, @theSec)
if theHour >= 12
    dialog.notify ("Good afternoon or evening.")
else
    dialog.notify ("Good morning.")
```

The script has six commands: seven lines, but one of them is a command split across two lines, using the backslash as the line continuation character. Notice that when there is line continuation like this, indentation is *not* meaningful; the line which begins with @theYear could just as well be at the same level as the line which begins with date.get instead of being indented under it—I have indented merely for legibility.

On the other hand, the indentation of the line which follows the if line, and that of the line which follows the else line, *is* meaningful. After an if line, the way that Frontier knows what commands to execute if the condition is true is by looking for the bundle indented under that if line. Here, that bundle consists of just a single line, but it could consist of any number of lines. The same is true for else. In a parallel way, looping constructs such as loop, for, and while show what commands they apply to by use of an indented bundle.

Since indentation is meaningful in script outlines (except where a line is being continued), it is an error to use it in any other way. For example, this script will not compile:

```
s = "Hello, World!"
    dialog.notify(s)
```

However, indentation is so useful for organization and legibility that the UserTalk language provides a construct, the **bundle** keyword, one of whose purposes is to allow arbitrary indentation of lines.* The following script, though rather pointlessly organized, is legal.

Example 4-4. The bundle construct

```
s = "Hello, World!"
bundle
    dialog.notify(s)
```

If there were some good reason for wanting to indent the `dialog.notify` line, the **bundle** construct would be the way to do it.

Having scripts be outlines also makes UserTalk user-friendly, because it helps one write, edit, and read a script. An outline is easy to view: subordinate bundles may be collapsed or expanded, and the indentation visually reflects a script's structure. An outline is easy to reorganize: lines can be moved up and down, promoted and demoted, with keyboard commands or by dragging, and when a line that has a subordinate bundle is moved, the bundle moves with it. An outline is easy to navigate. It is easy to delete lines, to create and type new lines—and so on.

Scripts Outside of Outlines

Frontier does also provide for a text-based rendering of scripts, for use in non-outline environments. In this alternate rendering, the rules are very like those of C: indentation is meaningless, and so is lineation, mostly;† simply put, every subordinated bundle must be surrounded by curly braces, and every command that is not followed by a curly brace must be followed by a semicolon if another command follows.‡

Frontier provides splendid automatic conversion between the two formats. This makes it easy to transfer material in either direction between script windows and wptext windows (not to mention documents belonging to other programs, such as email, word processors, and even outliners).

---

* The second chief purpose of the **bundle** construct appears in Chapter 7, *The Scope of Variables and Handlers*.

† Lineation is meaningful in the text rendering of a script in the following senses: you basically cannot break lines within a literal string; and line-end marks the end of a comment.

‡ A right curly brace is followed by a semicolon, if another command follows. `else` is not preceded by a semicolon.

For example, given Example 4-3, if you click on the triangle of the `local` line (to select the whole line) and choose **Select All** and then **Copy** from the **Edit** menu, then go into a wptext edit window and **Paste**, the result will look something like this.*

Example 4-5. A text-based rendering of a script

```
local (theDay, theMonth, theYear, theHour, theMin, theSec);
date.get (clock.now(), @theDay, @theMonth,
@theYear, @theHour, @theMin, @theSec);
if theHour >= 12 {
    dialog.notify ("Good afternoon or evening.")}
else {
    dialog.notify ("Good morning.")}
```

Appropriate semicolons and curly braces have been automatically inserted. So have tab characters and return characters, to give legible lineation and indentation, but you could alter this without changing the script's meaning; for example, this would be an equivalent (but, to a human, largely illegible) script in a wptext window:

```
local (theDay, theMonth, theYear,
theHour, theMin, theSec); date.get
(clock.now(), @theDay, @theMonth,
@theYear, @theHour, @theMin, @theSec); if
        theHour >= 12 {dialog.notify
("Good afternoon or evening.")} else {
dialog.notify ("Good morning.")}
```

Automatic conversion works the other way, too. Go back to Example 4-5, rejoin the two `date.get` lines into one, select the whole script and **Copy** it, and then **Paste** into a script edit window. You get a legal outline version of the script again: the semicolons and curly braces are removed, and the tabs (spaces would have done just as well) are used as cues to decide how to indent each line.†

It is also permitted, and is sometimes considered good programming style, to use semicolons and curly braces even in a script edit window. Semicolons allow more than one command to occupy a line; curly braces can bring a bundle onto the same line as the command to which it is subordinate. For example, this script:

Example 4-6. Asking your name and greeting you

```
local (theReply)
theReply = user.name
if not dialog.ask("What is your name?", @theReply)
    return
dialog.notify ("Hello, " + theReply + "!")
```

---

* Frontier joins the two parts of the continued `date.get` line into one, so I have split them again for legibility. Note that you are permitted to run this script even from this "alien" environment; select it all and choose **Run Selection** from the **Main** menu. In this way you can confirm that it is still a legal script.

† Frontier also would not mind if the semicolons and curly braces weren't there; it would still use the tabs (or spaces) to get the indentation right.

can be legally rewritten like this:

```
local (theReply); theReply = user.name
if not dialog.ask("What is your name?", @theReply) {return}
dialog.notify ("Hello, " + theReply + "!")
```

The second format is not uncommon. The first two lines of the first version are made one in the second, because they are short and conceptually very closely related. The content of the bundle subordinate to the `if` is extremely short, so it is convenient to bring it up onto the same line.*

It is permitted to use "otiose" semicolons at line-end or before a closing (but not an opening) curly brace, and to let curly braces take the place of indentation (the opening curly brace must appear on the same line as the material to which the bundle is subordinate); so, it is legal to rewrite Example 4-6 like this:

```
local (theReply); theReply = user.name;
if not dialog.ask("What is your name?", @theReply) {
return;};
dialog.notify ("Hello, " + theReply + "!");
```

But no one would actually ever use such hideous, pointless style.

How Scripts Are Represented in This Book

This book uses a hybrid formatting for scripts. The problem is to represent how a script looks in a script edit window. This could be done by means of a screenshot, as shown in Figure 4-1.

```
local (theDay, theMonth, theYear, theHour, theMin, theSec)
date.get (clock.now(), @theDay, @theMonth, \
    @theYear, @theHour, @theMin, @theSec)
if theHour >= 12
    dialog.notify ("Good afternoon or evening.")
else
    dialog.notify ("Good morning.")
```

Figure 4-1. How scripts really look

But the use of a graphic for every script in the book would be inconvenient for author, publisher, and reader. On the other hand, the text-based rendering:

```
local (theDay, theMonth, theYear, theHour, theMin, theSec);
date.get (clock.now(), @theDay, @theMonth, \
    @theYear, @theHour, @theMin, @theSec);
```

---

* It is not legal to bring the content of an `else` bundle onto the same line in curly braces unless both the `if` and the `else` are on the same line. This may be a bug.

```
if theHour >= 12 {
    dialog.notify ("Good afternoon or evening.")}
else {
    dialog.notify ("Good morning.")}
```

introduces semicolons and curly braces that are distracting and unnecessary.

So I have reproduced the outline's lineation and indentation *without* introducing semicolons and curly braces. This is not an automatic conversion; I wrote a utility script to do it. For the curious, and without explanation, here it is.[*]

Example 4-7. How the scripts in this book were copied

```
on obtainLineText()
    if script.isComment()
        return "« " + op.getLineText() + cr
    else
        return op.getLineText() + cr
op.firstsummit()
op.fullexpand()
local (s)
s = obtainLineText()
while op.go (flatdown, 1)
    local (t = op.level ())
    while --t {s = s + "    "}
    s = s + obtainLineText()
clipboard.putvalue (s)
```

Comments in Scripts

A *comment* is text meant to be seen by someone reading a script, but ignored by Frontier during execution. In a UserTalk script window, comments are of two types: either an entire line is a comment, or a line is a command but a comment is appended to it. The script editor handles these comment types in different ways.

To turn an existing command line into a comment, select the line or within its text, and choose **Toggle Comment** from the **Script** menu. This works the other way too: selecting a line or within a line that is a comment, and then choosing **Toggle Comment**, makes it a command.

To cause a new line to be a comment instead of a command as you're creating it, create the line by hitting Shift-Return instead of simply Return.

---

[*] This is actually a somewhat simplified version of the script that was used. It works for most of the scripts in this book, but would not work for absolutely any script, because of the way comments work. The real script is given as Example 18-4.

The status of a line is indicated by the symbol that precedes it: a line that is a command has a triangle; a line that is a comment has a "chevron" (strictly, a left guillemot). Figure 4-2 shows a screenshot illustrating both.

```
local (theReply)
default to the name stored in the database
theReply = user.name
present a dialog asking the user's name
   and quit if the user cancels
if not dialog.ask("What is your name?", @theReply)
   return
user said OK; put up a dialog saying hi, using the name given
dialog.notify ("Hello, " + theReply + "!")
```

Figure 4-2. Comments

Observe that one of the comments has a line subordinate to it, and that this line is also a comment. A line subordinate to a comment is automatically rendered a comment. If a command line is moved into subordination to a comment, or created in subordination to a comment, it will be changed into a comment. However, it is not a full-fledged comment; Frontier remembers that it is really a command line. This allows the line to be automatically changed back to a command line if it is moved out from comment subordination. This automatic comment subordination is a handy feature: multiple comment lines can be collapsed into one, and it is easy to comment out a bundle of commands temporarily by subordinating it to a comment.

To append a comment to a command, type Option-\ (backslash); this types, at the insertion point, the actual left guillemot character, as opposed to representing it graphically in place of the line's triangle.

One use of appended comments is to describe in English what happens in subordinate lines, so that if those lines are collapsed one knows what they do without expanding and reading them. Thus this:

```
if not dialog.ask("What is your name?", @theReply) « quit if cancel
   return
```

collapses to this:

```
► if not dialog.ask("What is your name?", @theReply) « quit if cancel
```

This book's hybrid notation deals with comment lines just like Frontier's own cut-and-paste conversion: the graphically produced guillemot is rendered by an ordinary guillemot character, as follows:

```
local (theReply)
« default to the name stored in the database
theReply = user.name
```

5

Handlers and Parameters

We have seen that creating a script object creates a UserTalk verb. If a script object at `workspace.sayHi` consists of the script presented as Example 4-3, then it is possible to call it as a verb from any other script:

```
workspace.sayHi()
```

If you execute a script that contains this line, then at this point the script at `workspace.sayHi` will execute, much as it would if you opened `workspace.sayHi`'s edit window and pressed its **Run** button.

But there is a complication. Scripts sometimes do not function correctly as verbs unless they contain an *eponymous handler*. And scripts called with a verb call do not always function the same way as when their **Run** button is pressed.

This chapter explains these important matters, and describes in detail the syntax of verb calls.

What Is an Eponymous Handler?

Let's examine this term one word at a time.

A *handler* is a verb definition: it tells the verb's name, how many parameters it expects, and what commands it consists of. Syntactically, a handler is a control structure introduced by the keyword on. This keyword is followed by the name of the verb that the handler defines, plus parentheses containing a comma-delimited list of variables to receive each of the verb's parameters (if the verb takes no parameters, the parentheses will be empty). The bundle subordinate to the on line is the sequence of commands to be executed when the verb is called.

For instance, here is our Example 4-3 routine rewritten as a handler taking no parameters.

Example 5-1. A handler

```
on sayHi ()
    local (theDay, theMonth, theYear, theHour, theMin, theSec)
    date.get (clock.now(), @theDay, @theMonth, \
        @theYear, @theHour, @theMin, @theSec)
    if theHour >= 12
        dialog.notify ("Good afternoon or evening.")
    else
        dialog.notify ("Good morning.")
```

An *eponymous* handler is a handler whose **on** line is at summit level, and whose name is the same as the last element of the name of the script object in which it lives. For instance, if the Example 5-1 handler lives in a script object at **workspace.sayHi** or at **people.man.sayHi** or in an entry called **sayHi** anywhere else in the database, and if the handler is not subordinate to anything else in the script, then it is that script object's eponymous handler.

The Handler Rule

We are now ready to enunciate the entirety of the "Handler Rule."

1. A script object, in order to be called as a verb from a script, must contain an eponymous handler.[*]

2. Despite rule 1, a script object that lacks an eponymous handler can still be called as a verb from a script, provided the verb takes no parameters.

3. When a script is called as a verb, then:

   a. If it contains an eponymous handler, the eponymous handler's bundle is executed—and anything else in the script is ignored.

   b. If it doesn't contain an eponymous handler, the commands at summit level are executed.

4. When a script edit window's **Run** button is pressed, the commands at its summit level are executed.

Let's consider some implications of this. First, compare Example 4-3 and Example 5-1. Imagine that one or the other of these is the complete content of **workspace.sayHi**. So in the former case, **workspace.sayHi** has no eponymous handler; in the latter, it does. Then, either way, we can call **workspace.sayHi()**—in the case of Example 4-3, this is because there are no

---

[*] For this reason, when you choose **New Script** from the **Table** menu, Frontier creates the beginning of an eponymous handler for you.

parameters (rule 2)—and it will execute. Example 4-3 will execute according to rule 3b. Example 5-1 will execute according to rule 3a.

Now suppose `workspace.sayHi` consists of Example 4-3. If we press its edit window's **Run** button, it executes just as if it had been called with a verb call (rule 4).

But suppose `workspace.sayHi` consists of Example 5-1. If we press its edit window's **Run** button, nothing happens!* That's because rule 4 says that Frontier will execute the script's summit-level commands; but the only summit-level command here is a handler, and a handler takes no action: it's just a definition. The commands in a handler's bundle execute only when the handler is called. So, if we want pushing the **Run** button to perform the action defined by the eponymous handler, we must include, after the eponymous handler (at summit level, not as part of the handler), a call to the handler.

We can write such a call in one of two ways. We can, if we wish, call the script object itself. `workspace.sayHi` would then look like the following example.

Example 5-2. Calling the eponymous handler as a script

```
on sayHi ()
    local (theDay, theMonth, theYear, theHour, theMin, theSec)
    date.get (clock.now(), @theDay, @theMonth, \
        @theYear, @theHour, @theMin, @theSec)
    if theHour >= 12
        dialog.notify ("Good afternoon or evening.")
    else
        dialog.notify ("Good morning.")
workspace.sayHi()
```

The second way to call the `sayHi()` handler is directly, treating `sayHi` not as a script object in the database but as a "local" handler (discussed in more detail in Chapter 7, *The Scope of Variables and Handlers*). This is done by using just the handler's name, not a path.

Example 5-3. Calling the eponymous handler as a local handler

```
on sayHi ()
    local (theDay, theMonth, theYear, theHour, theMin, theSec)
    date.get (clock.now(), @theDay, @theMonth, \
        @theYear, @theHour, @theMin, @theSec)
    if theHour >= 12
        dialog.notify ("Good afternoon or evening.")
    else
        dialog.notify ("Good morning.")
sayHi()
```

---

* Except that Frontier, figuring that our pressing the **Run** button in this situation probably means we don't understand the Handler Rule, will put up a warning dialog.

This is legal because the sayHi() handler is "visible" directly to the other parts of the same script, in accordance with the rules of scope. Such, in fact, is the most common way of constructing a script that contains an eponymous handler. Either way, the script now contains an eponymous handler and, nevertheless, can be tested with the **Run** button.

Notice that pressing the **Run** button in Example 5-2 or Example 5-3 does *not* cause some sort of vicious loop! Consider Example 5-2 first.

We press the **Run** button, and Frontier starts executing all summit-level commands in order (rule 4). The first of these, the handler, is just a definition, so Frontier takes note of the definition and goes on to the next command, which is the call work-space.sayHi(). Frontier now goes out to the database, finds the object workspace.sayHi, discovers that it is indeed a script object, discovers further that it possesses an eponymous handler, and (rule 3a) executes *only* the eponymous handler's bundle—the rest of the script, consisting of the call workspace.sayHi(), is ignored, and therefore execution comes to an end with no circularity.

Now consider Example 5-3. We press the **Run** button, and Frontier starts executing all summit-level commands in order (rule 4). The first of these, the handler, is just a definition, so Frontier takes note of the definition and goes on to the next command, which is the call sayHi(). Obedient to the call, Frontier executes the sayHi() handler's bundle. This completes the execution of the last (and only) action line of the script, and execution comes to an end.

To be sure, it is *possible* for a script to call itself in a vicious loop. If, for instance, workspace.sayHi contained no eponymous handler, but did contain a line calling workspace.sayHi()—which is legal, since sayHi() takes no parameters—we would loop forever, whether we pressed the **Run** button or called it from another script:[*]

```
local (theDay, theMonth, theYear, theHour, theMin, theSec)
date.get (clock.now(), @theDay, @theMonth, \
    @theYear, @theHour, @theMin, @theSec)
if theHour >= 12
    dialog.notify ("Good afternoon or evening.")
else
    dialog.notify ("Good morning.")
workspace.sayHi ()
```

The moral is: when in doubt, include an eponymous handler, to prevent such loops!

---

[*] Actually, we would not loop literally forever; we would loop 40 times and then hit Frontier's stack limit on recursion (see Chapter 7, *The Scope of Variables and Handlers*).

Stubs

Taken together, rules 3a and 4 of the Handler Rule enable the following technique for writing and testing scripts.

Suppose `sayHiTo()` is the eponymous handler of `workspace.sayHiTo`, and that it is like `sayHi()`, but takes a parameter—for instance, the parameter is to be someone's name, and our dialog will now greet that person by name. (Rule 2 does not apply, so, by rule 1, our script *must* contain an eponymous handler.) The script might go like the following example.

Example 5-4. A handler with a parameter

```
on sayHiTo (theName)
    local (theDay, theMonth, theYear, theHour, theMin, theSec)
    date.get (clock.now(), @theDay, @theMonth, \
        @theYear, @theHour, @theMin, @theSec)
    if theHour >= 12
        dialog.notify ("Good afternoon or evening, " + theName + ".")
    else
        dialog.notify ("Good morning, " + theName + ".")
```

By rule 3a, we can call this script from another script (providing a parameter in the call), like this:

```
workspace.sayHiTo ("Matt")
```

But, as we have already seen, we cannot run our script by pressing the **Run** button, because it contains no executable commands; to do so, we must introduce a call to `sayHiTo()`, after the handler, at summit level.

Well, as long as we are doing this, we may as well take the opportunity to *test* `sayHiTo()`. We might, for example, add several calls to `sayHiTo()`, with different parameters, just to make sure they all work. So, we might end up with something like the following example.

Example 5-5. An eponymous handler and a stub

```
on sayHiTo (theName)
    local (theDay, theMonth, theYear, theHour, theMin, theSec)
    date.get (clock.now(), @theDay, @theMonth, \
        @theYear, @theHour, @theMin, @theSec)
    if theHour >= 12
        dialog.notify ("Good afternoon or evening, " + theName + ".")
    else
        dialog.notify ("Good morning, " + theName + ".")
sayHiTo ("Santa Claus")
sayHiTo ("Totally B. Fuddled")
sayHiTo ("Mom")
```

If we press the **Run** button, we get the dialog three times in a row, with each of the names substituted (by rule 4). But if, as before, we call:

```
workspace.sayHiTo ("Matt")
```

from another script, we don't get those three dialogs: we just get one dialog, with the name "Matt" appearing in it (by rule 3a).

Our three calls to `sayHiTo()` within `workspace.sayHiTo` are a *stub*, executed during testing when we press the **Run** button, but not when we really *use* the verb by calling it. It is very common to write scripts consisting of an eponymous handler and a stub.

Call Results

A verb call doesn't just cause the verb to execute; it also has a value. Every verb generates a result as the final act of its execution. This result becomes the value of the verb call, within the calling script.

The script in which the verb call appears may or may not wish to use the verb call's value in some way. If it does, it must capture it somehow—for example, by assigning it to a variable. Otherwise, the value is simply thrown away, without penalty.*

Consider, for example, this simple script:

```
dialog.notify ("Hello, World!")
```

The verb `dialog.notify()` performs an action: it puts up a dialog on the screen. This really is all we want from it when we call it. But it also has a result. In this case, we are not concerned with the result, so we have done nothing to capture it. But if we *were* concerned with the result, we could capture it—by treating the call `dialog.notify()` as a value. For example, we could assign that value to a variable:

```
whatResult = dialog.notify ("Hello, World!")
```

This script runs `dialog.notify()` (causing the dialog to appear on the screen), and also gets its result, and assigns that result to the variable `whatResult`. Even so, *we* still don't know the result, because we don't know what `whatResult` is. To find out, let's display it—for example, by handing it as a parameter to `dialog.notify()`:

```
whatResult = dialog.notify ("Hello, World!")
dialog.notify (whatResult)
```

---

* Thus, as in C (or Lisp), a UserTalk verb can serve as what Pascal calls a procedure and/or as what Pascal calls a function.

The second dialog that appears says **true**; that's the result of the first call to `dialog.notify()`. This is not a particularly interesting result, to be sure. But sometimes, the result of a verb call *is* interesting; in fact, sometimes learning the result of a verb call is the whole point of calling that verb.

If a verb does not specify what its own result is to be, that result can be somewhat unpredictable; it is often the result of evaluating the last executed line of the script, but one cannot rely on this. So if a verb wants to specify its own result, it should do so explicitly. This is done with the `return` keyword. Actually, the `return` keyword has two effects: it states what the verb's result should be, and it terminates execution of the verb.* (UserTalk has no separate `halt` or `exit` keyword.)

The syntax of `return` is as follows. The value that the verb's result is to have appears after the word `return`, on the same line with it. This value is often surrounded by parentheses as a matter of style, but this is not required. If the word `return` appears by itself with no value, **true** or **false** is returned (probably **true** if we've gotten this far in the execution of the verb); this syntax is generally used simply in order to halt execution of the verb before the last command is reached.†

Thus, to make a verb that reports the hypotenuse of a right triangle given the two sides as parameters, you might say (supposing the script was at **workspace.hypotenuse**):

```
on hypotenuse (a, b)
    return trigCmd.sqrt ((a * a) + (b * b))
```

It would then be possible to use the value of a call to this verb as part of a calculation in some other script:

```
x = 3 + workspace.hypotenuse (8, 9)
« x is now 15.04159458
```

Line Evaluation

Given any line of UserTalk in a script edit window, it is possible to evaluate (and, if the line is an executable command, to execute) just that line. To do so, select the line or click within the line, and choose **Run Selection** from the **Main** menu. This causes Frontier to evaluate the line, and to insert its value as a comment subordinate to the line.

---

* "The verb" is the deepest handler in which the `return` appears, or the script itself if the `return` is not inside a local handler. If handler B appears as a local handler inside handler A, you cannot use a `return` inside handler B to stop execution of handler A as well. One way to accomplish this might be to pass an error up from B and trap it in A. Error-trapping is discussed in Chapter 12, *Control Structures.*

† Of course, when the last command is reached, the verb returns automatically; if it is not desired to return a particular value, there is generally no need to include a `return` as the last line of a handler or script. It can't hurt, though, to do so.

The technique is possible only if the line is understandable as a complete and independent script. For instance, in Example 5-5, selecting the line:

```
sayHiTo ("Mom")
```

and choosing **Run Selection** yields an error: Frontier has never heard of `sayHiTo`, because the handler defining it was never executed. On the other hand, if we change the line to:

```
workspace.sayHiTo ("Mom")
```

assuming the script lives in `workspace.sayHiTo`, selecting the line and choosing **Run Selection** does work—the dialog appears, and the value `true` is inserted as a comment subordinate to the line.

Line evaluation is a common way of learning the result of a verb call. For example, we could rewrite `workspace.hypotenuse` as an eponymous handler and a stub, as follows:

```
on hypotenuse (a, b)
    return trigCmd.sqrt ((a * a) + (b * b))
workspace.hypotenuse (3, 4)
```

Selecting the last line and choosing **Run Selection** causes the value `5.0` to be inserted as a subordinate comment; since this is what we expect `hypotenuse()` to return given 3 and 4 as parameters, it appears that the handler is correctly written.

In this book I shall often use the convention of a subordinate comment to present the result of a verb call.

Parameters

Example 5-4 illustrates the basic syntax for a handler defining a verb that takes a parameter. The name `theName` is completely arbitrary; any legal variable name will do. When a handler is actually called for execution, the variables named in its parameter list are created, and values are assigned them, based on evaluation of the corresponding expressions used in the call.

Example 5-6. Verb call with a parameter

```
myName = "Matt"
workspace.sayHiTo (myName)
```

When we execute this script, in the second line, the variable `myName` is evaluated; and its value, which is the string `"Matt"`, is handed to `sayHiTo()` to put into its variable called `theName`. Then, from `sayHiTo()`'s point of view, as it starts executing, it finds that a variable `theName` exists, whose value has been initialized to the string `"Matt"`.[*]

When execution of the `sayHiTo()` bundle is over, the variable `theName` that it was using goes out of existence. This is because parameter variables in a handler are local. (Local variables and scoping are dealt with in Chapter 7.)

It is an error to call a verb with a number of parameters different from the number of parameter variables in the handler definition. For example, given Example 5-4, it is an error to say:

```
workspace.sayHiTo ()
```

It is also an error to say:

```
workspace.sayHiTo ("Matt", "Mom")
```

Nonetheless, UserTalk provides a certain degree of flexibility as to the number of parameters in a verb call, as we shall see shortly.

Pass by Value

Let's consider further the relationship between the caller script of Example 5-6 and the handler in Example 5-4. The variable `theName` which appears in the latter is, while it exists, like any other variable. This means that its value can be changed, for example, by assigning it a new value. Nevertheless, such a change will have *no effect* upon the variable `myName` used in the calling script. This is because the value of `myName` was copied in order to initialize the value of `theName`. The two variables, `theName` and `myName`, are separate entities.

In technical terms: all parameters, without exception, are passed by value in User-Talk.[†] We shall see later on (Chapter 8, *Addresses*), though, that it is possible to pass a parameter in such a way that a variable belonging to the calling script *can* be changed by the called handler. We shall also learn (Chapter 7) that the scoping rules sometimes allow a variable belonging to the caller to be changed by the called handler, without being passed as a parameter at all.

---

[*] A distinction of nomenclature is sometimes drawn between parameters from the handler's point of view, which are variables created and given values when the handler is executed, and parameters from the caller's point of view, which are the values to be assigned to the handler's parameter variables. Sometimes "parameter" versus "argument" is used to make this distinction. But "parameter" has become common for both in UserTalk; when the context does not make the meaning obvious, I will speak of the handler's "parameter variables."

[†] Contrast C, where parameters are passed by value except for arrays, a feature which has always confused the heck out of me.

Multiple Parameters

If a verb takes more than one parameter, those parameters are separated by commas both in the handler definition and in the call.

Which parameter value in the call is assigned to which parameter variable in the handler? By default, values are assigned to the parameter variables based on the order in which they are provided in the call.

Thus, in this example, a is initialized to 1, b is initialized to 5, and c is initialized to 32 (as the result demonstrates).

Example 5-7. Default parameter passing

```
on addThreeAndMultiplyToo (a, b, c)
    return (a + (2 * b) + (3 * c))
addThreeAndMultiplyToo (1, 5, 32)
    « 107
```

Parameters can be provided in any order in a call, though, by specifying for each value what parameter variable it is to be assigned to. The parameter variable's name (from the handler definition) is followed by a colon and the value to be assigned to it. Thus, in this example, c is initialized to 1, a is initialized to 5, and b is initialized to 32.

Example 5-8. Named parameter passing

```
on addThreeAndMultiplyToo (a, b, c)
    return (a + (2 * b) + (3 * c))
addThreeAndMultiplyToo (c:1, a:5, b:32)
    « 72
```

The two syntaxes can be combined, with some parameter values named and others not. All the unnamed values must precede all the named values, and are assigned according to the default parameter passing rule. The following, therefore, is legal:

```
addThreeAndMultiplyToo (5, c:1, b:32)
    « 72
```

WARNING   Throughout this book, when the syntax of built-in UserTalk verbs is summarized, the parameter names used are *not* the actual names used in the verb's definition. Instead, descriptive names are given, to make the purpose of each parameter more evident. If you want to call a UserTalk verb by naming a parameter, you should look up the verb in the database to learn the parameter's actual name.

Omitted Parameters

A handler's bundle cannot execute without values for every parameter variable. However, some or all parameter variables can be assigned default values within the on line's parameter variable list. Values for any of those parameters can then be omitted from the call; the default values will be used for any parameters whose values are not specified in the call. But if a parameter's value *is* specified in the call, the specified value overrides the default value (if any).

Thanks to the calling syntax rules just described, Frontier will always be able to figure out which passed value goes to which parameter variable, and hence which values have been omitted.

An illustration is worth a thousand words.

Example 5-9. Omitted parameters

```
on addThree (a = 1, b = 2, c = 3)
    return (a + b + c)
addThree (10, 20, 30)
    « 60
addThree ()
    « 6
addThree (10)
    « 15
addThree (10, 20)
    « 33
addThree (c: 30)
    « 33
addThree (c:30, b:20)
    « 51
addThree (10, c:30)
    « 42
```

Indefinite Number of Parameters

UserTalk does not permit a handler to accept an indefinite number of parameters. There is an indirect mechanism for this, though, by way of the list datatype. Lists are discussed in Chapters 10 and 11; however, here is an illustration (a list is delimited by curly braces). The handler expects one parameter, but that parameter is a list which can consist of any number of items:

```
on addAny (theList)
    theTotal = 0
    for x in theList
        theTotal = theTotal + x
    return theTotal
addAny ({ 1, 5, 32, 4, 97 })
    « 139
```

Referring to Database Entries

The database is a major resource of Frontier, and just about any UserTalk script will likely refer to some entry in it.

An entry in the database is a lot like a variable. It has a name, and it has a value, and you are free to get or set that value. Operations that can be performed with variables can all be performed with database entries, and vice versa. Indeed, a database entry really *is* a kind of variable. Nevertheless, there are differences, and it is often useful to draw a distinction between the two. When we need to lump them together, we can say "database entry or variable," or simply *object*.

This chapter describes the differences between variables and database entries, and explains how to refer to database entries with UserTalk.

How Variables and Database Entries Differ

In Example 4-1, **s** is a variable:

```
s = "Hello, World!"
dialog.notify(s)
```

In the following script, **workspace.s** is an entry in the database.

Example 6-1. Assignment into the database

```
workspace.s = "Hello, World!"
dialog.notify(workspace.s)
```

Running each of these two scripts would appear to give just the same result—identical dialogs appear—but there is an important difference. After running the second script, examine the **workspace** table and you will find that an entry called s has been created within it, and that its value is the string `"Hello, World!"` This highlights one important difference between variables and database entries: Database entries are *persistent*, meaning that they continue to exist, maintaining their values, even when no script is running.* Variables, on the other hand, are temporary; in the first script in this chapter, the variable s is destroyed when the script finishes executing.

Another difference between variables and database entries is that database entries are *global*, meaning that they can be referred to from anywhere. There is only one `workspace.s` (if there is any at all), and any script can speak of it unambiguously. Hence it is possible for one script to assign a value to `workspace.s` and another to use that value subsequently. Indeed, this is an important technique for making a value available amongst scripts.

You might wonder: If variables and database entries are alike, how does Frontier know which is being referred to by a name? For one thing, in our previous examples, at least, the variable name contained no dot, and the database entry name did. Generally speaking, variable names cannot contain a dot (they can be forced to do so, but it's a poor idea), so there's a clue Frontier uses. But it is not the only clue. The question is a major one, and is discussed in detail, partly in this chapter, partly in the next.

Being Careful with the Database

Since a script is allowed to change the value of *any* entry in the database, you should proceed with great caution—the same caution with which you would consider changing a value in the database manually, but even more so, because a script, once set going, is out of the user's immediate control. (On the other hand, you may *get* values from the database without hesitating; we saw an example of this in Example 4-6, where the value stored at `user.name` was used to supply a default in a dialog.)

Overly drastic accidental damage to the database is checked, to some degree, by limits on what Frontier is willing to do to the database in response to an assignment command. In Example 6-1, a database entry `workspace.s` was created in response to an assignment statement; implicit creation is an important feature of assignment. But if we had said:

```
workspace.ourSubtable.s = "Hello, World!"
```

---

* This, of course, does not mean that a database entry, once created, lives forever: persistent does not mean permanent. You can delete a database entry manually, and the same action can be performed from within a script.

and if the `workspace` table did not contain a subtable called `ourSubtable`, a runtime error would have resulted: Frontier refuses to create a table implicitly like this. (Of course, there are verbs to create a table explicitly.)

Frontier will also balk at an attempt to replace an existing non-scalar database entry or variable (for instance, a script object, a wptext object, or a table) with a scalar value (such as 7 or `"Hello"`).

Still, Frontier will happily, in response to an assignment command, replace one scalar with another, or a scalar with a non-scalar, or a non-scalar with a non-scalar, altering the value and the datatype of the object to obey the assignment. So assignment has plenty of power to do damage. And later, we will meet verbs that can overcome all these safety limitations. So be careful what you say!

Calculated Names

In referring to a database entry or variable from a script, it is sometimes necessary to calculate part or all of its name at runtime. Consider, for instance, the business of referring to entries in your part of the `people` table. When Frontier is first obtained and started up, a dialog asks you for your initials. A subtable is then automatically created inside the `people` table; the name of the subtable is those initials.

Suppose, now, that you wish, from inside a script, to refer to this subtable. How is this to be done, when you don't know its name? Of course, I know that *my* subtable is called `people.man`; but I don't know what *yours* is called, so how can I write a script to give to you that can reliably refer to the corresponding subtable in *your* database? The solution is to use the fact that the user's initials are also stored as an entry in the database, at `user.initials`. On my machine, the value of `user.initials` is `"man"`; I don't know what it is on yours, but my script can find out.

But how is the script to make use of this knowledge? It cannot be right to speak of `people.user.initials`; this is a reference to an entry called `initials` in a table called `user` in the `people` table. What we want is to construct, *at runtime*, a database object reference, composed of `people` followed by the *value* of `user.initials`. This is signified by enclosing the part of the name to be calculated in square brackets, like this:

```
people.[user.initials]
```

This notation forces pre-evaluation of the expression in square brackets, and then uses that value in constructing an object reference. What goes into the square brackets can be *any* evaluatable UserTalk expression. To illustrate further, in this script:

```
s = "Hello"
workspace.[s] = "Hi"
```

the variable `s` is evaluated so that a database entry `workspace.Hello` is created and assigned the value `"Hi"`. In this script:

```
workspace.[date.daystring(date.dayofweek(clock.now()))] = 3
```

three built-in Frontier verbs are called, so that a database entry `workspace.Saturday` (because I'm writing this on a Saturday) is created and assigned the value 3.

For certain database entries, this square-bracket syntax is actually required in order to reference them at all. That's because these database entries have names that are in some way nonstandard. A standard database entry name consists entirely of alphanumeric characters (an underscore is also permitted), beginning with a letter. It is permissible to give a database entry a name that does not conform to this rule, but then one cannot refer to it normally; for example, one could manually create an entry in the `workspace` table called 1997, but referring to it as:

```
workspace.1997
```

generates a syntax error. The solution is a trick: specify 1997 in the reference by using the string literal `"1997"` and forcing its evaluation with square brackets, as follows:

```
workspace.["1997"]
```

Because a number will be coerced to a string in this kind of expression, it is also possible in this case to say `workspace.[1997]`, without the quotation marks.*

The same notation also handles the situation where a name contains spaces or other punctuation:

```
workspace.["my workspace entry"]
```

Also, this notation is the only way to refer to a name that is a UserTalk constant or keyword. UserTalk constants and keywords are evaluated before anything else; so if you tried to speak of an entry called `if` in the `workspace` table as `workspace.if`, the `if` would be seen as the keyword `if` and a syntax error would result. However, `workspace.["if"]` refers successfully to such an entry. This is discussed further in Chapter 9, *Special Evaluation*.

---

* It is also permissible to omit the preceding dot; thus, `workspace["1997"]` would work just as well. But one cannot use `workspace[1997]` in this case, because it has another meaning. These two notations are not object references, but array operations, which are dealt with in Chapter 11, *Arrays*.

All we have said here of database entry names is equally true of variable names. For instance, the following script assigns the value "Fun!" to a local variable route66 by constructing the variable's name:

```
local (route66)
["route" + (33 + 33)] = "Fun!"
```

There is also another, entirely different mechanism for calculating object names: this involves constructing and dereferencing a string, which actually belongs to the technique of referring to objects not by name, but by address. This is discussed in Chapter 8, *Addresses* and Chapter 10, *Datatypes*.

Partial References

To save programmers from the clumsy, error-prone tedium of having to use a full object reference to access any item in the database, UserTalk permits the use of *partial references* for entries in the database. In a partial reference, the first element or elements of the name are omitted. To understand how to use partial references, you need to know the rules by which Frontier resolves a database entry name.

In what follows we pretend that there are no variables—just database entries. The way the presence of variables complicates the picture is discussed in the next chapter.

The Search Order

When Frontier sees an object reference in a UserTalk expression, it tries to interpret it by looking for the object in a series of set locations in the database, as follows:

1. The database itself is called root, and so the name of every item in the database theoretically begins with root, but it is always permitted to omit this. If an object reference *does* begin with root, Frontier sees the object reference as *not* being a partial reference. If an object reference does *not* begin with root, Frontier initially tries to resolve it by prefixing root to it.

So, for example, a reference suites is taken to mean root.suites (which exists); a reference root.workspace is taken to mean root.workspace (which exists); a reference workspace.veeblefetzer is taken to mean root.workspace.veeblefetzer (which, I believe it is safe to assume, does not exist).

2. Frontier next tries to resolve the reference by consulting the list of *search paths*. This list is maintained at system.misc.paths; if you look at the table, which is reproduced as Table 6-1, you can interpret it by mentally removing

the initial @ from each of the values (not the names).* Frontier runs down the list from top to bottom, prefixing each search path in turn to the object reference (also prefixing `root` if the name does not now begin with `root`).

Table 6-1. Search paths

Name	Value
path00	@people.man
path01	@system.verbs.globals
path02	@system.macintosh.globals
path03	@system.verbs.builtins
path04	@system.compiler.["kernel"].lang
path05	@system.macintosh
path06	@system.compiler.["kernel"]
path07	@system.verbs.constants
path08	@system.macintosh.constants
path09	@root.suites
path10	@system.verbs.apps
path11	@root.system
path12	@system.extensions
path13	@system.verbs.colors

So, for example, a reference `long` is resolved as follows. First (step 1), Frontier finds that `root.long` does not exist; then (step 2), it finds that `root.people.[user.initials].long` does not exist; then (continuing with step 2) that `root.system.verbs.globals.long` does exist, and that's the end of the search.

How Frontier Behaves

We now understand the order in which Frontier searches. But what is Frontier's behavior during the search? When does it decide to stop the search? What constitutes failure to resolve the reference, and what constitutes success?

Distinguish two cases, having to do with the form in which the original object reference is given (that is, what *you* originally said, not what variations Frontier may construct on the name during the search process). Let's call them the `x` form and `x.y` form. The `x` form is where you provide only a single element, such as `long`. The `x.y` form is where you provide more than one element; everything but

---

* The value of the first entry in the table depends upon the user's initials, and was created when Frontier was first obtained and initialized, so mine reads @people.man, but yours probably doesn't. In any case, the path leads to the people.[user.initials] table.

the last element is x. So, if you say workspace.myTable.myEntry, then work-space.myTable is x, and myEntry is y.

1. If the original reference is of the x form, then Frontier is happy to *get* a value or *modify* an existing value, but under no circumstances will it *create* a new database entry. This is logical, because you haven't sufficiently specified where a new entry would go. Therefore, if the search process constructs the name of an object that exists, the search stops: the name has been resolved. Otherwise the search fails after all the search paths are exhausted.

2. If the original reference is of the x.y form, then Frontier takes the first element of x and looks in each search path for a table of that name. If it never finds one, the search fails after all the search paths are exhausted. If it does find one, *the search stops,*[*] the name is reassembled, and the following tests are applied until one of them is satisfied:

   a. If x.y exists, the name has been resolved.

   b. If the whole of x does not exist, the search fails.

   c. If x exists, and x is not a table, the search fails.

   d. If x exists and is a table, then if the original command was to assign a value to y, y is created and the value is assigned.

   e. If x exists and is a table, but the original command was to obtain a value from y, the search fails because y does not exist.

If the search fails, a runtime error is raised.

All of rules 2a through 2e make perfect sense given the premise in rule 2 itself; it is that premise which is, perhaps, somewhat surprising. If you were looking for the "pl" in "people", you would not stop at the first "p" and declare that there is no "pl" just because *this* "p" is not followed by "l". But what Frontier does is rather like this, and the results can be counterintuitive.

Suppose, for example, that we wish to know the value of system.misc. paths.path00. There is a search path to root.system; yet if we ask for misc.paths.path00, an error is generated. Why? Because, according to rule 2, Frontier begins by searching for just misc. There is no root.misc, so Frontier starts trying the search paths in turn. Well, before Frontier ever tries the search path root.system, it tries the search path system.macintosh. This table contains a subtable called misc! The search now stops. Since system.macin-tosh.misc turns out not to contain a paths, the search is declared a failure (the whole of x does not exist).

---

[*] Actually, this is not quite true. If the original reference is a verb call of the form x.y(), and if the first element of x is found but x.y does not exist, the search continues. But there are almost no circumstances where it is helpful to know this.

Being Careful About Search Paths

Since Frontier will very happily accept a partial reference to a database entry that you ask it to alter the value of, or even delete altogether, a certain amount of caution is clearly advisable. Surprises can crop up, especially when you believe you are talking about a variable and Frontier disagrees. Hitherto, we have been rather free and easy with creation of variables: in Example 4-1 we blithely said:

```
s = "Hello, World!"
```

But suppose we had said, instead:

```
cr = "Hello, World!"
```

This would not have created a variable `cr`; rather, it would have changed the value of an important UserTalk constant, `system.verbs.constants.cr`. That's because there is a search path to `system.verbs.constants`, so `cr` already exists along the search paths. By rule 1 above, Frontier would not have created `cr` in the database from this partial reference of form `x`; but it will gladly modify an existing `cr`. The way to avoid this type of misunderstanding is to declare your variables, as discussed in the next chapter.

Another implication is that it is possible to subvert the search process and break scripts accidentally. Suppose you were to create an entry called `long` in `people.[user.initials]`. Meanwhile, `system.verbs.globals.long` is an important built-in UserTalk verb; it converts its argument to an integer. For example, `long("34")` is 34, turning a string into a number. But if you say `long("34")` now, Frontier will encounter `people.[user.initials].long` first in the course of its search, and will stop, never finding out about `system.verbs.globals.long`. Therefore, the creation of an entry `people.[user.initials].long` causes all scripts that use the built-in UserTalk verb `long()` to break.

This is a pity, because `people.[user.initials]` is just where one would like to put utility scripts, and one would like to put them in tables with the same name as the verbs they supplement. For example, the `string` verbs lack a verb for getting the substring made up of the last *n* characters of a string. If you write one, calling it `rightN()`, you might wish you could put it in a table called `string` in `people.[user.initials]`, so that you can call it by saying `string.rightN()`. But the existence of such a table would cause the verb `string()` to break, and with it every script in the database that calls it—and there are many.

Fortunately, there are verbs to help you learn in advance how Frontier will resolve a name you are thinking of using. Their value, in this context, lies in the fact that when you give a partial reference to one of these verbs, Frontier (naturally) resolves it before the verb receives it.

For instance, `parentOf()` returns the name of the table, if any, where a named object resides. Therefore, handing to `parentOf()` the first element of a name you are thinking of using* tells whether it is in use along the search paths. For instance, `parentOf(long)` yields `"system.verbs.globals"` (which is a search path, or Frontier wouldn't have found it); so `long` is in use along the search paths, and it would be a bad idea to say:

```
long = 7
```

or to define `people.[user.initials].long`.

Another useful verb is `defined()`, which tells whether a name has a value. For example, `defined(misc.paths)` yields `false`, so this is not going to be a successful way to refer to `system.misc.paths`.†

Another important device is to Command-double-click, in an edit window, a name you are thinking of using. This causes Frontier to try to jump to a database entry by that name; again, the name is resolved using the search paths, though in this case other search methods are used as well (see the script at `system.misc.command2click` to learn more), so if the jump succeeds, the name may be in use among the search paths.

with

It often happens that, in order to save typing and improve the readability of a script, one would like to establish a temporary search path to a table which is not listed in `system.misc.paths`. Suppose, for example, we wish to obtain the square root of the sine of 13. Both square roots and sines are obtained through scripts in `system.extensions.trigCmd`. We already have a search path to `system.extensions`; so we could say:

```
x = trigCmd.sqrt ( trigCmd.sin ( 13 ))
```

But it is easier to read if we resort to UserTalk's **with** keyword, which lets us abbreviate object references. Our code would then look like this:

```
with trigCmd
    sqrt (sin (13))
```

A **with** statement is followed on the same line by a reference; this is called the **with**'s *domain*. When it encounters the **with**, Frontier attempts to resolve the

---

* A good way to do this is in the Quick Script window. Choose **Quick Script** from the **Open** menu, type `parentOf(`*whatever*`)` into the edit box, and click the **Run** button (or hit the Enter key). The answer comes back in the response box at the bottom.

† A paradox arises with verb calls, because of the rule mentioned a couple of footnotes back. To illustrate: if you create a table in the `people.[user.initials]` table called `string`, then `defined(string.length)` returns `false` but `string.length("hi")` works anyway.

domain, raising a runtime error if it cannot do so (or if it isn't a table). Having resolved the domain, Frontier proceeds to the with's indented bundle, and as it resolves each reference therein, it begins with an extra step: it looks to see whether that reference names an entry in the domain table. Only if it does not does Frontier pass on to the normal reference resolution procedure.

So, in the previous example, Frontier first tries to resolve trigCmd. This succeeds, and the resolved domain for the with is the table system.extensions.trigCmd. Then, when it comes to sqrt, Frontier looks first to see if systems.extensions.trigCmd.sqrt exists. It does; if it didn't, Frontier would then have tried to resolve sqrt in the normal way, looking for root.sqrt and then traversing the search paths.

There is essentially no penalty for using, inside a with bundle, partial references to things that don't live in the domain. If we say:

```
with trigCmd
    dialog.notify (sqrt (sin (13)))
```

the search for dialog.notify will begin by looking to see whether system.extensions.trigCmd.dialog exists. It doesn't, but this extra check only takes an instant and then we're off on the normal resolution process. You can't short-circuit the extra check anyway; even if you were to say:

```
with trigCmd
    root.system.verb.builtins.dialog.notify (sqrt (sin (13)))
```

Frontier would begin by checking for system.extensions.trigCmd.root.

On the other hand, you occasionally will need some extra specification to force Frontier to look outside the domain. For example, there is a built-in verb delete() which deletes database entries. If you were to say:

```
with Eudora
    « ... some stuff ...
    delete (myEntry)
    « ... some more stuff ...
```

there would be a problem; Eudora, which resolves to system.verbs.apps.Eudora, contains an entry called delete, which isn't the one you mean. Since the domain is checked first, you need to be more explicit:

```
with Eudora
    « ... some stuff ...
    system.verbs.globals.delete (myEntry)
    « ... some more stuff ...
```

The name you give inside the with bundle is what counts in deciding whether the original reference is of the form x or of the form x.y. Therefore, by rule 1 ("How Frontier Behaves," earlier in this chapter), a partial reference consisting of just one element in a `with` bundle can never cause a new database entry to be created. Saying:

```
with workspace
    x = 7
```

cannot create a new database entry, but only set an existing one. This provides a measure of security; you have not been sufficiently specific (you *might* mean to create `workspace.x`, but you might not), so Frontier is cautious.

It is possible to list more than one domain in a `with`, separating them by commas. This is really just a shorthand for nested `with`s; in other words:

```
with Eudora, eventInfo
    x = signature
```

is equivalent to saying:

```
with Eudora
    with eventInfo
        x = signature
```

Each subsequent domain is resolved by checking first to see if it is in an earlier domain. Here, for instance, after `Eudora` is resolved as `system.verbs.` `apps.Eudora`, the first step in resolving `eventInfo` is to see whether it is in that domain table. (It is.) References inside nested `with`s are resolved by checking each enclosing domain, starting with the innermost and working outward, before proceeding to normal resolution. So here in the case of `x`, Frontier looks first for `system.verbs.apps.Eudora.eventInfo.x`, then for `system.verbs.apps.` `Eudora.x`, and then goes on to `root.x` and the search paths.

There is more to know about `with`; see the next chapter.

7

The Scope of Variables and Handlers

In this chapter, when we speak of *variables* and *handlers*, we mean local variables and local handlers. The word *local* signifies that these variables and handlers are "owned" by the scripts in which they appear, as opposed to database entries, which are *global* (accessible to any script).*

We have met a local variable in Example 4-1:

```
s = "Hello, World!"
dialog.notify(s)
```

When we press this script's **Run** button, the script starts executing, and in the first line, the variable s is created. In the second line the same variable s is used. The script comes to an end, and s goes out of existence. No other script can access this s unless this script, while running, explicitly calls the other script in such a way as to provide such access (we shall learn in Chapter 8, *Addresses*, how this is done).

We have met a local handler in Example 5-3:

```
on sayHi ()
    local (theDay, theMonth, theYear, theHour, theMin, theSec)
    date.get (clock.now(), @theDay, @theMonth, \
        @theYear, @theHour, @theMin, @theSec)
    if theHour >= 12
        dialog.notify ("Good afternoon or evening.")
    else
        dialog.notify ("Good morning.")
sayHi ()
```

---

* Strictly speaking, the term "local" is probably redundant. All variables are local variables, as opposed to database entries, which are global. All handlers are local handlers, as opposed to script objects living in the database, which are global. However, from another point of view, database objects are a kind of variable, and eponymous handlers (which are global, in a sense) are a kind of handler; so the term "local" is still useful for emphasis and clarity.

When we press this script's **Run** button, the script starts executing, and in the first line, the verb sayHi() is created. In the second summit-level line the same verb sayHi() is called. The script comes to an end, and sayHi() goes out of existence. No other script can access sayHi() unless this script, while running, explicitly calls the other script in such a way as to provide such access.*

This chapter explains how Frontier distinguishes between references to local variables and references to database entries. It also discusses how and when a local variable or handler comes into existence and goes out of existence, and what areas of a script can "see" a given local variable or handler; such matters are termed the *scope* of the variable or handler.

local

Variables and database entries share the same namespace. This means that every time Frontier encounters an object reference, it must make a decision as to whether what is being referred to is a variable or a database entry. It is important to understand how Frontier makes this decision; otherwise, unexpected results can occur.

For instance, in Example 4-1, s is taken by Frontier to be a variable, not a database entry. But that's just because we were lucky in our choice of name. As we saw in the previous chapter, had we chosen to call our variable cr instead of s, executing the script would have changed the value of an important UserTalk constant, system.verbs.constants.cr, and scripts referring to this constant henceforth would have broken. There is thus a danger of accidentally harming the database, because variables and database entries share the same namespace.

How can the database be protected from such accidents? One way might be to memorize the name of every entry along the search paths, and avoid using any of those names as variable names; but this is impractical and unnecessary. UserTalk provides a means of specifying that a particular name is to be understood as referring to a local variable and not to a database entry. It is called "declaring a variable", and it is done with the **local** keyword. For instance, in the following example, cr is taken to be a variable throughout, and the database is untouched.

Example 7-1. Local declarations protect the database

```
local (cr)
cr = "Hello, World!"
dialog.notify (cr)
```

---

* Or unless sayHi() is this script's eponymous handler, but that's a special case.

What declaring a variable with `local` really does is to bring the variable into existence. I shall now explain why this makes such a difference.

In the previous chapter, we saw how Frontier resolves an object reference; but I omitted from the description of this process its very first step, and also its very last step.

The first step in the resolution of references is this: before ever trying to resolve a name as a reference to a database entry, Frontier first examines the name to see if a variable by that name exists. If so, the name is resolved as referring to that variable.

Now we understand how Example 7-1 works. In the first line, a local variable `cr` is brought into existence. In the second and third lines, Frontier begins the process of resolving the name `cr` by checking whether a local variable `cr` exists; it does, and so it is the local variable `cr`, not the database entry `cr`, that is affected.

The last step in the resolution of references is this: if an object reference is of the `x` form (a single element), and if it cannot be resolved as a database entry, and if the command was to set the value of the object, Frontier creates the object as a local variable and sets its value. It is as if Frontier declares the variable with `local` where the script has neglected to do so; and so the action is called an *implicit local.*

Now we understand how Example 4-1 works. In the first line, a name `s` is encountered, and Frontier attempts to resolve it. Frontier first looks for `s` among the existing variables, and fails to find it (no variables exist). Next it tries to resolve it as a partial reference to a database entry; but no `s` is found along the search paths, and Frontier will not create a database entry from a partial reference of the `x` form, as we know. So Frontier performs an implicit local, creating a variable `s` for us; and it sets the value of that variable. When we reach the second line of Example 4-1, `s` *does* exist as a variable.

Since implicit locals exist, it is not strictly necessary to declare your variables; but to do so is good programming practice and is strongly recommended, because it helps to protect the database from accidents.

Even when a variable name is declared local, one should still try to avoid having it match a database entry on the search paths—because one might want to refer to that database entry in the same script. Naming a local variable `cr` makes access to UserTalk's constant `cr` in the same script more difficult; it can no longer be done by saying `cr`—you have to say `system.verbs.constants.cr`, to show that you mean the database entry, not the local variable. Still, such a situation is merely inconvenient, not dangerous.

local Syntax

A `local` declaration can declare one or more names. There are two completely equivalent syntaxes: the name(s) can go into parentheses in the same line as the word `local`,[*] using a comma-delimited list if there is more than one name; or the name(s) can appear as individual lines of a bundle subordinated to the word `local`. Also, a `local` declaration may assign initial values to any of the variables. So this syntax:

```
local (theNumber = 4, theString = "howdy", anotherVar)
```

is equivalent to this syntax:

```
local
    theNumber = 4
    theString = "howdy"
    anotherVar
```

Since lines can be combined using semicolons, you could also write the second version like this:

```
local
    theNumber = 4; theString = "howdy"; anotherVar
```

A local variable created without explicit assignment of an initial value is implicitly assigned a value of `nil` and a type of `unknownType`. This value can be used in operations because it will be implicitly coerced (as discussed in Chapter 10, *Datatypes*) in a sensible way: as a string, it equates to the empty string; as a number, to 0; and so on.

There is a bug involving initial value assignment within a `local` declaration if what's on the right side of the equal sign is a non-scalar. For instance, suppose `workspace.myOutline` is an outline. Then it is bad to say:

```
local (o = workspace.myOutline) « don't do this
```

The bug is that the local copy of the outline is never created; instead, o becomes another name for `workspace.myOutline`. The workaround is to declare the local variable o in one statement, and assign it a value in another.

Another bug is that if you comment out the last line of a `local` bundle (such as the line containing `anotherVar` in our second syntax illustration earlier, the script will refuse to compile.

---

[*] A common beginner's error is forgetting the parentheses.

Variable Scope

The scope of a variable is the innermost bundle in which the variable's `local` declaration appears (look for the word `local`, not the variable name).* The following rules then apply:

1. A variable cannot be seen by a command outside its scope.

2. When a variable's scope has finished executing, the variable is destroyed (it "goes out of scope").

3. In case of a name conflict, the innermost scope for that name takes precedence.

Rule 3 is related to the way variable and database entry names are resolved: local scope is more "inner" than the database; the database is global, sitting outside the whole script.

If a variable will be needed in only a limited region of a script, it is common practice to declare the variable in the bundle where it is needed, restricting its scope. Indeed, it is not uncommon, where the program logic has not already required a bundle, to "bundleize" part of a script deliberately to restrict a variable's scope, using the `bundle` construct introduced in Example 4-4; to help with this, there is a **Bundleize** command in the **Script** menu.

Restricting the scope of a variable can help avoid accidental misuse of an already existing variable. For example, here is an impractical but illustrative script.

Example 7-2. Playing with scope

```
local (n = 1)
bundle
    local (n)
    for n = 1 to 3
        msg (n)
        clock.waitseconds (1)
msg (n)
```

The `msg()` verb displays its parameter in Frontier's Main Window.† The `for` construct denotes a loop that counts incrementally, so its subordinate bundle will be run three times, using the values 1, 2, and finally 3 for n.

When you run the script, you will see the Main Window count 1, 2, 3, then change back to 1. Why? The script twice declares a variable called n; the scope of the first

---

* How to identify the bundle of an implicit local declaration is discussed later in this chapter.

† It is very common to use `msg()` to display information during the course of a script's execution, to help track the script's activities, provide feedback during a lengthy operation, or debug. It has an advantage over the `dialog` verbs in that it doesn't pause execution or require any user response. It doesn't bring the Main Window to the front, though, so make sure that the Main Window (called **Frontier.root** in the **Window** menu) is visible before you run a script where you want to watch the results of `msg()`.

is the whole script, the scope of the second is the inside of the `bundle` construct. When we get to the `for` loop, the second n is innermost in scope, in comparison with the n declared at the start of the program; so it is this n, not the one at the start of the program, that is used in the `for` loop and displayed in the Main Window. Then, when we get to the last line of the script, we are outside the `bundle` construct, and the innermost n no longer exists. We are left with only the n created in the first line of the script. This n was initialized to 1 in the first line— and its value *was never changed thereafter*, because the n in the `for` loop was a different n! So when we display n now, we see **1**.

The technique is a common one. The programmer bundleizes the declaration of the counting variable and the `for` which uses it. The counting variable has no other purpose in the script; it comes into existence just in time for the `for` to use it, and goes out of existence when the `for` is over. Even if a variable by the same name already exists in the script, thanks to the deliberately restricted scope of the counting variable, the other variable of the same name is unharmed.

It is a runtime error to declare a local variable which already exists with identical scope. So, for example:

```
local (n = 1)
local (n)
    « error
```

Handler Scope

Handler scope is analogous to variable scope. The scope of a handler is the innermost bundle in which its on definition appears. One might not intuitively think of a handler as coming into existence and going out of existence, but it does. When an on line is encountered, a local variable is created: it has a name (as declared in the on line) and a value (the handler's code, compiled). This variable is available until its scope finishes executing, at which point it is destroyed.

Indeed, handler scope is the same as variable scope, since handlers and variables share the same subset of the namespace—the local namespace. Just as local variables are checked first before trying to resolve a name as a reference to a database entry, so are local handlers—it's the same namespace.

Since local handlers and local variables share the same namespace, a handler and a variable with the same scope cannot have the same name. So this is a runtime error:

```
local (theWord)
on theWord() « tries to declare theWord twice
    msg("hi")
theWord()
```

But the following script is legal, because there is only one `theWord`, which starts out as a handler and is changed to a number:

```
on theWord()
    return "Hello"
msg (theWord())
theWord = 6
msg (theWord)
```

And this is legal, because two different scopes are involved:

```
local (theWord = "hello")
bundle
    on theWord()
        msg("hi")
    theWord()
    clock.waitseconds(1)
msg (theWord)
```

The same considerations apply to the choice of handler names as for variable names. It would be foolish, though not harmful, to name a local handler `long()`, because within this handler's scope an attempt to call the UserTalk verb `long()` would call the local handler instead; you'd have to specify the UserTalk verb by calling it as `system.verbs.globals.long()`.

Uses of Local Handlers

The purpose of a local handler (as opposed to the eponymous handler) is usually to encapsulate a utility operation needed by commands within its scope. This is generally done in the interests of making the script clearer, more elegant, easier to write, and easier to understand.

For example, in Example 5-3, the two wings of our `if...else` construct are almost identical. We could encapsulate the similarities into a local handler, leaving only the differences, as in the following example.

Example 7-3. Abstracting with a local handler

```
on sayHi ()
    local (theDay, theMonth, theYear, theHour, theMin, theSec)
    date.get (clock.now(), @theDay, @theMonth, \
        @theYear, @theHour, @theMin, @theSec)
    on sayGood (s) « local handler, encapsulating common functionality
        dialog.notify ("Good " + s + ".")
    if theHour >= 12
        sayGood ("afternoon or evening")
    else
        sayGood ("morning")
sayHi ()
```

There is no gain here in efficiency; indeed, there is a slight efficiency loss, since we must now carry the overhead of an added level of depth. Nonetheless, this style of coding has a certain clarity that is much to the taste of UserTalk programmers.

Consider the local handler `obtainLineText()` used in Example 4-7:

```
on obtainLineText()
    if script.isComment()
        return "« " + op.getLineText() + cr
    else
        return op.getLineText() + cr
op.firstsummit()
op.fullexpand()
local (s)
s = obtainLineText()
while op.go (flatdown, 1)
    local (t = op.level ())
    while --t {s = s + "    "}
    s = s + obtainLineText()
clipboard.putvalue (s)
```

The script might have been written without the local handler, but the code would then have been harder to understand:

```
op.firstsummit()
op.fullexpand()
local (s)
if script.isComment()
    s = "« " + op.getLineText() + cr
else
    s = op.getLineText() + cr
while op.go (flatdown, 1)
    local (t = op.level ())
    while --t {s = s + "    "}
    if script.isComment()
        s = s + "« " + op.getLineText() + cr
    else
        s = s + op.getLineText() + cr
clipboard.putvalue (s)
```

The code of Example 4-7 is vastly more legible and logical. True, the simplification involved is very slight: `obtainLineText()` is very short, and is called in only two places. But the repeated material represents a specialized, meaningful action, and so encapsulating it into a local handler makes the program noticeably easier to understand.

with, again

The order in which a name inside a `with` is tested for resolution is:

1. The local namespace inside the `with`'s indented bundle
2. The `with`'s domain

3. The local namespace outside the `with`

4. The database

This order is the only one that makes sense—Frontier is simply proceeding outwards in its search—but a surprising possibility for error lurks here. This was enunciated so clearly in a now famous note on the Internet by Scott Lawton that it is usually referred to as "Lawton's Law." Scott's code went like this:

```
local (id, s = whatever)
with Eudora
    id = create(s)
```

The intention here was to set the local variable `id` to the result of `Eudora.create()`. Unfortunately, the `Eudora` table itself contained an entry called `id`, and this was what was actually changed (according to step 2). This was a very important value which should not have been touched, and all of Scott's commands involving `Eudora` broke from then on. It took him a long time to track down the problem.

The database would not have been tromped if Scott had written:

```
local (s = whatever)
with Eudora
    local (id)
    id = create(s)
```

But since Scott needed the value of `id` outside the `with` block, this wouldn't have worked for him: his only solution was to choose another name for `id`. The danger is always lurking, and is part of the price one pays for the tight integration of the database with the rest of the namespace.

Implicit Locals

When Frontier generates an implicit local, it has to decide what scope to give the local variable it is about to create. To make this decision, Frontier works its way upward through all bundles containing the line in which the undeclared variable was encountered, and to each bundle it applies following two rules, in order:

1. If the bundle contains a `local` declaration, Frontier declares the local variable at that level.

2. If the bundle is the top level of a handler, Frontier declares the local variable at that level.

The process stops as soon as the local variable is declared; if neither condition is ever met, Frontier declares the local at summit level.

This script doesn't do anything, but it illustrates these points:

```
on eponymousHandler()
    on localHandler()
        « x will be declared local here
        for x = 1 to 3
            local (n = 5)
            « y will be declared local here
            y = x * n
```

A handler's parameters are implicitly declared local at the handler's top level. Here is an impractical but illustrative script:

```
on countTo (n)
    « n is a local variable with its scope starting here
    local (i)
    for i = 1 to n
        msg (i)
        clock.waitseconds (1)
countTo (3)
```

Dynamic Scoping

The subordinate bundle of a local handler is considered, on any particular occasion when it is executing, to dwell physically at the point where the handler was called. To illustrate schematically, a script that looks like this:

```
on myHandler()
    « the statements of myHandler()
local (x = 6)
with Eudora
    myHandler()
```

is equivalent to this:

```
local (x = 6)
with Eudora
    « the statements of myHandler()
```

This is significant in two regards. First, any object references among the statements of myHandler() are now inside a with; resolving them, Frontier will look in the Eudora domain before in local namespace outside the with, and the database in general.

Second, it follows from the nature and rules of scope that the handler's subordinate bundle has access to variables and handlers within whose scope the handler was called. So, for instance, in this script:

```
on myHandler()
    « the statements of myHandler()
local (x = 6)
with Eudora
    on anotherHandler()
        « the statements of anotherHandler()
    myHandler()
```

the statements of `myHandler()` (when it is called in this particular way) are able to see and call `anotherHandler()`, and they are able to see `x`. Even more significant, perhaps, they are able to *change* `x`.

Since where a handler seems to be, and hence what variables and handlers its statements can see, depends upon where it is called on any particular occasion, UserTalk is said to have *dynamic scoping* of local handlers.

Dynamic scoping involves calls only to local handlers—it does *not* extend to calls to script objects in the database. In early versions of Frontier it did, but this was found to cause programmers too much confusion. UserTalk's limited dynamic scoping is not confusing, provided one remembers it; indeed, it seems quite natural. Consider this script:

```
local (y = 7)
on whatIsY()
    msg(y)
bundle
    local (y = 22)
    whatIsY()
```

When the script is run, **22** appears in the Main Window, not **7**. And this, one feels, is as it should be; `whatIsY()` is called from within a scope where `y` is 22, and it would be unnerving if the call had the power to jump outside of that scope.

Variables Global to a Handler

A variable to which the statements of a handler have access, and which is not local to the handler, may be described as *global* to the handler. Variables global to a handler provide a way to pass information to the handler directly, and not as a parameter. Here, for example, is Example 7-3 rewritten in such a way that `sayGood()` does not require any parameters:

```
on sayHi ()
    local (theDay, theMonth, theYear, theHour, theMin, theSec)
    date.get (clock.now(), @theDay, @theMonth, \
        @theYear, @theHour, @theMin, @theSec)
    local (what)
    on sayGood ()
        dialog.notify ("Good " + what + ".")
    if theHour >= 12
        what = "afternoon or evening"
    else
        what = "morning"
    sayGood ()
sayHi()
```

The handler `sayGood()`'s bundle can see `what` directly; `what` is global to `sayGood()`, and `sayGood()` can use `what` without being passed it as a parameter.

This sort of thing can save a great deal of overhead, because parameters are passed by value in UserTalk. For example, passing a string as a parameter means making a copy of that string; if the string is long, that's a lot of overhead, which builds up significantly if the handler is called many times. Taking advantage of the ability of local handlers to see variables directly can cut back on this overhead.

It is also common to take advantage of the fact that a handler has the power to change any variable that is global to it. Here, for example, is a pattern that is used frequently:

```
local (htmltext)
on add (s)
    htmltext = htmltext + s
if height != -1
    add (" height=" + height)
if width != -1
    add (" width=" + width)
if hspace != ""
    add (" hspace=" + hspace)
if align != ""
    add (" align=" + align)
```

The idea is to build up a string, concatenating pieces to it one at a time. The local handler `add()` is simply to save us from having to say:

```
htmltext = htmltext + ...
```

over and over again. The variable `htmltext` is global to `add()`, so `add()` is able to alter its value directly. This greatly reduces the number of times `htmltext` must be copied in the course of the routine. So, abstracting `add()` as a local handler threatens to generate overhead, but writing it so as to take advantage of global variables cancels the threat.

Naturally, with increased power comes increased responsibility and danger. Consider this script:

```
local (n = 1)
on countTo (what)
    for n = 1 to what « uh-oh!
        msg (n)
        clock.waitseconds (1)
countTo (3)
msg (n)
```

The programmer intended the n inside `countTo()` as a temporary counting variable, but has tromped on the value of the n declared in the first line. To avoid this, the second n should have been explicitly declared local in `countTo()`. Again we see the value of declaring one's variables.

Recursion

Since a call to a local handler is necessarily within the handler's own scope, statements within the handler can see the handler's name. This implies that local handlers can call themselves (are *recursive*) in UserTalk. The permitted depth of recursion is not tremendous, but it is sufficient for many purposes: 40 levels.

There is, however, an important caveat: an eponymous handler executing because its script was called by another script cannot call itself. That's because the on line declaring the handler was never executed; Frontier simply used it to discover that there was an eponymous handler and leapt into that handler to begin executing.

This is not the place for a full discussion of recursion.[*] Typically, a handler that calls itself will do so only based on some decision, and will often use the result of that call to decide whether to call itself again. In UserTalk, recursion is usually most appropriate when the increase of depth (the handler calling itself) mirrors an increase in depth in the entity on which it is to operate (such as file in folder in folder, or entry in table in table).

For example, suppose we wish to know whether a given table or any of its subtables or any of those subtables' subtables (and so on) contains an entry with a given name. The very statement of the problem suggests recursion immediately. The actual implementation involves arrays, datatypes, addresses and dereferencing, which we haven't come to yet; but the following pseudo-code shows how to structure the recursion.

Example 7-4. Pseudo-code illustrating recursion

```
on tableContains (theTable, entryName)
    on thisTableContains (theTable, entryName) « local handler for recursion
        for each entry of theTable
            if this entry's name is entryName
                return true
            else
                if this entry is a table « recurse into subtable
                    if thisTableContains (this entry, entryName)
                        return true
        return false
    return thisTableContains (theTable, entryName)
```

---

[*] Beginning programmers often fear recursion, as something dangerous or magical. The best way to get a sense for recursion is to learn Scheme (a simple Lisp dialect).

8

Addresses

A UserTalk object, as we know, is a variable, a handler, or a database entry. You refer to an object with an object reference: in effect, you use the object's name. But there are times when you might like to do things with object references other than use them. For instance, it would be nice to be able to make an object reference be the value of a variable.

Imagine, for instance, a handler to create, in any table in the database, an entry called **x** and assign it the value 3. Here is some pseudo-code that expresses what the handler does:

```
on makeX (theTable) « this is fake UserTalk, do not imitate
    theTable.x = 3
```

But what sort of variable can **theTable** be? We don't want it to be an actual table; that would mean we're making an entry **x** in a table which lives in a local variable—not in the database. And how can we call **makeX()**? Suppose we want **makeX()** to operate upon the **workspace** table. We cannot say **makeX(workspace)**, since this just copies the whole contents of the **workspace** table into the parameter. What we want to do is somehow package up the object reference **workspace**, hand that package to **makeX()**, and have **makeX()** open the package and find the object reference there, so that it knows what table to operate on.

UserTalk provides a way to make such a package: it's called an *address*. This chapter explains the syntax of addresses and discusses their primary uses—and some of their dangers.

Address Syntax

The *address* of an object is a pointer to that object. It's obtained and signified by preceding the object's name with @. For example, **workspace.notepad** names an object in the database; `@workspace.notepad` is the address of that object. We say that the `@` operator *takes the address* of the object whose name it is applied to.

It is a runtime error to try to take the address of an object if Frontier cannot possibly resolve the reference to that object. For example, `workspace` is a valid object reference, because Frontier can resolve it to `root.workspace`; so `@workspace` is legal. And `workspace.zzz` is a valid object reference, even though `workspace.zzz` does not exist, because Frontier *could* resolve it if you were to say this:

```
workspace.zzz = "hello"
```

So `@workspace.zzz` is legal. But if `workspace.zzz` does not exist, then `workspace.zzz.yyy` cannot possibly be resolved; so saying `@workspace.zzz.yyy` generates a runtime error.

An address becomes really useful only when it is *dereferenced*. An address is dereferenced by following a reference to it with `^`. For example, suppose `myAddress` is a variable whose value is the address `@workspace.notepad`. Saying `myAddress^` then dereferences that address.

A dereferenced address is a reference to the object pointed to by that address. In essence, taking an object's address packages up a reference to that object; dereferencing the address opens the package, and there's the object reference, ready to be used. In other words, *dereferencing an address is the same as using the name.*[*]

For example, suppose you have made the following declarations and assignments:

```
local (s = "Hello")
on myHi ()
    dialog.notify ("Hi!")
local
    addrWorkspace = @workspace
    addrS = @s
    addrHi = @myHi
```

Then you can say `addrWorkspace^` anywhere you would use the name `workspace`; you can say `addrS^` anywhere you would use the name `s`; and you can say `addrHi^` anywhere you would use the name `myHi`. For example, you can say:

```
addrWorkspace^.x = 7
    « like saying, workspace.x = 7
```

---

[*] This mantra is the UserTalk equivalent of "Port is left, starboard is right," or "Buy low, sell high." Repeating it several times into the mirror each morning will, in short order, make a beginning UserTalk programmer into an expert.

```
addrS^ = addrS^ + ", World!"
    « like saying, s = s + ", World!"
addrHi^()
    « like saying, myHi()
```

You can't use parentheses to modify the application of @ or ^. Only one @ can appear in connection with a name; it must precede the entire name, and it takes the address of the object referred to by the entire name. On the other hand, ^ can appear after any element; it dereferences the value of the object referred to by as much of the name as precedes it. If more than one ^ is applied to a name, they are evaluated from left to right. If both @ and ^ are applied to a name, ^ is evaluated first. So:

```
workspace.anAddr = @blue
workspace.anAddr^
    « "0000FF", because it's like saying blue
local
    addrWorkspace = @workspace
addrWorkspace^.anAddr^
    « "0000FF", because it's like saying workspace.anAddr^
addrWorkspace.anAddr^
    « error! causes Frontier to look for a table "addrWorkspace"
@workspace.anAddr^
    « @system.verbs.colors.blue
@addrWorkspace^.anAddr^
    « @system.verbs.colors.blue
```

The dereference operator can also be used to convert a string to an object reference; *dereferencing a string is the same as using the name.* So, for instance:

```
stringWorkspace = "workspace"
stringWorkspace^.x = 7
    « like saying, workspace.x = 7
```

This actually involves implicit coercion of the string to an address (see Chapter 10, *Datatypes*); dereferencing **"workspace"** turns it into **@workspace** first. Therefore, dereferencing a string causes a runtime error if the resulting address is illegal, as described previously.

Double-dereferencing (or more) is perfectly possible. Watch closely:

```
workspace.anAddr = @workspace.y
workspace.x = "workspace.anAddr"
workspace.x^^ = 8
    « like saying, workspace.y = 8
```

More exotic combinations are possible as well. For example, the verb **window.frontmost()** reports, as a string, the name of the object whose edit window is frontmost. The script for that verb lives in the database at window.frontmost (of course). Now suppose workspace's edit window is front-most. Then:

```
myString = "window.frontmost"
myString^()^.x = 7
    « like saying, workspace.x = 7
```

Believe it or not, I have had occasion to phrase things this way in my code.

Name Construction

A frequent use of addresses is to help with the construction of object references. For example, suppose there is a table at workspace.myTable.mySubtable and we wish to make several entries in it. To reduce clutter and the likelihood of a typing error, we might say:

```
theTable = @workspace.myTable.mySubtable
theTable^.x = "hey"
theTable^.y = "ho"
theTable^.z = "hey nonny no"
```

We now also see how to write our handler makeX():

```
on makeX(theTable)
    theTable^.x = 3
```

And to call it so that it operates on the workspace table, we would say makeX(@workspace).

Similarly, a frequent technique for calculating an object reference at runtime is to store or construct a string, and then dereference it to use the name; this comple-ments the square-bracket notation we studied in Chapter 6, *Referring to Database Entries*. So, for example:

```
theTable = "workspace"; theEntry = "x"
s = theTable + "." + theEntry
    « now s contains "workspace.x"
s^ = 3
    « like saying, workspace.x = 3
    « in this case we could have said [theTable].[theEntry] = 3 instead
```

Address Parameters

Suppose you need a verb to return more than one result value. Suppose, for example, we have a script at workspace.theAngles which calculates the three angles of a triangle given the three sides. A call to workspace.theAngles() can have only one value; how can three values be returned?

The solution is to pass as parameters to the verb the addresses of variables in which the verb is to store its results.[*] Such parameters are called *address parameters*. For example, we might define `theAngles()` like this.

Example 8-1. Handler accepting addresses to return multiple results

```
on theAngles (angle1, angle2, angle3, addrSide1, addrSide2, addrSide3)
    « ... do the calculation based on the three angles ...
    addrSide1^ = ...; addrSide2^ = ...; addrSide3^ = ...
```

And we might call it like this.

Example 8-2. Calling such a handler

```
local (angle1, angle2, angle3, side1, side2, side3)
« ... initialize the angles ...
workspace.theAngles (angle1, angle2, angle3, @side1, @side2, @side3)
```

Let's review Example 8-1 and Example 8-2 in detail. The caller uses the `@` operator to obtain the addresses of its local variables `side1`, `side2`, and `side3`; it passes those addresses to `theAngles()`, which uses the `^` operator to dereference the parameter variables holding the addresses, and assigns values to those dereferenced variables. This means that `theAngles()` has assigned values to the *caller's* local variables `side1`, `side2`, and `side3`!

The results of the calculation are sitting in the caller's `side1`, `side2`, and `side3`, even before `theAngles()` has finished executing. There is no need for `theAngles()` to return any other information, so `theAngles()` ends without a `return` statement. By the same token, the caller has no need to capture the value of the call to `workspace.theAngles()`; after the call, the variables `side1`, `side2`, and `side3` have been set with the information the caller desired.

I have already said (Chapter 5, *Handlers and Parameters*) that in UserTalk, parameters are passed by value, and so any changes that a verb makes to its parameter variables have no effect within the caller. Now we see how to circumvent this restriction. An address parameter gives a verb the ability to meddle with what the address points to.[†]

---

[*] There is another solution, using the list datatype. A list, signified by curly braces, can consist of several items, yet it is only one value. (See Chapter 11, *Arrays*.) So:

```
on theAngles (angle1, angle2, angle3)
    local (side1, side2, side3)
    « ... do the calculation, obtaining side1, side2, and side3 ...
    return {side1, side2, side3}
```

`theAngles()` returns a list consisting of the values of the three sides. However, there is a certain inconvenience to this approach; after a script calls `workspace.theAngles()`, it must parse the resulting list to get at its contents. In any case the method using addresses goes back to a time before Frontier had list datatypes, and remains very prevalent.

[†] Of course, the address itself is passed by value; but the new copy of the address points to the same object as the original address did.

As a matter of style, the programmer has chosen, for the three address parameter variables of `theAngles()`, names beginning with `addr`. This has no special meaning to the computer; it is simply a reminder to readers of the code. Such a coding practice is strongly recommended, especially because (as we shall see) misuse of addresses can lead to disastrous results.

We can now understand these lines from Example 5-3:

```
local (theDay, theMonth, theYear, theHour, theMin, theSec)
date.get (clock.now(), @theDay, @theMonth, \
    @theYear, @theHour, @theMin, @theSec)
```

The built-in verb `date.get()` analyzes a date into six values representing its day, month, year, and so on. It takes seven parameters, but only the first contains information the caller is asking to have analyzed; the other six are addresses of places to store the six results. In this case, we want to know the hour that it is right now; so the first parameter is the result of a call to `clock.now()`, which gets the current date-time value from the computer's internal clock. After the call to `date.get()`, the results are sitting in our variables `theDay`, `theMonth`, and so on.

Actually, only `theHour` is of interest to us, but we cannot call `date.get()` without handing it the required number of parameters, and since these are to be addresses, we should hand `date.get()` actual addresses of existing objects; hence we create five "dummy" variables whose values we don't care about. In reality, though, we need not have created five different dummy variables; we might have said:

```
local (theHour, dummy)
date.get (clock.now(), @dummy, @dummy, @dummy, \
    @theHour, @dummy, @dummy)
```

The one `dummy` variable receives each value in turn, changing five times as `date.get()` executes; it has then served its purpose.

Not infrequently, a handler will return information both by way of address parameters and as its result. That's how many of the dialog verbs work, for example. Recall Example 4-6:

```
local (theReply)
theReply = user.name
if not dialog.ask("What is your name?", @theReply)
    return
dialog.notify ("Hello, " + theReply + "!")
```

`Dialog.ask()` puts up a dialog containing a user-editable field, and then returns `true` if the user clicked **OK** and `false` if the user clicked **Cancel**. If the user did click **Cancel**, we want to abort the whole routine; so we test to see if the result is `false` and exit if so. But `dialog.ask()` also takes as its second parameter the address of a variable to which it returns the text that the user entered into the edit field; so after `dialog.ask()` has been called, we have a second piece of returned information sitting in `theReply`.*

Further Uses of Address Parameters

Handing a verb an address parameter can eliminate the overhead involved in passing a large value. Suppose, for instance, we want our verb to extract information from a string. Handing a verb a string value as a parameter copies the whole string, which may be very large; handing it an address copies only the address, which is tiny.† This is comparable to having a local handler operate upon a variable global to itself.

Another reason for handing a verb an address parameter is that the verb is to operate upon the "thing itself," not its value. A verb that deletes an object from the database, for example, doesn't want to know that the *value* of that object is `"Hello"`; it wants to know *what* object it is. If the object is at `workspace.myString`, saying `workspace.myString` as the parameter would pass `"Hello"`; it is `@workspace.myString` that should be the parameter. UserTalk has a great many verbs that operate upon objects *qua* objects, and they generally take address parameters.

Addresses of Verbs

The use of an address parameter pointing to a handler or script object allows a script to call a verb which is not specified until runtime. This increases the generality and value of the script.

Suppose, for instance, we have a script that is to help us make backups of recent files. It looks at all the files in a given folder, and in each case, adds the name of the file to a list, if the file is more recent than a certain date and time. Without worrying about the details, the structure of our script might look like this:

```
on listIfRecent()
    « ... some sort of loop to examine each file ...
        if file.modified(thisFile) > lastBackupDate
            « then add thisFile to the list
```

It then occurs to us that we might have many occasions for making selective lists of files in folders. We observe that at present we have a boolean test in the `if` line, which returns `true` or `false` for the particular file we are looking at. It seems wasteful to write the same routine over and over, differing only in the

---

* It happens that `dialog.ask()` also *uses* the value in the variable whose address it is handed in its second parameter, if that variable has a value, as the default text appearing in the dialog's edit field. That's why we initialize `theReply` with a reasonable default value before calling `dialog.ask()`.

† Nevertheless, most of UserTalk's built-in string verbs take strings, not addresses of string variables, as their parameters.

contents of this one line. If only we could specify *at runtime* what the test should be! By passing a verb's address as a parameter, we can do just that.

Example 8-3. Handler accepting a handler address

```
on listIfWhatever(addrTester)
    « ... some sort of loop to examine each file ...
        if addrTester^(thisFile)
            « then add thisFile to the list
```

To call `listIfWhatever()`, we hand it, as a parameter, the address of a handler or script. That handler or script must obviously be expecting one parameter, namely a file pathname. And it must return `true` or `false` reporting whether that file meets some criterion or not. Then `listIfWhatever()` calls that handler or script to perform the actual test.

For example, to call `listIfWhatever()` to the same effect as `listIfRecent()`, we could say the following.

Example 8-4. Telling a handler what handler to call

```
on isRecent(aFile)
    return (file.modified(aFile) > lastBackupDate)
workspace.listIfWhatever(@isRecent)
```

But now it would be just as easy to use `listIfWhatever()` to generate a list of files that are textfiles, or files that are larger than a certain size.

For a further example of this technique, see `samples.basicStuff.aliasGatherer`, and the way it is called by `samples.basicStuff.appAliases`. It is also used by most of the visit verbs; see Chapter 12, *Control Structures*.

Being Careful with Address Parameters

An all-too-easy coding accident is to hand something that is not an address as a parameter to a verb that expects an address. Typically, this is due to forgetting to put @ before a name. Sometimes this mistake will just cause a runtime error, but it can be a recipe for disaster; it can harm the database, and the trouble that results can be difficult to track down.

The danger often comes from the fact that it is possible to dereference a string. Suppose, for instance, one were to say the following.

Example 8-5. Bad way to call dialog.ask()

```
local (theReply)
theReply = "Last"
if not dialog.ask("What is your last name?", theReply) « no no NO!!!
    return
```

The dialog appears, the user types something—let's say **Neuburg**—and hits the **OK** button. Now, `dialog.ask()` was expecting the second parameter to be an address. It dereferences that address so that it can store, in the object pointed to, whatever the user typed in the dialog. If the program's third line had been correct, like this:

```
if not dialog.ask("What is your last name?", @theReply)
```

then `"Neuburg"` would have ended up stored in `theReply`. But as it is, Frontier will dereference the string `"Last"`, yielding an object reference, `last`: Frontier is going to try to store `"Neuburg"` in `last`. Unfortunately, this is possible: `last` resolves to `system.macintosh.constants.last`, and `"Neuburg"` is therefore stored there—replacing the value of an important constant. To top it all off, it could be a long time before the damage manifests itself in misbehavior; and even a diligent user who deduces what has happened may not know what value `last` *should* have, and may need to revert to a backup or a clean root.

Now let's turn up the heat. Consider the verb `new()`. It takes an address and creates a new variable or database entry there—overwriting anything that may be there already. It can create an object of any type. It is, for example, often used to create a new outline in the database. The following command creates a new outline at `workspace.s`, wiping out whatever may be at `workspace.s` already:

```
new (outlineType, @workspace.s)
```

Suppose the programmer forgets the `@`, and types the following instead.

Example 8-6. A really, really bad way to call new()

```
new (outlineType, workspace.s)
```

And suppose `workspace.s` exists, and that it is a string. And suppose that string is `"system"`. When `new()` dereferences `"system"` to get an object reference, it's going to be talking about `root.system`, where the heart of Frontier lives: constants, verbs, extensions, the UserTalk compiler, everything. Now Frontier is going to put an outline at `root.system`! Executing this line will replace the whole inner workings of Frontier with an empty outline!

These examples are not farfetched. They should suffice to instill caution when using addresses. You cannot avoid addresses, nor should you; they are tremendously valuable, and a clear understanding of them is part of any good UserTalk programmer's repertory. Still, you should be careful.

9

Special Evaluation

In general, Frontier evaluates names before using them. When you say:

```
x = y
```

it is the value of **y** that is assigned to **x**; when you say:

```
myScript(y)
```

it is the value of **y** that is handed as a parameter to **myScript()**. Under certain circumstances, however, Frontier behaves exceptionally with regard to evaluation. Sometimes it doesn't evaluate a name; sometimes it bypasses the normal evaluation rules. This chapter is about these and related matters.

Non-Evaluation

Four verbs—**defined()**, **parentOf()**, **sizeOf()**, and **nameOf()**—are *special forms*: they look like ordinary verbs, but they aren't. When they are used, their parameter is not evaluated. These verbs have in common that their parameter is an object reference—not an address—and it was this object reference itself, not its value, which passed to the verb.

To see why this is a good idea, consider **defined()**, which returns **true** or **false** depending on whether its parameter is the object reference of an existing object. Suppose **defined()** were not a special form. If we handed **defined()** an object reference and that reference was evaluated, we wouldn't be talking about the object itself. We could not hand **defined()** an address, though, because if the object in question didn't exist, it might be illegal to take its address in the first place.

The other special forms are parallel; one can speak to them of nonexistent objects without causing a runtime error.

If what you have is an address or string denoting the object reference, and you wish to hand that object reference as parameter to one of these verbs, you must dereference the address or string. For example:

```
defined (fakeTable.fakeEntry)
    « false
myString = "fakeTable.fakeEntry"
defined (myString^)
    « false, as expected; if we had said defined(myString)...
    « ...it would have returned true - the string variable myString exists
```

Because of the non-evaluation, it is permissible to dereference a string denoting an unresolvable reference.[*]

Pre-Evaluation

Like most computer languages, UserTalk has a number of *reserved words*. The user is not free to employ such reserved words to designate ordinary objects. For example, the keywords, such as `if`, `with`, `on`, `return`, and `local`, are reserved words. If you could call a variable `local`, then whenever you said `local`, Frontier would have a hard time deciding whether you were referring to your variable or starting a `local` declaration.

The way Frontier enforces the special status of its reserved words is by pre-evaluating them. When you use a reserved word, Frontier never even tries to resolve it as a name; it discovers beforehand that the word is reserved. The resolution process is, as it were, short-circuited in order for reserved words to work.

Pre-evaluation is implemented by way of three lists of reserved words, which are maintained inside `system.compiler`. The keywords are kept at `system.com-piler.language.keywords`. Certain verbs that need special treatment—including the special forms we have just mentioned—live at `system.compiler.language.builtins`. In addition, the UserTalk constants—in other words, names whose values are predefined at compiler level—are in `system.compiler.language.constants`.

It is best not to attempt to give an object a name that is a reserved word. Because of pre-evaluation, which can extend even to individual elements of an object reference, you will not be able to refer normally to such an object. Here are some examples of the kinds of knots you can tie yourself in if you try it:

• You cannot directly define a local variable called `infinity`; `infinity` is a compiler constant, and pre-evaluation means you are trying to define a local

---

[*] Certain other verbs, such as `delete()`, appear to be special forms in other ways, performing evaluation in an unusual fashion, but they are less significant.

variable called 2147483647. You would have to speak of ["infinity"] instead.

- You can define a local handler called pack, but you cannot call it. pack() is a built-in verb, and pre-evaluation means that when you say pack() you call the pack() in system.compiler.language.builtins.

- Similarly, you can define a handler in the workspace table called pack, and you can call it by saying workspace.pack(), but if you say:

```
with workspace
    pack()
```

it is the built-in pack() that is called, because the pre-evaluation precedes the application of the with.

Also, it is unwise to call a database entry by one of the names constants, keywords, or builtins.

Constants

The built-ins and reserved words are discussed individually later on, under the topic appropriate to the functionality of each. The UserTalk constants are as follows:

- The no-value constant, nil. This is the value of any variable or database entry that has never been explicitly assigned a value; you may assign it as a value if you wish.

- The largest positive integer UserTalk can represent, dubbed infinity.

- The eleven "directions," used in describing cursor movement when operating on such objects as wptexts and outlines: the most important are up, down, left, right, flatup, and flatdown.

- The two booleans, true and false.

- The datatype names. Their use is discussed in Chapter 10, *Datatypes*.

evaluate()

We have seen that you can obtain an extra round of evaluation for an element of an object reference that is a valid UserTalk expression, by enclosing the element in square brackets:

```
people.[user.initials]
```

By the same token, you can obtain an extra round of evaluation for any UserTalk expression. The expression must be constructed as a string; that string is then handed to evaluate() and the result of evaluating it is returned.

`evaluate()` is useful when an expression is unknown until runtime. This situation typically arises when a script, during the course of its execution, turns to the user to obtain a UserTalk expression. The communication takes place by way of a string. For example, `suites.samples.basicStuff.runClipboard()` is a script that lets the user copy text from another application for Frontier to evaluate in UserTalk; the user can then paste the result into the original application. And use of `evaluate()` in Frontier's Web site management facilities is what makes it possible to calculate pieces of a Web page at runtime; for instance, if `{clock.now()}` appears within the text of a Web page, the current time is substituted when the page is actually produced.

10

Datatypes

This chapter discusses the UserTalk datatypes: what sort of entities they represent, how they are represented as literals, and how they are transformed from one type to another.

UserTalk has 31 datatypes (plus an extra type, **unknownType**, for handling objects that have not yet been assigned a value). Every UserTalk value is of exactly one of these types. It is customary to speak of an object's type ("**s** is a string"), and I often speak that way too; but this is actually shorthand for speaking of the object's *value's* type. Objects themselves do not have datatypes; indeed, any object can be assigned a value of any type in UserTalk, even if it already has a value of a different type. We say that UserTalk *lacks strong typing.*[*]

Internally, Frontier designates the datatypes with string4 identifiers. (String4s are discussed further below.) However, these string4 identifiers are given English-like constant names, so you don't have to use the string4s yourself. In **system. compiler.language.constants**, English-like names for the datatypes are paired with the string4 values whereby Frontier designates the datatypes. So, for example, to make a new outline in the database, you can say:

```
new (outlineType, @workspace.myOutline)
    « you don't have to say, new ('optx', @workspace.myOutline)
    « (though you could if you wanted)
```

The English-like names for the datatypes are reserved words; see Chapter 9, *Special Evaluation.*

You can learn the datatype of a value by passing it as a parameter to the verb **typeOf()**; what is returned is the string4 designating that datatype.

---

[*] Contrast Pascal and C, where a specific type is associated with every object name. UserTalk's approach to types is more like Scheme's.

For most of the datatypes, Frontier also provides *coercion* verbs, which let you transform a value to an equivalent value of a different datatype.

A rather daunting feature of Frontier is that there are so many different names associated with types in various contexts. The English-like constant names are not the same as the names you see in the Kind column of a table edit window; these in turn are not the same as the names in the **Kind** popup at the bottom of a table edit window; and these are not the same as the names used in the **Table** menu to make a new object. The name of the coercion verb and the constant name of the type sometimes don't stand in a simple relationship. And the string4 names are just codes, and no facility is provided for translating them into English; when you receive a string4 from `typeOf()`, you may not know what datatype it means, and there is no automatic way to find out.[*]

Table 10-1 summarizes the situation.

Table 10-1. Types, Coercion Verbs, and Names Associated with Them

ID	Constant Name	New...	Kind Column	Kind Popup	Coercion Verb
'optx'	outlineType	**Outline**	outline	Outline	[N/A]
'wptx'	wptextType	**Text**	wp text	WP-Text	[N/A]
'tabl'	tableType	**Subtable**	table	Table	[N/A]
'scpt'	scriptType	**Script**	script	Script	[N/A]
'mbar'	menubarType	**MenuBar**	menu bar	MenuBar	[N/A]
'pict'	pictureType	**Picture**	picture	Picture	[N/A]
'shor'	intType, shortType	[N/A]	number	[N/A]	short()
'long'	longType	**Number**	number	Number	long(), number()
'exte'	doubleType	**Float**	double	Float	double()
'sing'	singleType	[N/A]	single	[N/A]	single()
'fixd'	fixedType	**Fixed**	fixed	[N/A]	fixed()
'TEXT'	stringType	**String**	string	String	string()
'char'	charType	**Character**	char	Character	char()
'type'	string4Type	**String4**	string4	String4	string4()
'date'	dateType	**Date**	date	Date	date()
'data'	binaryType	**Binary**	binary	Binary	binary()
'bool'	booleanType	**Boolean**	boolean	Boolean	boolean()

---

[*] Well, there *is* an automatic way, actually, but it's rather an elaborate trick. To find out, for instance, what the constant name for the 'qdrt' type is, execute the following script in the Quick Script window:

```
with system.compiler.language.constants {displaystring('qdrt')}
```

Table 10-1. Types, Coercion Verbs, and Names Associated with Them (continued)

ID	Constant Name	New...	Kind Column	Kind Popup	Coercion Verb
'addr'	addressType	**Address**	address	Address	address()
'QDpt'	pointType	**Point**	point	[N/A]	point()
'qdrt'	rectType	**Rectangle**	rect	[N/A]	rect()
'cRGB'	rgbType	**Color**	rgb	[N/A]	rgb()
'tptn'	patternType	**Pattern**	pattern	[N/A]	[N/A]
'obj '	objspecType	**Object Spec**	objspec	Object Specifier	objspec()
'fss '	filespecType	**File Spec**	filespec	File Specifier	filespec()
'alis'	aliasType	**Alias**	alias	Alias	alias()
'enum'	enumerator-Type	**Enumerator**	enumerator	Enumerator	enum()
'list'	listType	**List**	list	List	list()
'reco'	recordType	**Record**	record	Record	record()
'dir '	directionType	**Direction**	direction	Direction	direc-tion()
'tokn'	tokenType	[N/A]	token	[N/A]	[N/A]
'code'	codeType	[N/A]	compiled code	[N/A]	[N/A]

Summary of the Datatypes

The datatypes may be categorized as follows:

Non-scalar

These are types that have their own edit windows within Frontier: outline-Type, scriptType, wpTextType, tableType, menuBarType, pictureType. All other types are *scalars*, meaning that they are editable by typing directly into a table's Value column. A local variable can be of a non-scalar type; it is perfectly legal, for example, to declare a local variable and use new() or assignment to turn it into an outline. You can even open an edit window on such an outline!

Number

intType and longType are integers; intType is like a short in C, longType is like a long and is more commonly used. singleType and doubleType are single-precision and double-precision floating-point; doubleType is the more common. fixedType is fixed-point, and exists mostly for compatibility with the computer's toolbox routines (I have never seen it used).

Shorts range from −32768 to 32767; longs range from −2147483648 to 2147483647. Single-precision is accurate to about 9 digits, double-precision to about 19. A numeric literal is assumed to be a long, unless it contains a decimal point, in which case it is assumed to be a double. Frontier does not understand scientific notation. A hexadecimal literal may be represented by beginning it with `0x` (for example, `0xF` is 15).*

String

`stringType` is a string; a string literal is delimited either by double quotes (straight quotes) or by smart quotes (curly quotes). `charType` is a single character; a char literal is delimited by single quotes. `string4Type` is four characters packed into the same space as an unsigned `longType`; a string4 literal is four characters delimited by single quotes. `charType` and `string4Type` have certain affinities with strings, but also certain affinities with integers, as we shall see when we discuss coercion.

The literal representation of strings and chars is a large topic and is discussed further at the end of the chapter.

Date

Dates, embracing both dates and times, use the `dateType`, which is actually an unsigned long representing the number of seconds since some fixed date (on Mac OS, this is midnight on 1 January, 1904).

Binary

`binaryType` lets a scalar contain raw data of any kind; for example, a picture file such as a GIF can be imported into the database as a binary. A binary actually has two types: its own, which is `binaryType`, and that of its contents.

Boolean

`booleanType` is the type of the constants `true` and `false` and of the results of logical tests; what appears after `if` must be a boolean or be coercible to a boolean. For the operators that can be used in a boolean expression, see Chapter 44, *Operators*.

Address

`addressType` is a pointer to an object. Addresses are represented as literals in tables and scripts by @ followed by the name of the object pointed to; elsewhere, they appear coerced to a string naming the object.

Pattern

`patternType` is a representation of a 16-by-16 bitmap pattern as 16 hex digits; I have never seen this used.

---

* To convert a number to its hex representation, use `string.hex()`; see Chapter 14, *Strings and Chars*.

List

listType and recordType are ordered collections of items of any scalar type. As literals, they are delimited by curly braces and the items are separated by commas; so, for instance, this is a list, consisting of two items which are strings:

```
{"hi", "there"}
```

Records are lists where the individual items also have names, which can make it more convenient to specify an item; the names must be (or evaluate to) string4s. As literals, records look like lists except that each items is preceded by its name plus a colon. Thus, for example:

```
{'name':"Matt Neuburg", 'occu': "none", 'aged': 43}
```

Lists and records can be used to communicate with other programs, which sometimes expect or return these types, and they are also valuable as internal datatypes.

Set of shorts

pointType, rectType, and rgbType are ordered sets of two, four, and three shorts, respectively, which represent points (*x* and *y*), rectangles (top, left, bottom, and right), and RGB colors (red, green, and blue). As literals, they cannot be represented directly; they must be coerced from lists or strings of comma-separated numbers.

On Mac OS, the Cartesian plane is inverted top-to-bottom; the main screen has a top-left of 0,0 and both *x* and *y* values on the screen are positive.

Direction

directionType is used for the direction constants, which are passed as parameters to UserTalk verbs that manipulate the cursor in wptext and outline edit windows.

Token

tokenType is used for the tokens into which UserTalk keywords and built-ins are translated at compile time.

Compiled code

codeType is the type of the local "variable" generated by executing an on() declaration; the local handler is copied as compiled code into Frontier's compiler stack. Also, it is possible to strip the source from a compiled script for security purposes, leaving only the compiled code, which then can be called, but not read by a human being; a database entry containing such code is of type codeType.

Other

objspecType, filespecType, aliasType, and enumeratorType are mainly for communicating with other programs, which sometimes expect these types as parameters. It is only occasionally necessary to coerce explicitly to any of these types, chiefly in tweaking "glue" verbs (Chapter 32, *Driving Other Applications*). It often makes sense, when storing the pathname of a file in the database, to make it a fileSpec or alias.*

Coercion

Coercion is the transformation of a value into an equivalent value of a different datatype. To request coercion explicitly, one generally uses the coercion verbs, listed in Table 10-1.

No type-checking of parameters is performed when a verb is called. If, however, a value of incorrect type is passed as a parameter to a UserTalk verb that is genuinely built-in (i.e., implemented as a token somewhere in `system.compiler`), or if an object is subjected to an operator inappropriate to its type, Frontier will silently attempt to coerce the object to the correct type. If an operator that takes two operands is applied to objects of different types, Frontier will silently attempt to coerce at least one of them so that their types match. Such silent, automatic coercion is called *implicit coercion*. If Frontier cannot perform an attempted implicit coercion, a runtime error occurs.

We now discuss a couple of types for which coercion is performed by verbs not listed amongst the coercion verbs, and then describe the most commonly encountered coercions.

Binaries

A binary has two types—its own, which is **binaryType**, and that of the data it contains, called its *internal type*. The internal type can be any type whatsoever. In fact, it does not even have to be a Frontier datatype; it is merely a string4 attached to the binary, and this can be any string4 at all.

The ordinary coercion verb **binary()** may be used to coerce a value of any type to a binary; the binary's internal datatype is automatically set to the type of the original value. Going the other way, if a binary's internal type is one of the scalar types, that scalar type's coercion verb can be used to extract the data from the binary. For example, if you have said:

```
workspace.mybinary = binary("Hello, World!")
```

then `string(workspace.mybinary)` evaluates to `"Hello, World!"`

---

* An alias stored in the database works just as an alias file does: if the file to which it points is moved, it changes its value and continues to point to it. A fileSpec, on the other hand, is frozen and cannot be counted on to be valid the next time Frontier is started up. In the case of fileSpec database entries, Frontier compensates, to some extent, by silently coercing a fileSpec database entry to an alias when Frontier is shut down, and back to a fileSpec when Frontier is next started up; but on the whole, fileSpec values should be regarded as temporary and confined to a particular script execution.

However, transformation to and from a binary can also be performed with the verbs `pack()` and `unpack()`. `pack()` is identical in effect to `binary()`, differing only in the syntax of the call; you provide as second parameter the destination address. So these are equivalent:

```
workspace.mybinary = binary("Hello, World!")
pack ("Hello, World!", @workspace.mybinary)
```

The reverse of `pack()`, namely `unpack()`, is the only way to extract a non-scalar from a binary. It also has a major advantage over coercion as a way to extract what is in a binary: you don't have to know in advance the type to which the binary is to be coerced—the internal type is used automatically. `unpack()` takes two addresses, that of the source (the binary) and that of the destination. So the following:

```
unpack (@workspace.mybinary, @workspace.whatWasInMyBinary)
```

would place the string `"Hello, World!"` into `workspace.whatWasInMyBinary`.

To learn a binary's internal datatype, call `getBinaryType(theBinary)`. To set a binary's internal datatype, call `setBinaryType(addrBinary, theType)`. In practice, these verbs are used chiefly to manipulate the internal datatype as a marker—to create your own custom types, for instance, or to allow the data to be handed to another program that expects a particular type. For example, when Frontier's Web management features are used to read a GIF or JPEG image file into the database, the resulting binary is explicitly given an internal type of `'GIFf'` or `'JPEG'`; when the binary is to be written out to disk as a file later, the internal datatype is checked in order to determine the suffix to be appended to its name.

Sets of Shorts

Points, rects, and rgbs can be set by coercing a comma-delimited string or list consisting of the correct number of shorts; for example, these both work:

```
myPoint = point ("1, 2")
myPoint = point ({1, 2})
```

To obtain the value of a component of a point, rect, or rgb, one approach is to coerce it to a list, because it is easy to get a list's *n*th element (discussed in Chapter 11, *Arrays*):

```
myPointAsList = list (myPoint); x = myPointAsList[1]
```

But it is often more convenient to use verbs which set or get all the component values at once. The set verbs have the desired entity as their result; the get verbs take as parameters the entity and addresses of objects into which to put its parts. Thus:

```
point.set (x, y)
point.get (pt, addrX, addrY)
rectangle.set (top, left, bottom, right)
rectangle.get (rect, addrTop, addrLeft, addrBottom, addrRight)
rgb.set (red, green, blue)
rgb.get (rgb, addrRed, addrGreen, addrBlue)
```

A Mac OS color is represented by a set of three unsigned shorts; but Frontier has no such datatype. Thus white, which is represented by {65535,65535,65535} in Mac OS, is represented as {-1,-1,-1} in Frontier.

There is a search path to **system.verbs.colors**, which contains 140 named rgb colors, ready to use. Thus, to set a color, one can simply say:

```
myColor = aquamarine
```

However, these colors are not rgbs: they are hex triples, in Web style. To convert one of them to an rgb, use the following utility.

Example 10-1. tripletToRGB()

```
on tripletToRGB(trip)
    on sign(n)
        if n > 32767 {n = n - 65536}; return n
    on extract(n)
        return number("0x" + trip[n] + trip[n+1]) * 257
    local
        theRed = sign(extract(1))
        theGreen = sign(extract(3))
        theBlue = sign(extract(5))
    return rgb.set(theRed, theGreen, theBlue)
```

Coercion Results

On the whole, the various datatypes coerce from one type to another in an intuitive and intelligent manner. The following discussion is not complete, but in it I have tried to cover the most interesting and important cases.

TIP   Frontier helps you know what coercions are possible amongst scalars, in any table edit window; when you select a table entry, the **Kind** popup enables only those datatypes to which the selected entry (given its kind and value) can be coerced. (And choosing from the popup performs the coercion.)

Non-scalars

The non-scalars consisting of textual matter—script, outline, menubar, table, wptext—can be coerced to strings.* There are no verbs for coercing back in the other direction; however, pasting into a script, outline, or wptext can be a substitute.

Scripts and outlines are coerced to strings identical to the result of copying the script or outline and pasting it into a textual environment.† Coercion of a wptext to a string loses formatting information, but the text itself remains intact.

Menubars are an odd case, because a menubar is an outline in which an item can have a script associated with it. When a menubar is coerced to a string, the result is like that for a outline, with each script appearing indented below its menu item. Curly braces and semicolons are *not* added to the script parts, because the menubar is itself more like an outline than a script:

```
Add to Glossary...
    local (name = webBrowser.getFrontWindowTitle ())
    if dialog.ask ("Name:", @name)
        local (s = "<a href=\"" + webBrowser.getFrontWindowURL ())
        s = s + "\">" + name + "</a>"
        user.html.glossary [name] = s
        filemenu.save ()
Open Glossary
    Frontier.bringToFront ()
    edit (@user.html.glossary)
```

Tables are coerced to strings so as to simulate the layout of a table, using a three-column tab-delimited structure. Each line contains the object's name, its type (followed, oddly, by a colon), and its value (coerced to a string). The conversion does not work perfectly, and chokes on tables of any real size, but it is useful for small tables of simple data.

```
myshort      number:   6
mysingle     single:   6.0
newAlias     alias:    Desire:AppleScript™ Utilities:Frontier 4.1:
newBinary    binary:   0xAF963122
newBoolean   boolean:  false
```

For conversion between outlines and lists, see on `toys.outlineToList ()`, Chapter 18, *Non-Scalars.*

---

* A picture can be coerced to a string, but it just says `"<<picture>>\r"`.

† We have already discussed how Frontier pastes a copied script into a textual environment. Outlines are just the same, except that no curly braces or semicolons are inserted.

Scalars

Strings are the most universal type: they can be coerced to and from all other scalar types. Since coercions involving strings are generally obvious, what follows is mostly a summary of the commonly encountered scalar coercions *not* involving strings:

Numbers

Number types coerce to one another as expected. Coercing a floating-point to an integer truncates its value.

Chars and string4s

Numbers, chars, string4s, and strings are interrelated. A string is made up of chars; but, from another point of view, a char is a short and a string4 is a long.

Coercing a char to a string yields a one-character string, and a one-character string can be coerced to a char. Coercing a char to a number yields its ASCII numeric value, and vice versa (UserTalk has no `asc()` function). This can feel surprising; one expects `number('3')` to be 3, but of course to get that you'd have to say `number(string('3'))`. A string4 can be coerced to a long, and vice versa.

Booleans

The boolean values `false` and `true` can be coerced to numbers, yielding 0 and 1, respectively. In the reverse operation, any number except 0 yields `true`; 0 yields `false`. Any string except `"false"` (not case-sensitive!) or the empty string coerces to `true`.

Points, rects, and rgbs

A point can be coerced to a long and vice versa; the *x*-component of a point is stored as the lo-byte of the long, the *y*-component as the hi-byte. Points, rects, and rgbs coerce to strings as comma-delimited series of numbers.

Lists

A list or record coerces to a string which looks just like the list's or record's literal representation, curly braces and all.* For example:

```
string( {"hello", 3} )
    « "{\"hello\", 3}"
```

Any scalar can be coerced to a list; the list contains one element, namely, the original scalar. Just the other way, a one-element list can be coerced to a scalar (though this would be an odd way to code; it is simpler just to extract the element using array notation).

Nothing except a string can be coerced to a record.

---

* When a list is coerced to a string, there's a bug: if any one item of the list becomes a string longer than 256 characters, the excess is truncated. For example, if x is a string of 200 characters and y is a string of 200 characters, then `string({x,y})` will work, because each individual item is less than 256 characters long; but `string ({1,{x,y}})` won't work, because one item, {x,y}, is some 400 characters long.

Directions

The directions can be coerced to and from numbers (nodirection becomes 0, up becomes 1, down becomes 2, and so on).

Files

FileSpecs and aliases are coerced to and from strings that are pathnames, describing paths to files in standard format. On the Mac, a file called *myFile* on my hard disk which is called *HD* can be aliased as alias("HD:myFile"); my hard disk would be alias("HD:") (notice the final colon after volumes and folders). Coercing an alias to a fileSpec resolves the alias (if the alias is valid; otherwise, an error results). One cannot coerce to a fileSpec unless the folder containing the file actually exists; but an alias can be made from any pathname string.

Dates

A date coerced to a string is formatted like this on my machine: "5/7/97; 11:35:04 PM". But your results may be different, because the coercion calls the system, which consults the date format settings in the *Date & Time* control panel.

A string can be coerced to a date even if it doesn't quite hold to this format; various separators can be used, and the name of the month can be written out or abbreviated. Thus, Frontier is flexible enough to handle:

```
date("7 May 1997")
date("may-7-97")
date("7 may, 1997 11:43 PM")
```

and many other variations. What isn't so easy is getting a simple time to be understood; date("11:36 AM") works because there is no 36th of the month, but date("11:30 AM") does not—you have to use a date verb (discussed in Chapter 16, *Dates*).

Addresses

An address is coerced to a string which, if dereferenced, would yield the same object reference as the address; for example:

```
string (@workspace)
    « "root.workspace"
string (@scheduler)
    « "suites.scheduler"
```

Strings can be coerced to addresses if they can be resolved as object references at the moment the coercion is made; thus, address("glubglub") fails because there is no object in the search paths called glubglub, but it would succeed if you said it within the scope of a local variable called glubglub. As mentioned in Chapter 8, *Addresses*, a dereferenced string is implicitly coerced to an address first.

Type-Dependent Operations

Where an operation takes two operands, if these are of the same type, the operation generally yields that same type and is performed according to the rules for that type. This seems sensible enough, but many beginners are surprised by the results.

Consider addition. The + operator is overloaded: adding numbers adds them arithmetically; adding strings concatenates them. Thus:

```
50 + 1
    « 51
"User" + "Talk"
    « "UserTalk"
"50" + "1"
    « "501", because they're strings
```

So far, so good, but when adding variables instead of literals one may be unaware of their datatype. Even very experienced UserTalk programmers confess to being occasionally caught out by this. This script asks the user for two numbers, adds them, and displays the result; but it displays the wrong result, because a dialog's edit field always contains a string, so the operands in the last line are strings and are concatenated, not added:

```
local (firstnum, secondnum)
if dialog.ask ("Give me a number.", @firstnum)
    if dialog.ask ("Give me another number.", @secondnum)
        dialog.notify (firstnum + secondnum) « oops
```

Integer operations sometimes pose surprises, too. If both operands are integers, the result will be an integer; so, if the result is out of range, it wraps around from the other end:

```
infinity + 2 == -infinity
    « true! This should give Georg Cantor something to think about
```

An integer divided by an integer yields an integer; the decimal part of the result, if any, is thus thrown away. Beginners are confounded by this sort of thing:

```
5 / 2
    « 2
```

If they know about coercion, they may try to compensate like this:

```
double (5 / 2)
    « 2.0
```

But of course the coercion comes too late; the division has already been performed as an integer division; having obtained the wrong result, we are merely expressing it as a double. Similarly, the following does not give the right answer:

```
double (2147483600 * 2)
```

The multiplication is integer multiplication, which goes out of range and wraps around to a negative result before the `double()` coercion ever takes place.

It is sufficient in cases like these (for reasons we shall explain immediately in the next section) to coerce one operand to the type of the desired result before the operation, like this:

```
double (5) / 2
    « 2.5
double(2147483600) * 2
    « 4294967200.0
```

Mixed Operands

Where an operation takes two operands, and these are of different types, or not of the right type, a rather tricky form of implicit coercion takes place. In general, as each operation is handled, there is coercion of its operands to the type of the "higher" operand. The precedence of types is roughly arranged as follows, from lowest to highest: boolean; short; point, long, char, string4, and direction; date; single and fixed; double; rect, rgb, alias, and address; string; binary; list; record.

So, taking addition as an example:

```
"50" + 1
    « "501", because they're both strings after the coercion
```

Evaluation is pairwise, so if you add several numbers with a string mixed in, the operation won't turn into concatenation until we reach the string. Thus:

```
3 + 4 + "5"
    « "75", because 3+4 is 7 and then "7"+"5" is "75"
```

When an operation involves two different types with the same precedence, the coercion is to the type of the left operand. Here, for example, is what can happen when integers and chars are mixed in an arithmetic calculation:

```
1 + '3'
    « 52, because '3' is ASCII 51
'3' + 1
    « '4', the char whose ASCII value is 52
```

Things are similar if you add a string4 and an integer.

The second example is particularly tricky. Because one operand is an integer, the operands are added as integers, but the result is a char. (Compare also the arithmetic treatment of dates; they become numbers for the purposes of the operation, but the result is presented as a date.) Contrast what happens when two chars are added:

```
'3' + '1'
    « "31"
```

Here, the operands are treated as strings (and concatenated), though neither is a string. It is not always easy to know what Frontier will do; experimentation is helpful.

The fault, though, is one of virtue: instead of complaining, Frontier tries to obey your commands gracefully. What on earth, for example, can the following possibly mean?

```
'helo' + true
```

Yet Frontier will permit it: `true` is converted to a string4 in accordance with the precedence rules, and then the two string4s are added, which means treating them as integers but representing the result as a string4 (the answer turns out to be `'help'`). We may think of booleans, dates, chars, string4s, together with actual numeric types, as "numeric equivalents"; if they are involved in numeric operations, they may be treated as numbers.

Runtime Coercion

There is no built-in facility to perform coercion to a datatype which will not be known until runtime.[*] Yet one does occasionally need such a facility. Suppose, for example, we want to present a scalar database entry in a dialog so that the user can edit it, and then replace the original value with the edited value. The edited value is always going to be a string, so we need to coerce it back to the type of the original scalar—whatever that may be.

Here is a utility verb, `coerceTo()`, which takes both something to coerce and a datatype to which to coerce it. It's simple: we make a record of all of the datatype-coercion verb pairs and perform a lookup; then we just construct and perform the coercion command. The script looks perfectly horrid, but it is worth having on hand.

Example 10-2. coerceTo()

```
on coerceTo (whatToCoerce, whatType)
    theRec = {charType:"char", longType:"long", binaryType:"binary", \
        dateType:"date", booleanType:"boolean", addressType:"address", \
        doubleType:"double", stringType:"string", directionType:"direction", \
        string4type:"string4", pointType:"point", rectType:"rect", \
        rgbType:"rgb", \
        fixedType:"fixed", singleType:"single", filespecType:"filespec", \
        aliasType:"alias", listType:"list", recordType:"record"}
    return theRec[whatType]^(whatToCoerce)
```

---

[*] On Mac OS there is a verb `coerceValue()`, but it isn't really a Frontier verb; it simply hands the arguments to the system's `AECoerceDesc()` function. This doesn't work for every Frontier datatype.

String and Char Literals

Character literals in UserTalk are delimited by single quotes. What appears between the single quotes must represent exactly one character, so `'h'` is legal, but `'hi'` will raise a compiler error.

String literals in UserTalk may be delimited either by ordinary double quotes, sometimes called straight quotes (`""`), or curly quotes, sometimes called smart quotes (`""`).* This provides an elegant way to express strings within strings. Suppose, for example, you wish to assign to a variable a literal string whose value is:

```
"I've dropped my toothpaste," said Tom crestfallenly.
```

You cannot do this by simply enclosing the value in straight quotes, like this:

```
x = ""I've dropped my toothpaste," said Tom crestfallenly." « won't compile
```

The problem is that Frontier sees what it thinks is an empty string literal:

```
x = ""
```

and then can't figure out what the rest of the line can possibly be. The solution is to enclose the value in curly quotes instead:

```
x = ""I've dropped my toothpaste," said Tom crestfallenly."
```

You can write it the other way round, too (use straight quotes for the outer pair and curly quotes for the inner pair), but the result is not the same; in the actual value of `x`, the quotes around what Tom says will be curly.

A second way to render a double quote inside a string is to "escape" it. "Escape" means that a special character is used within a string to change the value of what follows in some way. The basic escape character is the backslash. Note that all the backslash escape combinations can be used in either a string literal or a char literal. They are as follows:

`\"` Indicates a double quote. (Not needed in a char, but not illegal.)

`\'` Indicates a single quote. (Not needed in a string, but not illegal.)

`\r` Indicates a return character (char constant equivalent:† `cr`).

`\t` Indicates a tab character (char constant equivalent: `tab`).

`\n` Indicates a linefeed character (char constant equivalent: `lf`).

---

* On Mac OS, curly quotes are typed with Option-[and Option-Shift-[.

† These constant equivalents are at `system.verbs.constants`. They cannot be used inside a literal string, obviously, but are useful shorthand alternatives to char literals.

\x*nn*

> Indicates the character whose hexadecimal numeric equivalent is the two-digit short *nn*. It's okay to omit one of the digits (or both, in which case this is interpreted as \x00).

**** Indicates a backslash; necessary because the backslash is otherwise understood as escaping the next character, not as a literal backslash.

It is not an error to escape a character other than those listed here, but it won't have any effect, either.

Escape-notation sometimes confuses beginners. The key is to remember that it's just a way of writing; when you say "\r" in UserTalk you have used a string only one character long.

String literals are limited in UserTalk to 255 characters. However, string values (of database entries and variables) may be of any length, and one may use concatenation (or any other string operation) to generate them. Thus:

```
myString = "[ 500 characters ]" « fails
myString = "[ 250 characters ]" + "[ 250 more characters ]" « fine
```

How Frontier Displays String and Char Literals

The way Frontier displays string and char literals varies with the context; this can be confusing.

In reporting the value of a char or string in the Quick Script result box, or as a result of choosing **Run Selection** within a script, or in a table, Frontier will surround the literal with any required delimiters (such as double quotes or single quotes), and will use escape notation—for example, if the string contains a return character, you will see \r.* This is because the value is handed to the verb **displayString()** first, which generates the "external representation" of a string or char, the literal you would type if you were talking to Frontier. This is hard to read, because you have to understand escape-notation, but it's more precise because you really see what each character is, and it's useful because it shows how to represent the string as a literal.

In a dialog, or in the Main Window, a string or char value is interpreted. If the string contains a return character, then in a dialog you will see a new line at that point; the Main Window can't display interpreted return characters at all, because it's only one line. By the same token, 'a' displays 'a' in Quick Script, but **msg('a')** displays a in the Main Window, because the value is coerced to a string.

---

* There is a difference between the representation of literal strings in these situations, though: in the Quick Script result box, or as a result of choosing **Run Selection**, there are straight quotes around the string; in a table, there are not. Therefore, in a table, double quotes in a string literal are not escaped.

You can call `displayString()` yourself. For example, saying:

```
msg(displayString("Hello\rThere"))
```

causes `"Hello\rThere"` to be displayed, just as it would be if you said `"Hello\rThere"` in Quick Script. It gets tricky if you display the result of a call to `displayString()` in Quick Script, because `displayString()` is really being called twice. For example:

```
displayString("Hello\rThere")
    « "\"Hello\\rThere\""
```

This tells you the string literal you would have to type in order to make a dialog display "**Hello\rThere**"—which may or may not be what you wanted to know.

For a very different use of `displayString()`, see Chapter 32, *Driving Other Applications.*

11

Arrays

Tables, lists, records, strings, and binaries are all *arrays*. Arrays are ordered collections of *items*. A table's items are its entries, which can be of any type; list and record items can be of any scalar type; a string's items are chars. The items of a binary are treated as shorts or chars, depending on the context. As arrays, these datatypes have certain features in common; we will discuss these, and then proceed to some features that tables do not share.

Array Notation

The chief thing that all arrays have in common is that an individual item can be specified by means of an array specifier. Array specifier syntax, or *array notation*, is theArray[*theIndex*], where *theIndex* can be a number, or, in the case of tables and records, a name. The number is 1-based: that is, the first item is item 1. The name index for a table is a string; the name index for a record is a string4.

Array specifiers may be used to get or set the value of an item of an array. For example, if the first entry in **workspace** is a string object called **aLittleString**, whose value is **"hi"**, then:

```
workspace [1]
    « "hi"
workspace ["aLittleString"]
    « "hi"
```

Array specifier syntax using a name index on a table should not be confused with the syntax for constructing an element of an object reference at runtime. It is true that you can also access **workspace.aLittleString** by saying:

```
workspace.["aLittleString"]
```

This accesses the same entry, but in a different way. The name array specifier asks for that element of **workspace** whose name is **aLittleString**. The dot form constructs an entire object name, workspace.aLittleString, and uses it. The difference is clearer if the name of an entry in workspace is a number, say 2000: workspace.[2000] will access it, because we are forming the normally illegal name workspace.2000, but workspace[2000] will not, because there is (I presume) no 2000th entry in workspace.

Similarly for lists, records, and strings:

```
local
    myList = {"hello", "there"}
    myRecord = {'firs': "hello", 'seco': "there"}
    myString = "hello there"
myList [1]
    « "hello"
myRecord [1]
    « "hello"
myRecord ['firs']
    « "hello"
myString [1]
    « 'h'
```

Of course, the index does not have to be a literal; it is sufficient that it should evaluate to the desired value, or be coercible to the desired value. Thus:

```
local (x = 7 - 5)
myList [x]
    « "there"
myRecord ["firs"] « okay to use a string coercible to a string4
    « "hello"
myRecord ["1"] « okay to use a string coercible to a number, too
    « "hello"
```

Indexes can be "chained": the first specifies an item of an array, so if that's an array too, you can specify an item in it, and so on:

```
myList [1] [5]
    « 'o'
```

When using array notation to set the value of an array item, the change is made "in place": you aren't making a copy with a substitution performed, you're reaching right in and changing an item. In the case of a list or record you can use numeric array notation to create an entirely new item; the number must be exactly one more than the size of the list or record. In the case of a record, you can use name array notation to create a new item by name:

```
myString [1] = 'm'; myString
    « "mello there"
myRecord [2] = "and goodbye"; myRecord
    « {'firs': "hello", 'seco': "and goodbye"}
myList [3] = "you"; myList
    « {"hello", "there", "you"}
myRecord ['thir'] = "for now"; myRecord
    « {'firs': "hello", 'seco': "and goodbye", 'thir': "for now"}
```

With lists only, it is permissible to create a new item by setting the (fictitious) zeroth item of the list; that is:

```
myList = {"hello", "there"}
myList [0] = "you"; myList
    « {"hello", "there", "you"}, just like setting myList [3]
```

This saves one the trouble of determining beforehand how many items the list contains.

Array Verbs

Items of tables and records specified by number can be handed to the verb **nameOf()** to find out their name, which is returned as a string:

```
nameOf ( myRecord [1] )
    « "firs"
nameOf ( workspace [1] )
    « "aLittleString"
```

The number of items of an array can be determined by handing the name of the array to **sizeOf()**:*

```
sizeOf (myList)
    « 3
sizeOf (myString)
    « 11
```

A specified item of an array is removed with **delete()**. This doesn't make a copy lacking the specified item; it acts directly on the array:

```
myString = "Hello there"
delete ( myString [7] ); myString
    « "Hello here"
```

Some languages permit negative indices, letting you say such things as **myArray[-1]** to mean the last item of **myArray**, but UserTalk does not; you have to use **sizeOf()** and calculate the correct index. Here is a utility verb which returns the value of an array item specified by right-index. The parameters are the address of the array (to avoid copying a table or other large array) and a right-index value which can be positive or negative.

Example 11-1. rIndex()

```
on rIndex (addrArray, index)
    return (addrArray^[sizeOf(addrArray^)+1-abs(index)])
```

This ends our discussion of tables *qua* arrays; we shall have much more to say about working with tables through UserTalk in later chapters, especially Chapter 18, *Non-Scalars.*

---

* sizeOf() applied to a non-array reveals the amount of storage space it occupies, in bytes. This results in such wonders as sizeOf (infinity) == 4.

Lists and Records; Strings and Binaries

Several operators are defined on strings, binaries, lists, and records in common but not on tables. Note that there is normal implicit coercion of the operands (see Chapter 10, *Datatypes*).

Strings can be tested for equality with strings; so can lists with lists, and records with records. The operator is == (or **equals**). In the case of strings, the match is case-sensitive. In the case of lists and records, every corresponding pair of items must pass its own equality test in order for the whole test to return **true**; in this equality test, there is no implicit coercion—the types of each pair of compared items must match (except that a char can match a one-character string, and an address can match a string):[*]

```
{0} == 0
    « true, because the integer is coerced to a list first
{'h', 'i'} == {"h", "i"}
    « true, because a char "is" still a one-character string
{0} == {false}
    « false, even though 0 == false is true
```

Strings can be concatenated to strings, lists to lists, and records to records, using the plus-sign (+). The items of the second operand come after the items of the first operand in the result:

```
myString = "hello" + "there"
    « "hellothere"
myList = {"hello"} + {"there"}
    « {"hello", "there"}
myList = myList + "you"
    « {"hello", "there", "you"}, after "you" is coerced to {"you"}
myRecord = {'firs': "hello"} + {'seco': "there"}; myRecord
    « {'firs': "hello", 'seco': "there"}
```

In the case of a list, creating a new item by setting it directly is faster than "addition," and also doesn't involve implicit coercion to a list:

```
myList = {"hey", "ho"}
myList + {"hey", "nonny", "no"}
    « {"hey", "ho", "hey", "nonny", "no"}
myList + "{\"hey\", \"nonny\", \"no\"}"
    « {"hey", "ho", "hey", "nonny", "no"}
myList[0] = {"hey", "nonny", "no"}; myList
    « {"hey", "ho", {"hey", "nonny", "no"}}
myList[0] = "{\"hey\", \"nonny\", \"no\"}"; myList
    « {"hey", "ho", "{\"hey\", \"nonny\", \"no\"}"}
```

---

[*] This is because, owing to a feature of the operating system, chars and addresses are coerced to strings *before* they are stored in lists.

A substring can be removed from a string, or items from a list or record, using the minus sign (–). What is removed from the first operand is the first instance of all of the items of the second operand together and in the same order. This involves an implicit equality test between pairs of items, which works as described for == above. If the second operand isn't contained in the first, there is no error; the first operand is returned unchanged:

```
myString = "Madam, I'm Adam"
myString - "adam"
    « "M, I'm Adam"
myString - "Adam"
    « "Madam, I'm "
myList = {"h", "e", "l", "l", "o"}
myList - {"e", "o"}
    « {"h", "e", "l", "l", "o"}, the sub-list wasn't found
myList - {"e", "l"}
    « {"h", "l", "o"}
```

Three boolean tests, `beginsWith`, `endsWith`, and `contains`, tell whether the first operand contains all the items of the second together and in the same order —starting at the first operand's first item (`beginsWith`), or ending at its last item (`endsWith`), or anywhere (`contains`).* Again, the item equality test works as for ==, discussed earlier:

```
"Madam, I'm Adam" beginsWith "Mad"
    « true
{45, 63, 21, "hike"} beginsWith 45
    « true, after 45 is coerced to {45}
{45, 63, 21, "hike"} contains {45, 21}
    « false
{45, 63, 21, "hike"} contains "k"
    « false, because none of its elements is "k"
```

Because of implicit coercion, it is not an error to apply these operators to things that are not lists, records, or strings, but some very strange results can be incurred:

```
2001 beginsWith 20
    « true, because "2001" beginsWith "20" is true
workspace beginsWith "t"
    « true, because workspace is coerced to "table: 26 items"!
```

Lists and records have a keyword unique to them, `for...in`, which loops over all the items. This is discussed in Chapter 12, *Control Structures.*

---

* In the case of strings, an old verb, `string.hasSuffix()`, is superseded by `endsWith` and is not discussed further in this book.

Empty Strings, Lists, and Records

The empty string can be expressed as a literal as `""`, and the empty list can be expressed as a literal as `{}`. As for the empty record, though, while Frontier can portray an empty record as `{}`, you can't use this as a literal representation, because it will be seen as the empty *list*. The way to create an empty record is to use **new()**. Here, for instance, we assign the empty record to a local variable:

```
local (x)
new (recordType, @x)
```

Binaries

The `==` test returns **true** only if the internal datatypes match, as well as every "character." "Addition" is supported, returning a binary with internal datatype `'????'`, but "subtraction" is not supported. **beginsWith**, **endsWith**, and **contains** work, but they are a little tricky: for example, if **x** is **binary("hello")**, then **x[1]** is reported as a short, 104, but:

```
x beginsWith 104
    « false
x beginsWith char(104)
    « true
```

12

Control Structures

This chapter describes the UserTalk control structures that have not already been described. They are categorized as looping constructs, visit constructs, conditional constructs, and errors.

For on and **return**, see Chapter 5, *Handlers and Parameters*. For **with**, see Chapter 6, *Referring to Database Entries*. For **local**, see Chapter 7, *The Scope of Variables and Handlers*. For **bundle**, see Chapter 4, *What a UserTalk Script Is Like*, and Chapter 7.

Control structures all involve indented bundles, as explained in Chapter 4; for the significance of this with respect to scope, see Chapter 7.

Looping Constructs

A looping construct provides a way to perform the same commands over and over. Repetition is the essence of computer programming; most programming tasks are solved by finding a way to express them as a loop. From a theoretical point of view, all looping constructs are equivalent; a variety is provided merely for convenience of expression.

Syntactically, the pattern is that the construct is introduced by a keyword, usually with further information; the commands to be repeated then appear as a bundle indented from the keyword's line. For example:

```
for x = 1 to 10
    msg(x)
    clock.waitseconds(1)
msg("All done!")
```

The repeated task here is to display the value of **x** in the Main Window and then wait for one second. The nature of the repetition is dictated in the **for** line: here, the task is to be performed ten times, with **x** taking on a new value each time—first 1, then 2, and so on until it has been performed with **x** being 10. Only then will execution proceed to the next command, which displays **All done!** in the Main Window.

Two keywords provide ways of short-circuiting a loop. The `continue` statement skips all subsequent commands within the loop and proceeds to the next repetition (if any). For example:

```
for x = 1 to 10
    msg(x)
    if x > 5
        continue
    clock.waitseconds(1)
msg("All done!")
```

All ten iterations are performed, but in the last five, where **x** is more than 5, the `clock.waitseconds()` line is skipped. So the routine counts from 1 to 5 slowly and then from 6 to 10 very quickly, and then says **All done!**.

The `break` statement jumps out of the loop altogether. Thus:

```
for x = 1 to 10
    msg(x)
    if x > 5
        break
    clock.waitseconds(1)
msg("All done!")
```

counts from 1 to 6 slowly before the `if` test succeeds. Then **All done!** is displayed.

The `continue` and `break` statements are as close as UserTalk comes to having a `goto`. They apply only to loops; it is not an error to execute a `continue` or `break` outside of any loop, but the result is that execution is completely aborted. They apply only to the innermost loop containing them; they cannot leap out to a higher level. Use of local handlers and `try...else` (discussed later in this chapter) can provide that sort of control.

for

There are two forms:

```
for x = num1 to num2
for x = num1 downto num2
```

Any variable or database entry name can be used for the iteration variable on the left side of the assignment. Both *num1* and *num2* must be integers (positive, negative, or zero); if not, they will be coerced to integers (if possible).

Before any action is taken, *num1* and *num2* are compared. If *num1* is less than or equal to *num2* (in the first form), or greater than or equal to *num2* (in the second form), the iteration variable is assigned the value of *num1* and the indented bundle is performed. Subsequently, *num1* is maintained in Frontier's internal memory: its value is incremented by 1 (in the first form) or decremented by 1 (in the second form) and compared to *num2* in the same way as before, and if the test succeeds, then the iteration variable is assigned that new value and the indented bundle is performed again— and so on, until the comparison fails, whereupon the value of the iteration variable is left untouched and the indented bundle is skipped, and we go on with the script.

Thus it is perfectly possible that the loop will never be performed, and that the iteration variable will never be assigned a value by the for line. This suggests a useful trick for ascertaining whether the loop was ever performed: give the iteration variable some telltale value beforehand, and test for it afterwards:

```
local (x = "never")
for x = value1 to value2
    « whatever
if x == "never"
    « then we know the loop was never performed
```

It is perfectly legal, within the loop, to make changes to the value of the iteration variable. This does no harm to the iteration process, because it is not really the iteration variable that is incremented or decremented before each repetition: it is the most recent value of *num1* in Frontier's internal memory. This value will be assigned to the iteration variable if the loop is going to be performed again, over- riding any change to the iteration variable made within the loop.

Since, after the last iteration, the iteration variable's value is left untouched, it can be used after the loop. This is useful, among other things, for finding out how many iterations were actually performed.

Here, for example, is a utility which, given an array, reports the number of its first item having a given value. This is not the cleverest way to write such a utility, but it illustrates the use of the iteration variable after the loop is over.

Example 12-1. whichEntry()

```
on whichEntry (addrArray, match)
    local (i, found = false)
    for i = 1 to sizeof(addrArray^)
        if typeOf(addrArray^[i]) == typeOf(match)
            if addrArray^[i] == match
                found = true; break
    if found {return i}
    return 0
```

for...in

The syntax is:

```
for x in myList
```

where *myList* must be a record, a list, or something coercible to a list. Again, any variable or database entry name can be used for **x**. The construct loops through the indented commands, assigning to **x** before each iteration the value of each item of *myList*, in order. The result is exactly like saying:

```
for i = 1 to the sizeof(myList)
    x = myList[i]
    « ...
```

The shorthand construct is very useful, but ordinary **for** is sometimes necessary (as in Example 12-1).

while

The keyword is followed by an expression which evaluates to (or can be coerced to) a boolean. If the boolean is **true**, the indented bundle is performed; this process is repeated until the boolean is **false**, at which point the indented bundle is skipped and execution proceeds. Note that the boolean is evaluated anew before each repetition, and that any variables referred to within it may have new values as a result of the previous repetition; this is in contrast to **for**, where *num1* and *num2* are evaluated only once and the iteration variable is set by Frontier before each iteration.

A typical use is to initialize some boolean or counting variable beforehand, check it in the **while** line, and change it during the loop. Recall these lines from Example 4-7:

```
local (t = op.level ())
while --t {s = s + "    "}
```

The idea here is to reflect, using spaces, the depth of the current line of a script or outline. We use the pre-decrement operator (see Chapter 15, *Math*) to take care of the subtraction for us right in the **while** line; we can use a number as a boolean because every number except 0 is coerced to **true**.

loop()

This construct has two syntaxes. In the first, there are no parentheses, nothing follows the word **loop** on its line, and we just loop forever; it is up to the commands inside the loop to escape, typically with **break** or **return**. The same thing can be accomplished by saying **while true**:

```
local (x = 1)
loop « could have said, while true
    msg(x++)
    if x > 10 {break}
```

In the second syntax, the parentheses contain three items, separated by semi-colons. The first item is a command to be performed before the first iteration of the loop; the second item is a boolean expression to be evaluated before each iteration (the indented bundle will be executed only if this is true); the third item is a command to be performed before each iteration except the first. This will be clearer if we call the three items *p1*, *p2*, and *p3*; then, on the first pass, we perform *p1* and check *p2*; on subsequent passes we perform *p3* and check *p2*; if *p2* is false we skip the indented bundle and go on with the script.

There is nothing here that could not be done with a while loop; but the syntax is very convenient. Suppose, for example, we want to do what a for loop does but we want to increment by our iteration variable by .1 instead of 1. A for loop doesn't allow this. So we might use a loop() construct instead, saying, for example:

```
loop (local (x = 1.0); x <= 2.5; x = x + 0.1)
   « ...
```

As with while, any variable used in the loop() line can be altered in the indented commands, and it is the altered value that will be used as preparations are made to perform the loop again. For example, what does this script do?

```
loop (local (x = 1.0); x < 2.5; x = x + 0.1)
   msg(x)
   clock.waitseconds(1)
   x = x + 4
msg(x)
```

It displays 1.0, then 5.1, and stops.

fileloop()

This construct is for looping over files in a folder on disk. The basic syntax is:

```
fileloop (f in folderPathname)
```

The name f is just by way of example; it can be any variable or database entry name. On each iteration of the loop, f is assigned the full pathname of a different file or folder in the folder designated by *folderPathname*. The indented bundle uses this information as desired. For example, here is a script that lets the user select a folder and then reveals whether that folder contains any invisible files or folders:

```
local (f, whatFolder, foundAny = false, s)
if not file.getFolderDialog ("Please choose a folder.", @whatFolder)
   return
fileloop (f in whatFolder)
   if not file.isVisible(f)
      foundAny = true
      break
```

```
s = "The folder " + whatFolder + " contains "
if foundAny
    s = s + "at least one invisible file or folder."
else
    s = s + "no invisible files or folders."
dialog.notify(s)
```

TIP   The pathname of the absolute top-level "conceptual" folder is the empty string. Saying `fileloop (f in "")` will loop through the names of your mounted volumes.

There is a second syntax. The parentheses contain a second item, separated from the first by a comma; the second item is a positive integer indicating the depth to which the search on disk should be recursively performed. If the depth is 1, we will look only within the folder named by *folderPathname*. If the depth is 2, we will look within the folder named by *folderPathname*, and also within any folders it contains. And so on. If the depth is `infinity`, we will look as deep as there are files within folders. It is not an error to assign too large a depth.

With this second form of `fileloop()`, there's a catch. Only names of files—not folders—will be assigned to `f`. Thus, these two statements do not quite do the same thing:

```
fileloop (f in folderPathname)
fileloop (f in folderPathname, 1)
```

They both assign `f` pathnames of just the immediate contents of `folderPathname`; but in the first case, these will be the pathnames both of folders and of files, whereas in the second case, they will be the pathnames of files only.

Here, for example, is a script that lets the user select a folder and then tells the name of the first invisible file (but not folder!) encountered at *any* depth within it:

```
local (f, whatFolder, foundAny = false, s)
if not file.getFolderDialog ("Please choose a folder.", @whatFolder)
    return
fileloop (f in whatFolder, infinity)
    if not file.isVisible(f)
        foundAny = true
        break
s = "The folder " + whatFolder + " contains "
if foundAny
    s = s + "at least one invisible file, " + f + "."
else
    s = s + "no invisible files at any depth."
dialog.notify(s)
```

The question then arises of how to get f to receive the names of both files and folders, yet pursue the search to a depth greater than 1. One answer is to use the first syntax of fileloop() but write a recursive handler; that is, we perform the recursion ourselves rather than asking fileloop() to do it. The technique is so common that its basic structure is worth memorizing; it is then easy to build any particular functionality around it.

Example 12-2. traverse()

```
on traverse (path)
    fileloop (f in path)
        if file.isFolder (f) « dive into the folder
            traverse (f) « recurse
```

For example, here is a routine to let the user select a folder and then tell the name of the first invisible file *or folder* encountered at any depth within it.

Example 12-3. Using traverse()

```
local (f, whatFolder, foundAny = false, s, whatF)
if not file.getFolderDialog ("Please choose a folder.", @whatFolder)
    return
on traverse (path)
    fileloop (f in path)
        if foundAny {return} « if we're unwinding the recursion, keep unwinding
        if not file.isVisible(f)
            foundAny = true; whatF = f « retain the file name
            return
        if file.isFolder (f) «dive into the folder
            traverse (f) «recurse
traverse (whatFolder)
s = "The folder " + whatFolder + " contains "
if foundAny
    s = s + "at least one invisible file or folder, " + whatF + "."
else
    s = s + "no invisible files or folders at any depth."
dialog.notify(s)
```

Here we've made some slight adjustments to take account of the recursive nature of things. We want to stop as soon as we hit the first invisible file or folder, but we can't just break out of the loop: we have to unwind the recursion all the way up to the surface. And, to retain the name of the invisible file or folder, we store it in a global, whatF, because f after traverse() unwinds completely will be the f of surface level, not the f that caused us to stop.

A particularly beautiful and educational example of traverse() (and UserTalk recursion in general) is suites.samples.basicStuff.buildFolderOutline(), which constructs an outline whose hierarchy reflects the hierarchy of folders and files within a user-designated folder.

If writing your own recursive handler scares you, or you just can't be bothered, you can use `file.visitFolder()` instead (see later in this chapter).

Visit Constructs

A visit construct is a looping construct rather like `for...in` or `fileloop()`: we start with a set of similar entities, and Frontier loops over them for us, handing us each entity in turn. The calling syntax is a bit tricky. A visit verb is an ordinary script,* not a keyword, so the commands to be executed during each loop cannot be indented below it. Instead, those commands are encapsulated in a handler or script whose address is handed as a parameter to the visit verb (see Chapter 8, *Addresses*).

In the discussion that follows, I shall refer to the handler whose address is passed as a parameter as "the proc"; it is often referred to in Frontier circles as a "callback," but I regard this as an incorrect use of the term. I shall refer to the set of similar entities over which a visit verb loops as its "domain."

The proc must return a boolean, or something that can be coerced to a boolean. The visit verb examines the proc's result after each iteration, and proceeds to the next iteration only if it is not `false`.

Further, the proc must take exactly one parameter in which to receive each entity in turn. The only exception is the proc for `op.visit()`; it takes no parameters.

There is something of an art to writing a proc, because it isn't itself a loop: it is called anew for each iteration. Thus, it cannot maintain state within itself. How, for example, might the proc determine how many times it has been called so far? A common solution is to employ variables global to the proc, where state can be maintained; we shall see an example of this immediately.

window.visit()

This construct takes one parameter, the address of the proc. The domain is the set of all open windows within the Frontier application. The parameter passed to the proc each time is a reference to an open window; the windows are traversed in front-to-back order (window references are discussed in more detail in Chapter 24, *Windows*).

---

* The scripts of the visit verbs make interesting and educational reading.

For example, here is a script that arranges all open Frontier windows into a cascade.

Example 12-4. cascadeWindows()

```
local(cascH = 27, cascV = 66)
on cascadeOneWindow (addrWindow)
    if addrWindow == "Frontier.root"
        window.setposition("Frontier.root", 5, 22)
        window.sendToBack("Frontier.root")
    else
        window.setPosition(addrWindow, cascH, cascV)
        window.bringToFront (addrWindow)
        window.getPosition(addrWindow, @cascH, @cascV)
        cascV = cascV + 22
        cascH = cascH + 22
    return true « not really needed
window.visit(@cascadeOneWindow)
```

The script defines the proc, and then just hands it to **window.visit()**. Notice the use of variables **cascH** and **cascV** global to the proc; this allows their values to be maintained between calls to the proc.

In this routine we treat the Main Window specially: we want it located first; it has no titlebar, so its "position" is its actual upper left-hand corner, whereas a normal window's position is the upper left-hand corner of its content region. We check where each window was actually placed, because Frontier will override our instructions if we try to use **window.setPosition()** to move the right edge of a window off the screen.

stack.visit()

This construct takes two parameters: the address of the stack and the address of the proc. The domain is all the values in the stack pointed to by first parameter. Thus, the proc has no opportunity to alter any values in the stack; it is receiving values, not addresses. Stacks are described in Chapter 22, *Stacks*.

op.visit()

This construct takes one parameter, the address of the proc; the proc should take no parameters. It is assumed that the "target" (the outline to be worked on) has already been set, and that a desired line of that outline has already been selected; **op.visit()** selects each subhead of that line and calls the proc, which can then use op verbs that act upon the currently selected line. (The "target" and the op verbs are discussed in Chapter 18, *Non-Scalars*.)

Not only subheads of the originally selected line are traversed, but also subheads of subheads. But the expansion state is not changed. So the domain of the verb is all the subheads of the originally selected line, to any depth at which they are currently expanded. A line is selected, and the proc is called, before the first of that line's expanded subheads (if any) is selected.

The originally selected line is not among those for which the proc is called. The original position of the cursor is not restored at the end; this is up to the calling routine, if desired.

In this dramatic and somewhat useless example, we make an outline containing the names of the 50 states of the United States and then delete any whose name contains the letter "e":

```
on deleteWithE()
    if op.getLineText() contains "e"
        op.deleteLine()
bundle « create the outline
    new (outlineType, @workspace.theStates)
    edit(@workspace.theStates)
    op.wipe(); op.setLineText("The States")
    loop (local (x = 1); x <= 50; x++)
        op.insert(states.nthState(x), down)
    op.go(up, infinity); op.demote()
op.visit(@deleteWithE)
```

The **bundle** material just makes the outline for us to work on; the only interesting part (for our purposes) is the proc, deleteWithE(). Notice that op.visit() is clever enough to deal with the possibility that the currently selected line might be deleted by the proc.

table.visit()

This construct takes two parameters: the address of a table and the address of the proc. Its domain is first, the address of the original table, and second, the address of each entry in that table, including the address of each entry in any subtables, and so on. The proc always receives a table just before its entries.

Often what one wants to do is process a table's contents but not the table itself. It is therefore useful to be able to avoid having the proc process the first value handed to it. Use of a variable global to the proc can help accomplish this.

In this example we count the number of scripts in the **suites** table, at any depth. For illustrative purposes, we show how to avoid processing the **suites** table itself, though there is no actual need to do so. (The technique slows down processing, since an extra check must be made on each call to the proc; but this is unavoidable.)

```
local (ct = 0, firstTime = true)
on isScript(addrThing)
    if firstTime
        firstTime = false
        return true
    msg (addrThing) « provide some feedback
    if typeOf(addrThing^) == scriptType
        ct++
    return true
table.visit(@suites, @isScript)
dialog.notify ("The Suites table contains " + ct + " scripts.")
```

file.visitFolder()

This verb is an alternative to a `fileloop()`. It takes three parameters: the path-name of a folder, an integer representing the depth of recursion to be used, and the address of the proc. Unlike `fileloop()` used with a depth specifier, the proc receives the names of both files and folders. Files in a folder are received just before that folder; this is the reverse of `table.visit()`.

Here is a rewrite of Example 12-3 using `file.visitFolder()`:

```
local (whatFolder, whatF = false)
if not file.getFolderDialog ("Please choose a folder.", @whatFolder)
    return
on checkVisibility (f)
    if not file.isVisible (f)
        whatF = f
        return false « signal to stop
    return true « otherwise keep looking
file.visitFolder(whatFolder, infinity, @checkVisibility)
s = "The folder " + whatFolder + " contains "
if whatF
    s = s + "at least one invisible file or folder, " + whatF + "."
else
    s = s + "no invisible files or folders at any depth."
dialog.notify(s)
```

Conditional Constructs

Conditional constructs allow a bifurcation in the sequence of command execution: depending on a test performed at runtime, this set of commands or that set of commands will be performed.

if...else

There are two forms: `if` may be used alone, or it may be used with `else`. The rest of the `if` line must evaluate to, or be coercible to, a boolean. For operators that can be used in boolean expressions, see Chapter 44, *Operators*.

The first form looks like this:

```
if boolean
    « what to do if so
« later stuff
```

The indented bundle will be executed only if the boolean is `true`.

In the second form, `else` appears as the first line following the `if` line at the same level, like this:*

```
if boolean
    « what to do if so
else
    « what to do if not
« later stuff
```

In this form, if the boolean is `true`, we execute the `if`'s indented bundle, then skip the `else` (and its indented bundle) and proceed to the "later stuff"; but if the boolean is `false`, we execute the `else`'s indented bundle before proceeding to the "later stuff."

case

The `case` construct is a way of specifying different actions depending upon various discrete values that an expression might have at runtime. This example, although by no means how one would optimally accomplish the particular task, illustrates the basic syntax clearly:

```
local
    howMany = dialog.threeWay ("How many times would you like me to beep?", \
    "Once", "Twice", "Three times")
case howMany
    1
        speaker.beep()
    2
        speaker.beep()
        clock.waitSixtieths(10)
        speaker.beep()
    3
        speaker.beep()
        clock.waitSixtieths(10)
        speaker.beep()
        clock.waitSixtieths(10)
        speaker.beep()
```

As you can see, what follows `case` on its own line is an expression to be evaluated; then, one level indented, there are possible values for that expression, and,

---

* It is a compile-time error for `else` to appear anywhere but in this relationship to an `if` (or `case`, or `try`).

indented from those, what to do if the expression matches that value.* If more than one discrete value is to have the same action, you can list multiple values on one line separated by semicolons, or on successive lines like this:

```
local
    howMany = dialog.threeWay ("How many times would you like me to beep?", \
    "Twice", "Two Times", "Three times")
case howMany
    1
    2
        speaker.beep()
        clock.waitSixtieths(10)
        speaker.beep()
    3
        speaker.beep()
        clock.waitSixtieths(10)
        speaker.beep()
        clock.waitSixtieths(10)
        speaker.beep()
```

An optional `else` may follow immediately at the same level as the **case** line; its indented bundle will be executed if none of the listed values was matched.

case will not accept ranges as values: the match between the expression and a value must be one of equality. However, a trick lets you take advantage of **case** syntax to build what amounts to a multiple `if...else`: use **case true**, and then list booleans as values. Since the booleans are tested in the order they appear, this is the equivalent of the `if...else if...else if...else` construct supported by many languages:

```
local (whatNum)
if not dialog.ask("Pick an integer from 1 to 10.", @whatNum)
    return
whatNum = double(whatNum)
case true
    whatNum < 1
        dialog.alert ("That's too small.")
    whatNum > 10
        dialog.alert ("That's too large.")
    whatNum ≠ long(whatNum)
        dialog.alert ("That isn't an integer.")
else
    dialog.notify ("Good, you can follow directions.")
```

If we reach the third case, the number is between 1 and 10 inclusive.

Because of a bug, a script will refuse to compile if it contains a **case** construct whose last value is commented out.

---

* Unlike C, it is not true that one "falls" from one case into the next unless prevented by a **break**.

Errors

An error during the execution of a script means that the script cannot continue, but UserTalk makes it possible to keep going anyway, or at least to come to a halt in an orderly fashion. In fact, we shall see that your script can deliberately cause an error as a means of determining what happens next.

try...else

If a runtime error occurs, Frontier puts up an error dialog and execution comes to a halt. Sometimes, though, an error is not worth stopping your whole script for. You can prevent a halt in execution by *trapping* for errors; your script remains in control even if an error occurs. This is done by placing any commands that you think might generate an error in a `try` bundle.

There may or may not be an `else` immediately following at the same level as the `try`. If there is no `else`, then if a command in the `try` bundle generates an error, the rest of the `try` bundle is skipped, and we proceed to the first command after the `try`. If there is an `else`, then if a command in the `try` bundle generates an error, the rest of the `try` bundle is skipped and we execute the `else` bundle, and then proceed; if there is no error generated in the `try` bundle, the `else` bundle is never executed.

A typical place for using `try` alone is in a loop where a non-fatal error can result—you don't care about the error, you just want to go on to the next iteration of the loop. For instance, the script at **samples.basicStuff.reallyEmptyTrash** uses a `fileloop()` to go through everything in the *Trash* folder and erase it. If an error occurs, it's just because the file is somehow in use; we don't mind leaving that file alone, but we'd like to continue with the loop. So the attempt at deletion is in a `try` bundle:

```
local (vol, f, trashfolder)
fileloop (vol in "")
    trashfolder = vol + "Trash:"
    try
        fileloop (f in trashfolder)
            try
                if file.isFolder
                    file.deleteFolder (f)
                else
                    file.delete (f)
```

Another common use of `try` alone is where you don't want to check for the existence of something whose nonexistence is unimportant but would cause an error.

For example, in this line from **toys.listToOutline**:

```
try {delete (@scratchpad.xxx)}
```

we want to delete the database entry **scratchpad.xxx** if it exists. If it doesn't exist, asking to delete it will raise an error, but we don't care because then there's nothing to delete anyway.

The **else** option might be used in case the error means that execution can't proceed without mopping something up, or to allow smooth reporting or analysis of the error. We'll look at some examples in a moment.*

scriptError() and tryError

A call to **scriptError()** actually causes an error. The verb's parameter is a string; if the error it causes is not trapped by a **try**, Frontier puts up its usual error dialog, containing that string.

Why would anyone want to do something like this? One reason might be simply that we have in fact ended up in a fatal situation, and need to abort. It isn't sufficient merely to **return**; that will just put the caller of the present handler back in charge. What we want to do is stop *all* activity, and supply an informative error message as we do so.

For example, in this snippet from **backups.backuproot()**, we try to create a folder using **file.sureFolder()**; we know if this fails because the call returns **false**. So we test for that, and throw an error if it does:

```
if not file.sureFolder (backupfolder) «couldn't create the folder
    scriptError ("Couldn't create the backup folder" + \
        "or a file named 'backups' already exists.")
```

Here is another example snippet, simplified from **html.data.standard-Macros.renderObject()**. The script is trying to set the value of **addrScript**; it has been given something called **name**, which is a kind of partial reference, and makes several guesses as to where in the database **name** might live. If none of them works, the whole operation needs to come to a halt:

```
addrScript = address (name)
if defined (addrScript^)
    return
addrScript = @html.data.page.tools^.[name]
if defined (addrScript^)
    return
addrScript = @user.html.renderers.[name]
if defined (addrScript^)
    return
scriptError ("Can't find a rendering script named \"" + name + "\".")
```

---

* Because of a bug, the **try...else** mechanism does not work when scripts such as **system.misc.command2click()** and **system.misc.closeWindow()** run automatically (i.e., in response to the user Command-double-clicking text or closing a window).

Another reason for throwing our own error is that a fatal error has already occurred, but we think we can make a more informative error dialog than the original error does. The original error is trapped in a `try`, and we substitute our own error message in the corresponding `else`.

In this example snippet, again simplified from `renderObject()`, we have been trying, in a `case` construct, to deal with an entity (pointed at by `addr`) in various ways, depending upon its type. Finally, in the `case`'s corresponding `else`, we make one last-ditch effort: we coerce `addr^` to a string. If *that* fails, we trap the error and substitute our own error message for the one Frontier proposes to generate:

```
try
    s = string (addr^)
else
    theError = "Can't render \"" + string (addr) + "\" because " + \
        " its type is \"" + objecttype + "\""
    scriptError (theError)
```

It frequently happens that we wish to know what the error message would have been for an error that has been trapped within a `try` (this error message will have been generated by Frontier or some application it was talking to, or by a call to `scriptError()`) if we hadn't trapped it in the `try`. Access to the error message generated in a `try` bundle is available in the corresponding `else` bundle through a global system variable `tryError`.[*]

In this example snippet from `html.ftpText()` we attempt to upload a file via FTP; if this fails, we examine the error message generated. If it has to do with memory, the situation is fatal, and we just pass the error along unchanged; otherwise, we assume a needed directory doesn't exist, so we create it and try again:

```
try
    ftpClient.store (f, domain, path, account, password)
else
    if tryError contains "memory"
        scriptError (tryError)
    else
        ftpClient.sureFilePath (domain, path, account, password)
        ftpClient.store (f, domain, path, account, password)
```

syscrash()

This verb causes your system to, well, crash. If you have a system-level debugger, such as MacsBug, Frontier drops into the debugger, and, if `syscrash()` was

---

[*] It is a runtime error for the name `tryError` to be used except in the `else` of a `try`.

given a string parameter, that string is displayed. This can be used for last-ditch debugging if Frontier itself is having some kind of problem (for example, I once used it to detect a bug in the `try...else` mechanism); but generally it is of no interest to UserTalk scripters, since Frontier's own debugging facilities will suffice. These are discussed in Chapter 13, *Running and Debugging Scripts*.

13

Running and Debugging Scripts

A script is just a series of instructions until it actualizes its potential by being executed, also known as *running* the script. Furthermore, before a UserTalk script can be run, it is always compiled. This chapter describes how compilation and running of scripts takes place. And, since bugs in your script are proverbially as inevitable as death and taxes, you're going to want to isolate and mend those bugs as they crop up; this you can do through Frontier's excellent debugging environment, which this chapter also describes. Finally, we discuss some additional ways in which Frontier helps you figure out what your script does and how to make it do what you want it to.

Compiling

Before a script can be run or debugged, it must be compiled. Compilation involves translation of the script into tokenized form (p-code) by way of the tables in `system.compiler`. Compilation is extremely rapid; on most machines, the time involved is typically too small to measure accurately (one- or two-sixtieths of a second).

You compile a script by pressing the **Compile** button in the script's edit window. If you make changes in the window and then press the edit window's **Run** button or **Debug** button without pressing **Compile**, Frontier compiles implicitly and proceeds. However, the compiled version generated by such implicit compilation is not the same as the compiled version generated by explicit compilation by way of the **Compile** button.

It is the explicit version that is run when a script is called as a verb. For this reason, if you make changes in a script window and then close the window, Frontier will offer to perform explicit compilation.[*]

The chief reason for this behavior is that a script can be executed without the user initiating it directly—for example, agent scripts run automatically in the background, and scripts can be triggered by commands from other applications. Scripts can call other scripts, so it is perfectly possible that a script object could be called while you are editing it. The script at that moment might be unfinished, buggy or dangerous. Therefore, it is the explicitly compiled version that runs; thus, there will be no danger as long as the script's current state has never been explicitly compiled.

So the existence of an implicitly and an explicitly compiled version of a script is important and useful. But it can also be confusing. If you modify a script, test it with the **Run** button, and then execute another script which calls it, the modifications may mysteriously fail to work—because you've forgotten to compile the script explicitly. A particularly insidious variant can arise when you press **Run** in a script that calls itself: initially, the implicitly compiled version runs, but when it calls itself, the script calls the explicitly compiled version. Then, when you try to track down the problem in Debug mode, debugging itself appears to have broken; the text version of the script is out of synch with the explicitly compiled version that is actually being executed.

The moral is: press the **Compile** button! On the other hand, if the script is one which might be called automatically, *don't* press the **Compile** button until the script is in acceptable shape (see Chapter 27, *Agents and Hooks*).

NOTE   Menu item scripts have no **Compile** button, because the implicitly compiled version is the only version needed—such a script will never be called in an unexpected automatic way.

The Kernel

The token tables that are located at `system.compiler.kernel` and `system.compiler.language` are called the *kernel*. A verb is truly "built in" to Frontier if it has a token here, because it is implemented in the Frontier application. All other verbs are implemented as scripts.

---

[*] When Frontier starts up, a script has no explicit compiled version, so Frontier will make one automatically the first time the script is called. Therefore, if you make changes to a script just after startup and close the script window, no compile prompt appears.

The programmer should never call the kernel explicitly. Pre-evaluation takes care of routing calls to the verbs in `system.compiler.language.builtins` (see Chapter 9, *Special Evaluation*), and scripts are provided which route calls to the other kernel verbs. So, for example, when you use `pack()`, pre-evaluation translates it to `system.compiler.language.builtins.pack`; when you use `msg()`, the reference is resolved to `system.verbs.globals.msg` and the script there calls the kernel.

Calls to the kernel happen primarily by means of the `kernel()` verb. The programmer should never use this verb! It works in unusual ways. It must be always the only command in its script. And it has the remarkable property that the parameters to its caller are passed on automatically. For instance, if you examine `system.verbs.globals.msg()`, you see that it receives a string parameter, but apparently fails to pass it on when it calls the kernel; that's because it's passing it on by a different mechanism.[*]

Compile Errors

When Frontier compiles a script, it puts up an error dialog if a syntax error prevents compilation. There is a **Go To** button in this dialog, and if you press it (or hit Enter), Frontier will place the insertion point after the error in the script.

When Frontier does this, it is thinking like a machine, so its idea of where the error is may differ from yours. You may have to hunt back a little to find what a human being would think of as the culprit. For example, if you try to compile the following:

```
locals (x = 6)
```

when the error message appears and you press **Go To**, Frontier places the insertion point after the =, because it is illegal to use this symbol in the parameter list of a verb call. The fact is, of course, that you never intended a verb call: the real problem is that you typed `locals` for `local`. Or, if you try to compile:

```
local (x = target.get()
msg(x)
```

then when the error message appears and you press **Go To**, Frontier places the insertion point in the second line, after the left parenthesis; but the problem is in the first line, where a right parenthesis has been omitted.

---

[*] Also known as "magic."

Script Object Verbs

Several verbs do compilation-related things to script objects as a whole.

To compile a script:

```
script.compile (addrScriptObject)
```

> Identical to pressing the **Compile** button in the script object's edit window.

To clear the memory occupied by a script's explicitly compiled version:

```
script.uncompile (addrScriptObject)
```

> The script will take longer to start executing when next called, but some memory
> is freed up. Chiefly a method of housekeeping; for an example, see
> `suites.samples.basicStuff.windowFormatter()` (discussed in Chapter 24,
> *Windows*).

To clear a script's text version:

```
script.removeSource (addrScriptObject)
```

> The script object becomes henceforward permanently uneditable—indeed, it
> ceases to be a script object, becoming instead a "compiled code" object—but it
> can still be called. Useful chiefly for security purposes, as a way to distribute a
> script no one can read. Since *you* can't read it either, be sure to keep a normal
> copy. Called by **Remove Source Code** in the **Script** menu.

To learn or specify a script's OSA dialect:

```
script.getLanguage (addrScriptObject)
script.setLanguage (addrScriptObject, languageString)
```

> Identical to using the language popup menu at the bottom of the script object's
> edit window. OSA dialects are typically `"UserTalk"` or `"AppleScript"`. On
> scripts in non-UserTalk dialects, see Chapter 33, *AppleScript*.

Running a Script

The following are the ways in which script execution may be initiated:*

*   The user presses the **Run** button in a script edit window.

    This causes evaluation of the surface-level commands in the script, which, as
    we saw in Chapter 5, *Handlers and Parameters*, may have a different effect
    from calling the script as a verb.

*   The user enters a UserTalk expression into the Quick Script window and
    presses its **Run** button (or hits Enter).

    The Quick Script window is summoned by choosing **Quick Script** from the
    Open menu. It can be resized, but the result area cannot be made to contain
    more than one line.

---

* Calling a script object from another script is not included, because by definition the caller is already
somehow running; the question is how that might ultimately have been caused.

Quick Script serves as a convenient command-line interface to Frontier. Its **Run** button causes evaluation of its contents as a UserTalk expression; thus, if the window contains a verb call, the verb will be executed. Another possibility is to enter a script directly into Quick Script; but note that if more than one command is to be entered, the semicolon-and-curly-braces syntax appropriate to a text environment must be used (Chapter 4, *What a UserTalk Script Is Like*).

The result of evaluating the contents of the Quick Script window appears automatically in the result field at the bottom of the window (after first being passed through displayString()—see Chapter 10, *Datatypes*). If you wish to capture the result of a Quick Script you must include commands to do so in the code itself.

- The user chooses a menu item (or types its keyboard shortcut) to which a menu item script is attached.

  This is how all Frontier menus (except for **File**, **Edit**, and **Window**) are implemented. Every menu item has a script; choosing the menu item runs that script. The script runs as when its **Run** button is pressed, executing its top-level commands; menu item scripts don't have eponymous handlers. The creation and editing of menu item scripts is discussed in Chapter 26, *Menus and Suites*.

- The user chooses **Run Selection** from the **Main** menu.

  What happens when the user does this depends upon what is selected at the time.[*]

  In a *table*, if a *string* object is selected, the string is evaluated; if a *script* object is selected, the script is called with no parameters.[†] The result is displayed both in the result field at the bottom of the table window and in the Main Window.[‡]

  In a *wptext*, the currently selected text is evaluated as a UserTalk expression. The result is displayed in the Main Window.[§]

  In a *menubar*, the selected menu item's script is run, just as if the menu item had been chosen from the menu.

---

[*] The behavior of **Run Selection** is implemented at system.verbs.globals.runSelection.

[†] If the script requires parameters, an error results.

[‡] This overwrites anything displayed in the Main Window during the evaluation; you may wish to modify this behavior. A MacBird card (a binary object of type 'CARD') can also be run by selecting it in a table and choosing **Run Selection**.

[§] This overwrites anything displayed in the Main Window during the evaluation; you may wish to modify this behavior.

In a *script* or *outline*, the entirety of the current line is evaluated as a User-Talk expression. This works even if the current line is itself a comment. The result of the evaluation is displayed as a comment indented below the originally selected line. If there was already a comment in that position, it is overwritten. The database is full of "one-liners," lines in scripts or outlines intended to be selected and evaluated with **Run Selection** for testing purposes; see, for example, `suites.webBrowser.examples.oneLiners`, or the last line of `suites.toys.commentDelete`.

- Frontier automatically calls an agent script.

  Agent scripts are called automatically at intervals which they themselves dictate. See Chapter 27, *Agents and Hooks*.

- Frontier automatically calls a "hook" script.

  Hook scripts are called automatically in response to certain events. See Chapter 27.

- A desktop script is opened from the Finder.

  See Chapter 29, *Import/Export*.

- Frontier receives a command from another application.

  See Chapter 34, *Driving Frontier From Outside*.

Debug Mode

Debug mode is entered by pressing a script edit window's **Debug** button. Thereupon, Frontier will propose to treat the script as if you had pressed its **Run** button, but it pauses with the first executable surface-level command selected and awaits further instructions.

A new set of buttons has now appeared at the top of the window, and you proceed by pressing one of them. (I'll describe what they do in a moment.) Frontier remains in Debug mode until either the script terminates naturally or you exit it with the **Kill** button. The buttons then return to their normal state.

In the heat of debugging, you might start making changes to a script. You should not continue debugging a script which you have accidentally changed, because the script's text may no longer correspond to the code Frontier is executing, and you'll be confused. Exit Debug mode and start debugging again.

During debugging, there may be several script edit windows, and you might press a button in a window that is not active. It might appear that nothing has happened, which can be confusing. What's going on is that the first click in a non-active window just makes the window active; you thought you were pressing the button, but you weren't.

Breakpoints and the Execution Path

A *breakpoint* is a line of a script that is specially marked: it starts with a "stop" hand icon instead of a triangle. The *execution path* is the complete train of commands that Frontier actually executes during a debugging session as one script calls another.*

In Debug mode, any time the execution path passes through a breakpointed line, Frontier will pause, open, and bring frontmost the script's edit window, and select the breakpointed line, without yet having executed it.

The presence of a breakpoint has no effect when you are not in Debug mode. The database is full of breakpoints left there by the developers of the scripts, but they do not affect normal execution.

To toggle a line's status between normal command and breakpoint, Command-click its triangle or hand, or choose **Toggle Breakpoint** from the **Script** menu. You can set a breakpoint at any time, not just in Debug mode. In Debug mode, it's fine to set and remove breakpoints "on the fly"; for example, if you set a breakpoint in a loop after a couple of iterations of the loop, Frontier will pause there on the next iteration (and then you can remove it if you like).

WARNING   An `on` line, an `else` line, a `bundle` line, or a line indicating one of the possible values in a `case` is not an executable line. Neither is a `local` or a line subordinate to a `local`, unless assignment takes place there. You can breakpoint such a line, but Frontier will never pause there. You can breakpoint a comment, but this will have no effect, and the hand will not be visible, until the comment is turned to a command line.

The Debug Mode Buttons

We now describe the behavior of the Debug mode buttons. The following two rules should be borne in mind:

1. Whenever Debug mode pauses, it selects the command it is about to execute.

2. Breakpoints take priority over everything else; in Debug mode, Frontier always pauses before executing a breakpointed line.

The **Go** button causes the script to continue execution without pausing until either all execution terminates (at which point Frontier leaves Debug mode) or a

---

* For purposes of debugging, the execution path is never considered to pass through any script which consists solely of a call to `kernel()`.

breakpoint is encountered. The **Go** button changes to **Stop** during execution; the **Stop** button causes a pause.

The **Step** button causes execution of the currently selected line, pausing before the next line to be executed. If the current line contains a verb call, the execution path enters and returns from the called script or handler without pausing (unless a breakpoint is encountered).

The **In** button causes the script to continue execution without pausing (unless a breakpoint is encountered) until a verb call is executed (Frontier pauses at the first executable line of the called handler or script) or until the current handler or script terminates.

The **Out** button causes the script to continue execution without pausing (unless a breakpoint is encountered) until the current handler or script terminates.

The **Follow** button causes the script to continue execution without pausing (unless a breakpoint is encountered), and each line of any open script through which the execution path passes is selected as it is executed. This makes it possible to watch the execution path, but no windows are opened, nor are windows brought to the front as the execution path passes through them; also, the speed is rather fast, and cannot be changed. The **Go** button changes to **Stop** during execution; the **Stop** button causes a pause.

The **Kill** button aborts execution and brings Frontier out of Debug mode. You can use Command-period instead.

Variable Values

In Debug mode, the value of local variables can be consulted. This is actually just a consequence of the fact that it is possible to pause. During execution, as Frontier encounters each new scope, it creates a subtable in `system.compiler.stack` and maintains there the names and values of any local variables created in that scope; when that scope goes out of existence, so does the corresponding subtable. In Debug mode, you can examine these subtables.*

The most convenient way to examine local variable values in Debug mode is to push the **Lookup** button. If the **Lookup** button is pressed while a line as a whole is selected, the subtable for that scope opens. If the **Lookup** button is pressed while the name of a variable is selected, the subtable containing that variable opens; or you can get the same effect by Command-double-clicking the name.

---

* You can change the values of variables in these tables, but you shouldn't: it's dangerous. As Doug Baron has said, "If Frontier supported locked windows these would be the first ones we'd lock."

Navigating the subtables by hand is also possible. Jump to `system.compiler.stack` and find the subtable containing the variable you wish to examine. The name of a subtable tells which scope within what script it represents. It may take some hunting to find a desired variable.

You can take advantage of a pause to examine other parts of the database as well. If the **Lookup** button is pressed while the name of a database object is selected, that object opens or is revealed; or you can get the same effect by Command-double-clicking the name.

Runtime Errors

When a runtime error is not trapped by a `try`, Frontier puts up an error dialog.

Even when Frontier is not in Debug mode, this error dialog can be used to help track down the error. Clicking the **Go To** button (or hitting Enter) opens the script window and puts the selection point at the line that caused the error. Or, holding down the **Go To** button pops up a menu which shows the calling chain; you can use this information to help figure out what was going on when the error occurred, as well as to navigate to the various handlers and scripts named in the menu.

If Frontier *is* in Debug mode, exactly the same thing happens, but the compiler's variable stack is not emptied, so you can also examine variable values. However, you can't continue debugging; after all, an error has occurred. So after you've finished examining things, you must exit Debug mode.

Since the compiler stack is emptied when a runtime error occurs not in Debug mode, a useful strategy is then to run the same routine in Debug mode so that variables can be examined when the error occurs again.

TIP   The text of the error dialog can be copied to the clipboard by choosing **Copy** when it is frontmost.

Runtime errors are by nature often pesky to track down, even with all the help Frontier gives you. It can be useful to modify the code temporarily to be more informative. A judicious sprinkling of `try...else` structures and `msg()` or `dialog.notify()` calls can help you work out what was happening when the error occurred.

Desperation

To abort a running operation, hit Command-period. You may have to hit it several times or hold it down.

If you think the database may have been damaged, choose **Revert** from the **File** menu, or quit without saving the database. If you're still worried about its integrity, you can check whether Frontier has lost the internal consistency of any tables by choosing **UserLand Testing** from the **Suites** menu and then **Validate Tables** from the **Test** menu that appears. But this won't tell you if you have overwritten or deleted some important value.

The best policy is to back up the database from time to time. Personally, I also keep a spare copy of a completely clean root downloaded from UserLand's Web site. That way, if I suspect the integrity of the database, I can migrate my personal stuff into the clean root. These matters are all discussed in Chapter 29.

Getting Help from DocServer

DocServer is an online help program included with the Frontier distribution. It contains documentation for two kinds of UserTalk features: (1) punctuation, operators, and keywords; and (2) verbs. It's very useful, though, as of this writing, the information it contains has some slight drawbacks. Not every built-in verb is documented, and important utility verbs and UCMDs are omitted. The documentation for some keywords fails to describe their full syntax. The documentation is sometimes unnecessarily obscure, and occasionally just plain wrong.

Also, the DocServer application is hard to navigate: the **Index** menu does not include every category of verb; no menu lists other verbs in the present category, and if you don't know the exact name of a verb, there are no good facilities for finding it.

Nonetheless, Frontier and DocServer work tightly together to provide a handy reference, and it is extremely useful to be conversant with the ways in which they do so:

*   In Frontier, if you are looking at the name of a verb in an edit window, Control-double-clicking that name takes you to the DocServer documentation for that verb.[*]

---

[*] This feature, and the next, are implemented by `system.misc.control2click`, which is automatically called when you Control-double-click in Frontier.

- In Frontier, if you are looking at the name of a verb in a script or outline edit window, Control-Option-double-clicking that name copies the verb's parameter information from DocServer to just after the name. Now you have placeholders for the expected parameters and can substitute actual values.

- In Frontier, choosing **DocServer** from the **Main** menu (or hitting Command-=) brings up a dialog into which you can type the name of a verb to jump to its documentation. (You can do the same thing in DocServer by choosing **Jump To Verb** from the **File** menu.)

- In DocServer, choosing **Paste Into Frontier** from the **Frontier** menu switches to Frontier and pastes at the insertion point of the frontmost window the name and parameter information of the verb you were just looking at.

- In DocServer, choosing **Open Glue Script** from the **Frontier** menu switches to Frontier and opens the script edit window of the verb you were just viewing.

- In Frontier, an outline at `suites.docs.verbList` lists the verbs (along with their parameters) in DocServer. This outline is a splendid way to browse DocServer; remember, Control-double-clicking a verb name takes you to its DocServer documentation. To see the outline, jump to it, or, from DocServer, choose **Open Verb List** from the **Frontier** menu. To rebuild the list, so that it reflects the current state of DocServer, choose **DocServer Verb List** from the **Suites** menu, and choose **Build Whole List** from the **Frontier** menu that appears.

Getting Help from the Database

Exploration of the database is a primary debugging (and learning!) technique. If a verb is implemented as a script, you can study that script. You can jump to any database entry whose name you see in an edit window by Command-double-clicking that name, and you can jump to any database entry whose name you know by choosing **Jump** from the **Open** menu. In both cases, partial object references are resolved for you.

III

Data Manipulation

This section continues the exposition of the UserTalk language to an advanced level, explaining how UserTalk manipulates and operates upon various kinds of data. There is a description of how UserTalk verbs and operators work upon strings and characters, numbers and other mathematical entities, and dates.

The basic Frontier entities such as the database, tables and table entries, outlines, wptexts, and so on, are also grist for UserTalk's mill: UserTalk can manipulate these like any other kind of data, creating and destroying database entries, rearranging outlines, editing wptexts, and so forth. UserTalk can read and write data in files on disk, both the data fork and the resource fork. Frontier is multithreaded, and threads too are a kind of "data" on which UserTalk can operate.

Finally, I describe an "artifical" datatype (stacks) implemented entirely with scripts; and there is a sketch of how UserTalk is extended through compiled code fragments (with further technical details in Part VII, *Reference*).

Readers aiming to take advantage of the full power of UserTalk will wish to study this section. On the other hand, those who are experiencing Frontier for the first time and who presently wish only to acquire an initial overview may prefer to proceed to Part IV, *Interface*.

14

Strings and Chars

This chapter discusses the verbs which perform utility operations on strings and chars. For information on the nature of strings and chars, the rules of string and char literal construction and display, and several key verbs and operators that work on strings, see Chapter 10, *Datatypes*, and Chapter 11, *Arrays*. On the importance of strings in coercion, see Chapter 10 (under "Coercion") and Chapter 8, *Addresses*.

Strings are surely the most important datatype in UserTalk, both because they are the most universally coercible and because of Frontier's aptitude for batch processing and construction of text files. Considering this, the repertoire of string verbs seems none too excessive. Many utility functions need to be built by the user—several are suggested or provided in the discussion that follows—and it has taken the addition of supplements such as the **regex** suite and the conversion of some script-based functions to a kernel implementation to render UserTalk's string manipulation truly suitable for Web- and network-related functionalities.

All the verbs described here create a new entity as their result. Thus, when we describe the purpose of a verb as "to remove a stretch of a string," we really mean, "to return a new string equivalent to the original string with a certain stretch removed." Position indices are 1-based. A char can be fed to a verb that expects a string and be implicitly coerced; the reverse is true only if the string is one character long. String matching is case-sensitive: see Chapter 45, *Punctuation*.

Substrings

To insert one string into another:
```
string.insert (substring, originalString, startIndex)
```

To remove a stretch of a string:

```
string.delete (string, startIndex, count)
```

To remove the end of a string, starting at the last occurrence of a given character value:

```
toys.popStringSuffix (string, character)
```

> Intended for removing delimited suffixes, such as ".html". See also the field verbs, later in this chapter.

To obtain a substring of a string:

```
string.mid (string, startIndex, count)
```

A nice feature of `string.delete()` and `string.mid()` is that *count* or *startIndex* can be too large, or 0, without generating an error. With `string.mid()` (but, oddly, not with `string.delete()`), *count* can even be negative. See Chapter 46, *Verbs*, for details.

UserTalk has no `string.left()` or `string.right()`, to return the leftmost or rightmost *n* characters of a string. No `string.left()` is really needed, since `string.mid()` with a *startIndex* of 1 will do; but it is useful to have `rightN()` as a utility script (you can't call it `right` because that's a reserved word).

Example 14-1. rightN()

```
on rightN (s, n)
    on pos(n)
        return n * (n >= 0)
    return string.delete (s, 1, pos(sizeof(s) - n))
```

Also, it is often useful to remove the rightmost *n* characters of a string. The following is a utility script, `rDelete()`, which does it.

Example 14-2. rDelete()

```
on rDelete (s, n)
    return string.mid (s, 1, sizeof(s) - n)
```

To obtain the nth character of a string:

```
string.nthChar (string, index)
```

Unlike array notation, it is not an error to supply to `string.nthChar()` an *index* which is 0 or too large.

To obtain a string's length:

```
string.length (string)
```

> Otiose, since `sizeOf()` is completely equivalent.

Case and Character Type

To change a string's case to uppercase or lowercase:
```
string.upper (string)
string.lower (string)
```
> The system is consulted, so diacritics are correctly handled.

It is useful to supplement these with boolean tests `isUpper()` and `isLower()`; here is an example.

Example 14-3. isUpper()

```
on isUpper (s)
    return s == string.upper(s)
```

To test whether a character is standard alphanumeric (such as can appear in a standard variable name):
```
toys.alphaChar (character)
```
To test whether a character is alphabetic:
```
string.isAlpha (character)
```
> Consults the system for its notion of an alphabetic character, so includes diacritics.

To test whether a character is a digit:
```
string.isNumeric (character)
```
> The routine `toys.isNumeric()` is otiose.

To test whether a character is punctuation:
```
string.isPunctuation (character)
toys.puncChar (character)
```
> The `toys` version takes a narrow view of what constitutes punctuation; the `string` version consults the system. I don't know what the purpose is of the `toys` version not seeing `'@'` as punctuation.

To strip out non-alphanumeric characters from a string:
```
toys.dropNonAlphas (string)
```
> Chiefly intended for generating universally legal filenames. Oddly, does not use quite the same definition of an "alpha" as `toys.alphaChar()` (underscores are stripped).

To turn multiple words into one word with capitalized elements:
```
toys.innerCaseName (string)
```
> Chiefly intended for generating universally legal filenames. For example:
> ```
> toys.innerCaseName ("inner case name")
>     « "innerCaseName"
> ```

It is easy to write a generalized verb that tests whether every character in a string meets a given test, thanks to UserTalk's ability to use a script address as a parameter (see Chapter 8).

Example 14-4. testEachChar()

```
on testEachChar (s, addrTest)
    local (x)
    for x = 1 to sizeOf(s)
        if not addrTest^(s[x])
            return false
    return true
```

You can write your own test or use one of those already included. For example, to test whether every character in a string meets the **string.isAlpha()** test:

```
testEachChar ("thisHasNoSpaces", @string.isAlpha)
    « true
```

Similarly, the following is a generalized verb that strips a string of all characters failing to meet a given test.

Example 14-5. stripTestFailers()

```
on stripTestFailers (s, addrTest)
    local (x, t = "")
    for x = 1 to sizeof(s)
        if addrTest^(s[x])
            t = t + s[x]
    return t
```

For example:

```
stripTestFailers ("The rain, in Spain", @string.isAlpha)
    « "TheraininSpain"
```

Find and Replace

To determine the index of the first location of a substring in a string:

```
string.patternMatch (substring, string)
```

> Returns 0 if *substring* doesn't occur.

Recall that 0 is coerced to **false** and all other numbers to **true** if used where a boolean is expected; thus, the result of **string.patternMatch()** can be used as a boolean test to see whether *substring* occurs at all.

Not infrequently, it is desired to determine a substring's first location in a string while ignoring some portion of the beginning of the string.

Example 14-6. patternMatchAfter()

```
on patternMatchAfter (pattern, s, index)
    local
        where = string.patternMatch (pattern, \
            string.mid (s, index, infinity))
    return (index + where - 1) * boolean(where)
```

Based on this, it is easy to write a utility to obtain the *n*th location in a string of a substring.

Example 14-7. findNth()

```
on findNth (pattern, s, count)
    local (index = 1, theLength = sizeOf(pattern))
    loop
        index = patternMatchAfter (pattern, s, index)
        if index and --count
            index = index + theLength
        else
            return index
```

See also the discussion of the string parsing verbs, later in this chapter.

To replace the first or all occurrences of one substring in a string with another substring:

```
string.replace (string, oldSubstring, newSubstring)
string.replaceAll (string, oldSubstring, newSubstring)
```

To delete the first or all occurrences of a substring in a string, call **string.replace()** or **string.replaceAll()** with the empty string as the third parameter. We can use this device to count the occurrences of a substring.

Example 14-8. countInString()

```
on countInString (pattern, s)
    local (t = string.replaceAll(s, pattern, ""))
    return (sizeof(s) - sizeof(t)) / sizeof(pattern)
```

To perform grep search and replace, the **regex** suite is provided. This is a port of the GNU **regex** library to a UCMD. There is undeniable value in being able to use grep, especially in parsing HTML and other structured text.

This is not the place for a full discussion of grep.[*] The suite includes a ReadMe wptext which explains both grep and the regex verbs. Admittedly, grep is probably not everyone's cup of tea; and writing a grep expression in UserTalk is not made any easier by the rules of string literal construction. Here, for instance, is a grep pattern that finds stretches of quoted material containing an escaped double quote (\"):

```
"\"[^\"]*(\\\\(\"[^\"]*\\\\)*\"[^\"]*)?\""
```

This looks more formidable than it is; it boils down to:

```
"[^"]*(\\("[^"]*\\)*"[^"]*)?"
```

---

[*] See Jeffrey E. F. Friedl, *Mastering Regular Expressions* (O'Reilly & Associates, Inc., 1997).

surrounded by quotes and rendered with escape characters.* Even so, it's fairly opaque. Such is the price of being able to take advantage of the power of grep. This power is well demonstrated by the included **regex.toys** examples, such as the utility which parses a browser bookmark file from HTML into an outline of title-URL pairs. Tasks of this class would be very difficult in UserTalk without **regex.**†

Here, we content ourselves with sketching the basic behavior of the three most important **regex** verbs. Consult the documentation in the suite for full information. Notice that the target string must be passed by reference (an address).

To perform a grep search and replace:
```
regex.subst (searchFor, replaceWith, addrString, caseSensitive?,
             maxSubstitutions)
```

To obtain a list of all substrings matching a grep pattern:
```
regex.extract (searchFor, addrString, addrList, groups, caseSensitive?)
```

To parse a string according to a grep pattern:
```
regex.split (searchFor, addrString, caseSensitive?)
```

> The result is a list of strings; these are what is left of the string at *addrString* when all substrings matching *searchFor* are removed, interspersed with the matches on any group expressions in *searchFor.*

Suppose, for example, we wish to put into lowercase everything not enclosed in double quotes. We can use **regex.split()** to break the string into pieces, run through the list lowering the case of those items that don't start with a double quote, and reassemble the string. (An extra verb, **regex.join()**, is supplied for just this purpose.) The following code is an illustration:

```
local
    myString = "This Is An \"Amazing\" Demonstration"
    theList = regex.split ("(\".*\")", @myString)
    x
for x = 1 to sizeOf(theList)
    if not (theList[x] beginsWith "\"")
        theList[x] = string.lower(theList[x])
return regex.join("", @theList)
    « "this is an \"Amazing\" demonstration"
```

---

* Since a backslash is already an escape character in grep, it must be escaped in a grep expression to be understood as a literal: to look for \ one must say \\. But now each of those must be escaped to be understood literally in a UserTalk string literal; whence the horrid \\\\ just to look for one backslash.

† A commercial application, TextMachine, performs grep search and replace through English-like codes. For example, the grep pattern in the main text could be rendered as:
```
"[quote][zeroOrMore not quote][backslash]" + \
"[zeroOrMore (quote, (zeroOrMore not quote), backslash)]" + \
"[quote][zeroOrMore not quote][quote]"
```
which is pretty easy to code, to debug, and to read. See *http://www.prefab.com/textmachine.html.*

The group-making parentheses in the first parameter to `regex.split()` are crucial; otherwise, `theList` will be:

```
{"This Is An ", " Demonstration"}
```

Pad and Trim

To generate a string made up of one character, repeated:
```
string.filledString (character, length)
```

To attach leading zeros to a number:
```
toys.padWithZeros (theNumber, length)
```

If we are making ten files, for example, and we call them *file1, file2, ... file10,* then the system will show their names sorted in this order: *file1, file10, file2, ...* To prevent this, we call the files *file01, file02, ... file10;* `toys.padWithZeros()` lets us create the suffixes using a counting variable.

To remove all leading or trailing instances of a character:
```
string.popLeading (string, character)
string.popTrailing (string, character)
```

It is useful to write a combined utility verb that removes a character from both ends of a string; `trim()` might be a good name for it. And one of the first User-Talk scripts I ever wrote was a utility verb **popAllTrailing()**, which accepts a string of characters, any of which are to be removed from the end of a string.

Example 14-9. popAllTrailing

```
on popAllTrailing(theString, theChars)
    while theChars contains theString[sizeof(theString)]
        delete(theString[sizeof(theString)])
    return theString
```

Parsing

To "parse" here means to recognize divisions of a string based on some delimiter character. UserTalk has various ways of doing this.

Fields

The fields of a string are the substrings before, between, and after all occurrences of the delimiter character, except that if the last character of a string is the delimiter, there is no field after it.

For example, suppose . is the delimiter. Then in `"root.system"`, there are two fields, `root` and `system.` In `".root.system."` there are three fields: the empty string before the opening delimiter; then `root`; then `system`—there is no field after the final delimiter.

To count the number of fields:
```
string.countFields (string, delimiterChar)
```

To obtain a particular field:
```
string.nthField (string, delimiterChar, n)
```

Often what you really want to know is where the *n*th field is, rather than its value; the utility `findNth()` in Example 14-7 will help with this. Another useful utility to write is one that deletes the *n*th field.

Note that to deal with paragraphs (or lines) you just use the field verbs with `cr` as the delimiter character.

To remove a UserTalk trailing comment:
```
toys.commentDelete (string)
string.commentDelete (string)
```

`toys.commentDelete()` is implemented by using « as the delimiter character and returning the first field. It works around the fact that `string.commentDelete()` is limited to 255 characters. But so is a line of a script, and `string.comment-Delete()` is smarter in a different way: it ignores « in a string literal.

Words

The term "words" is a little misleading; any character can serve as the delimiter, just as with fields. The real difference between fields and words is this: All runs of word-delimiters are counted as a single word-delimiter, and both leading and trailing word-delimiters are ignored.

For example, suppose `.` is the delimiter. Then `"..root..system.."` has six fields: empty, empty, `root`, empty, `system`, and empty. But it has only two words: `"root"` and `"system"`.

Instead of supplying the word-delimiter as a parameter to each verb, like the field verbs, the word-delimiter is a "hidden" global. Once changed it retains its value until changed again or until Frontier is shut down, at which point it reverts to the space character. Typically, if you want the word-delimiter to be something other than a space, you'll set it, do some word operations, and restore it.*

To set or obtain the word-delimiter:
```
string.setWordChar (character)
string.getWordChar ()
```

---

* The word-delimiter is truly global, not confined to the current thread only; this means that you may need to take care with a script running in a multithreaded context, lest it clash with some other thread over the word-delimiter's behavior. See Chapter 21, *Yielding, Pausing, Threads, and Semaphores.*

To count the number of words:
```
string.countWords (string)
```

To obtain a particular word:
```
string.nthWord (string, n)
string.firstWord (string)
string.lastWord (string)
```

Obviously `string.firstWord()` and `string.lastWord()` could easily be written as utility scripts using `string.nthWord()` and `string.countWords()`. Given that they do exist, though, it's a pity there are no `string.firstField()` and `string.lastField()` to parallel them.

As with fields, often what one wants to know is where the *n*th word is. This is slightly harder to write as a utility than Example 14-7 was, because by the definition of a word, it is not sufficient to know where the *n*th word-delimiter is. It is implemented for you as `wordInfo.getNthWordOffset()` (in `system.extensions`). For example, here is a routine to capitalize the first letter of every word.

Example 14-10. titleCase()
```
on titlecase(s)
    local (x, n)
    for x = 1 to string.countwords(s)
        n = wordInfo.getNthWordOffset(s,x)
        s[n] = string.upper(s[n])
    return s
```

To obtain the first sentence:
```
string.firstSentence (string)
```

This one's an oddity. There are no other "sentence" verbs, and the definition of a sentence is primitive. If you really need to work with genuine sentences from within Frontier, you're probably better off writing a grep pattern and using **regex**.

Pathnames

Pathnames specify a file, folder, or volume. Some character (on Mac OS it is a colon) delimits elements of pathnames.

To obtain just the name of a file:
```
file.fileFromPath (pathname)
```

To obtain the pathname of the enclosing folder:
```
file.folderFromPath (pathname)
```

To obtain the pathname of the enclosing volume:
```
file.volumeFromPath (pathname)
```

If a volume or folder is returned, the string includes a final colon. For example, if *pathname* is `"HD:someFolder:someFile"`, these verbs return, respectively, `"someFile"`, `"HD:someFolder:"`, and `"HD:"`. The idea is that you can confidently concatenate strings returned from these verbs to form new pathnames.

Formatted Numbers

To insert commas into the string representation of a number:

    string.addCommas (number)

For example, `string.addCommas(1234)` is `"1,234"`.

HTML-Related Conversions

The incorporation of Web site management features into Frontier has brought with it a number of utilities for performing HTML-related munging of strings. Some of these started life as scripts, were subsequently reimplemented as UCMDs, and finally were built into the kernel, for speed. Utilities optimized in this way remain in the database so that old scripts calling them don't break; for instance, `toys.iso8859filter()`, the original version, now simply calls `string.iso8859encode()`. Only the optimized version is listed here.

To pass a string's high-ASCII characters through a translate table:

    string.iso8859encode (string)

string is typically the text of a Web page, and the idea is to treat all high-ASCII characters so that this text becomes universally legible over the Web, by way of ISO 8859-1 (Latin-1) encoding. The default translate table used is an internal copy of one that sits at `suites.html.utilities.iso8859.table`. As you can see, entry names are numeric ASCII values; these are paired with the string to which they are to be translated.

The default table tries to deal with the fact that the Mac character set does not completely translate to the ISO 8859 standard. So, the ampersand named-entity reference is used when it exists (for example, ASCII 138, ä, becomes `"&auml;"`), and otherwise a description in square brackets is employed (for example, ASCII 170, the trademark symbol, is rendered as `"[trademark]"`). A few characters are translated directly to low-ASCII approximations: curly quotes become straight, en- and em-dashes become hyphens, bullets become o, and the like.

The verb is more flexible than its name suggests: you are allowed to substitute your own translate table. The rule is that if there is a table at `user.html.prefs.iso8859map`, it will be used instead of the default. The best way to create this table is to copy `suites.html.utilities.iso8859.table` into `user.`

`html.prefs`, rename it `iso8859map`, and alter the values to suit. What I do, actually, is to call the table something else, such as `iso8859map1`; that way, I can keep it on hand with its interfering with the default operation of `string.iso8859encode()`. Then, when I want a script to use my table, I have my script rename the table to `iso8859map`, call `string.iso8859encode()`, and then restore the name.

You may omit high-ASCII values from the table and the omitted values will simply translate to themselves. The routine does just one pass, so what a high-ASCII character is translated to can involve high-ASCII characters, including itself. You cannot use this verb to translate low-ASCII characters; low-ASCII numbers used as names in the translate table will be ignored.

To translate between a normal string and %-encoding:
```
string.URLencode (string)
string.URLdecode (string)
```

URLs can contain only a limited character set (as suggested, for instance, by RFC 1738); other characters are escaped using % followed by their hex ASCII value (octet), except that a space may be represented by +. These verbs perform the translation respectively to and from this encoding.

To parse CGI arguments:
```
string.parseHTTPargs (string)
```

This verb handles form data such as a browser generates as part of a POST method when the Submit button is pressed in an HTML form. Form data typically consists of **name=value** pairs delimited by ampersand and URL-encoded; this verb URL-decodes and parses the input into a list of strings, name followed by value. For example:

```
string.parseHTTPArgs ("name=Martin+M%9Fller&address=456+Main")
    {"name", "Martin Müller", "address", "456 Main"}
```

The chief function of **string.parseHTTPargs()** is to be called by **suites. webserver.parseArgs()**, so you probably won't call it directly; it is more likely that you will call **suites.webserver.parseArgs()**, particularly if you pass search arguments to a CGI with a GET method. See Chapter 40, *CGIs*.

To split an absolute URL into its components:
```
toys.URLsplit (string)
```

> Components are scheme, domain, and path; the domain does not end with a slash, so your use of this verb may need to supply one. For example:
>
> ```
> toys.URLsplit ("http://www.tidbits.com/matt")
>     « {"http://", "www.tidbits.com", "matt"}
> ```

To convert a relative URL to an absolute URL with respect to a given base:
```
pbs.utilities.parseRelativeURL (baseURL, relativeURL)
```

To extract a list of URLs:

    pbs.getLinks (*string*)

To remove all HTML tags:

    pbs.stripHTML (*string*)

The functionality of both **pbs.getLinks()** and **pbs.stripHTML()** might also be duplicated using **regex**, but the **pbs** versions are simpler and faster.

15

Math

Numeric datatypes and coercions, and how these affect the results of arithmetic operations, are discussed Chapter 10, *Datatypes.* Recall that if both operands are of the same type, the result will be that type; hence, when integers are divided, the result is an integer, and when integers are added, subtracted, or multiplied, the result may "wrap around" in order to stay within the range of permissible integer values.

Basic arithmetic is performed using familiar operators: addition (+), subtraction (−), multiplication (*), and division (/). If an expression contains more than one arithmetic operator, the pairs are evaluated in left-to-right order, except that multiplication and division are evaluated before addition and subtraction. To override this order of evaluation, enclose in parentheses any expressions to be evaluated first. So:

```
2 + 1 * 3
    « 5
(2 + 1) * 3
    « 9
```

The modulus or remainder operator (%) operates on integers, yielding the remainder when the first operand is divided by the second. The verb **mod()** may be used instead. Thus:

```
7 % 2
    « 1
mod (7, 2)
    « 1
```

Absolute values are obtained with the verb **abs()**.

Like C, UserTalk has the unary increment and decrement operators (++ and −−). They do nothing that cannot be accomplished more verbosely without them, but they are extremely convenient. Used with the name of a variable or database entry, they respectively add 1 to and subtract 1 from its value, and this is done either before or after the context as a whole is evaluated, depending on whether the operator appears before or after the name. In Example 4-7, the line:

```
while --t {s = s + "    "}
```

decrements t and then evaluates it for the while test. In this example it is crucial that t be decremented before we check it, because if its value before we start is 1 we don't want to append any spaces to s.

Many mathematical verbs are provided through a UCMD, trigCmd, living in system.extensions. Since there is a search path to system.extensions, it suffices to call one of these by prefixing trigCmd to its name; for example, trigCmd.sqrt(10).

The trigCmd verbs are: abs(); floor(), and ceil(), which round downwards or upwards, respectively; sqrt(); sin(), cos(), tan(); asin(), acos(), atan(); sinh(), cosh(), tanh(); exp(), log(), log10(). Several other functions can be derived from these; for example, one could write a utility pow() that raises its first parameter to the power of its second parameter, thus:

```
on pow (x, y)
    with trigCmd
        return exp(log(x)*y)
```

and similarly for finding *n*th roots. Other utilities, such as max() and min(), may be worth writing as well.

To obtain a decimal number's hexadecimal representation, use string.hex(). For example, string.hex(1234) is "0x04D2".

To convert from hexadecimal to decimal, just use number(). Since hexadecimals are strings starting with "0x", if you have a pure hex string such as "F1" you can convert it to decimal as follows:

```
hexstring = "F1" « or whatever
decimalValue = number ("0x" + hexstring)
```

You can convert bytes to kilobytes using string.kBytes(); the result is a string representing an integral number of kilobytes and ending in "K". For example, string.kBytes(1234) is "2K". See also samples.basicstuff.megabyte-String() for a similar conversion to megabytes.

The verb random() returns a pseudo-random integer.

Three bit commands operate on individual bits of a long. A long may be considered as made up of 8 hex digits, 32 binary digits; so its bits are numbered 0 to 31 (where 0 is the low-order bit). See on `bit.set()`, `bit.clear()`, and `bit.get()` in Chapter 46, *Verbs*.

Three `rectangle` commands are utility scripts to make some very basic manipulation of rects more convenient. See Chapter 46 on `rectangle.random()`, `rectangle.inset()` and `rectangle.outset()`. The last two are complementary, giving identical results if their parameters are the additive inverse of one another.

16

Dates

On internal storage of the date type, and on coercion of dates to and from strings, see Chapter 10, *Datatypes*.

Recall that on Mac OS, a date[*] is stored as an unsigned long representing the number of seconds since the start of 1904. Since an unsigned long is a numeric type, no special date operators are needed in order to perform date arithmetic; math operators can be applied directly to date values. Also, a long and a date can be combined in a math operation, because the long will be implicitly coerced to a date. For example, to find out the number of days between two dates, just subtract and divide by (60*60*24), like this:

```
long((date("Feb 3 1997") - date("Jan 28 1997")) / (60*60*24))
    « 6
```

The tricky part is that dates cannot be externally represented without coercion. Coercing a date with **double**() shows its numeric value more readably than coercing it with **long**(), because a date is an *unsigned* long; coercing with **long**() causes wraparound to a negative number for dates after early 1972. However, a double cannot be coerced with **date**(). The following utility converts a double back to a date.

Example 16-1. doubleToDate()

```
on doubleToDate(d)
    if d > infinity
        d = d - infinity - infinity - 2
    return (date(long(d)))
```

---

[*] By "date" I always mean "date-time." Times do not exist in the abstract, of course, but only as part of a date-time complex.

An unsigned long measuring seconds starting in 1904 can handle only up through early 2040. On the Mac, dates outside this range* are handled as a pair of longs, called a LongDateTime. Certain foresighted applications have already adopted this format for their internal dates; in order to communicate about dates with such applications, it may be necessary to convert between Frontier's date type and this LongDateTime type. The following scripts (by John Baxter) show how to do this; note that these scripts do not give Frontier itself the power to work with dates outside its range.

Example 16-2. ldtToDate()

```
on ldtToDate (ldt)
    local
        high = number (binary (string.mid (ldt, 1, 4)))
        low = number (binary (string.mid (ldt, 5, 8)))
    if high != 0
        scriptError ("Long date value out of range for conversion to date")
    return (date (low))
```

Example 16-3. dateToLDT()

```
on dateToLDT (aDate)
    local
        ldt = binary (string (binary (0)) + string (binary (aDate)))
    setBinaryType (@ldt, 'ldt ')
    return (ldt)
```

Sometimes coercion happens behind the scenes; don't be misled. If you say clock.now() in Quick Script, the response is legible; but this is only because the result, which is actually a date, has been coerced to a string for display. Such a string would be useless for further numeric calculation in a script. On the other hand, the *real* result of clock.now() could be used in numeric calculation.

Beware of making assumptions about the structure of string representations of dates. For example, on my computer, date.shortString(myBirth) yields "8/10/54"; when I lived in New Zealand, it yielded "10/8/54", with the day-number preceding the month-number. These results come from system calls which examine settings in the *Date & Time* control panel; therefore, they vary from computer to computer.

Nevertheless, Frontier's string-to-date coercion is remarkably intelligent, and can accomplish a great deal on its own. The following, for instance, is a (simplified) utility that converts the date line of an email header to GMT.

---

* Between 30081 BC and AD 29940, if you must know.

Example 16-4. convertToGMT()

```
on convertToGMT (s)
    local (theString = "", x, theOffset)
    on add(t)
        theString = theString + t + " "
    for x = 3 to 6
        add(string.nthWord(s,x))
    theOffset = string.nthWord(s,7)
    if theOffset[1] == '+' {delete(theOffset[1])}
    return date(s) - (number(theOffset)/100*60*60)
convertToGMT("Date: Sat, 7 Jun 1997 22:12:55 -0400")
    « "6/8/97; 2:12:55 AM"
```

To analyze a date into its numerical components:

```
date.get (theDate, addrDay, addrMonth, addrYear, addrHour,
         addrMin, addrSec)
```

To construct a date from its numerical components:

```
date.set (day, month, year, hour, min, sec)
```

It is wise, in specifying a *year* value for **date.set()**, to supply a four-digit number (e.g., 1997), because the result of supplying a two-digit number can be unexpected if Frontier guesses wrong about what century you mean.

It is not an error to supply **date.set()** with values that are out of range for their category, and indeed this can be a useful way to perform date calculation. For example, what is the date 40 days from today?

Example 16-5. Manipulating a date component

```
local (day, mon, yr, hr, min, sec)
date.get(clock.now(), @day, @mon, @yr, @hr, @min, @sec)
msg (date.set (day+40, mon, yr, hr, min, sec))
```

Frontier returns a legal date 40 days from now.

The technique is particularly handy when dealing with months and years, which vary in length. Month manipulations are a little tricky; Frontier's algorithm is sensible, but it may not give quite the results you expected. Increasing or decreasing a date's month component returns a date with the same day-number as the first—provided the new month possesses such a day-number. If not, we lapse over into the start of the following month. So, for example:

```
local (day, mon, yr, hr, min, sec)
date.get("march 31, 1997", @day, @mon, @yr, @hr, @min, @sec)
msg (date.set (day, mon-1, yr, hr, min, sec))
    « "3/3/1997; 12:00:00 AM"
```

Since there is no February 31, we lapse three days into March to get our result.

Given a date, to add or subtract one unit:

```
date.tomorrow (date)
date.yesterday (date)
date.nextWeek (date)
date.prevWeek (date)
date.nextMonth (date)
date.prevMonth (date)
date.nextYear (date)
date.prevYear (date)
```

These are just shortcuts, giving exactly the same results as the technique used in Example 16-5. For example, **date.tomorrow()** could have been implemented as:

```
on tomorrow(theDate)
    local (day, mon, yr, hr, min, sec)
    date.get(theDate, @day, @mon, @yr, @hr, @min, @sec)
    return date.set (day+1, mon, yr, hr, min, sec)
```

To obtain month information:

```
date.firstOfMonth (date)
date.lastOfMonth (date)
date.daysInMonth (date)
date.weeksInMonth (date)
```

To learn the day of the week:

```
date.dayOfWeek (date)
```

Returns a day-of-week number from 1 to 7, where 1 represents Sunday.

To convert from a day-of-week number to a day name:

```
date.dayString (number)
```

Implemented as a script which simply uses a **case** statement.

To obtain formatted date string representations from the system:

```
date.shortString (date)
date.abbrevString (date)
date.longString (date)
```

It is easier to list than to describe the results (on my machine):

```
x=date.set(10,8,1954,3,0,0)
date.shortString(x)
    « "8/10/54"
date.abbrevString(x)
    « "Tue, Aug 10, 1954"
date.longString(x)
    « "Tuesday, August 10, 1954"
```

To obtain formatted string representations of the current date and time from the system:

```
string.dateString ()
string.timeString ()
```

On my machine these yield:

```
string.dateString()
    « "Sun, Jun 8, 1997"
string.timeString()
    « "2:03:20 PM"
```

I believe that saying `string.dateString()` is equivalent to saying `date.abbrevString(clock.now())`.

There are constants (at **system.verbs.constants**) that allow you to type the full English name of a month instead of its month number, as follows:

```
x=date.set(10,August,1954,3,0,0)
    «"8/10/54; 3:00:00 AM"
```

17

Objects

An *object* is a database entry or variable—a storage location with a name and a value. This chapter discusses UserTalk verbs that manipulate objects as objects, creating them, deleting them, copying them, and so forth.

The reader should be thoroughly familiar with ways of referring to objects, and the circumstances under which an object reference will suffice to permit creation of an object. See Chapter 6, *Referring to Database Entries*, Chapter 8, *Addresses*, Chapter 9, *Special Evaluation*, Chapter 10, *Datatypes*, and Chapter 11, *Arrays*.

Verbs that operate on objects often take an address parameter that specifies the object. We have already pointed out (Chapter 8) the dangers of accidentally handing to a verb that expects an address parameter something that is not an address. The verbs studied in this chapter can destroy or alter the value of database entries, so those dangers are particularly acute.

WARNING   Some of these verbs can be dangerous! Verbs such as `new()`, `table.assign()`, and `delete()` wipe out existing values and cannot be undone. Be careful what parameter you hand such verbs.

In a table edit window, table entry names are limited to 31 characters in length. A UserTalk command can give to a table entry a much longer name, up to 255 characters (and can then operate upon that entry); this is usually unwise, though, since it can cause problems when one encounters the entry in a table edit window.

Creation and Destruction

To alter the value of an object, possibly creating the object if necessary:

```
=
table.assign (addrObject, value)
```

For details on the assignment operator (=), see Chapter 44, *Operators*. The difference between these two methods of assignment is that the assignment operator will balk at setting an existing non-scalar object to a scalar value, whereas `table.assign()` will replace anything with anything.

To create an "empty" object of specified type:

```
new (datatype, addrObject)
```

When using `new()` on a local variable, it is common practice to declare the local variable first, just to make quite certain that `new()` interprets the address correctly. So, for example:

```
local (myOutline) ·
new (outlineType, @myOutline)
```

To create all database tables needed to create a database entry:

```
table.surePath (string)
```

`table.surePath()` is a clever utility script, well worth studying, which solves the problem that it is impossible to refer to or take the address of a database entry whose parent table doesn't exist. (For this reason, *string* must be a string, not an address, and it must be a full path, not a partial reference.) `table.sure-Path()` ignores the last element of the path if the path doesn't end with a period, so you can use the same value both as the parameter to `table.surePath()` and to create the object. For example:

```
whatToCreate = "workspace.table1.table2.someString"
table.surePath (whatToCreate) « now table1 and table2 certainly exist
new (stringType, whatToCreate)
    « or: whatToCreate^ = "hello there", or whatever
```

To obtain a name guaranteed unique in its table:

```
table.uniqueName (startOfName, addrTable)
toys.uniqueTableName (startOfName, addrTable, numberOfDigits)
```

`table.uniqueName()` can be used to guarantee that creating a new table entry won't overwrite an existing table entry; it does this by attaching a serial number to the end of the name. `toys.uniqueTableName()` is just the same, except that it pads the serial number with zeros so that the resulting names sort correctly. What these verbs return is actually an address, which is suitable for creating the entry:

```
local (whatName = table.uniqueName("testing", @workspace))
new (stringType, whatName)
    « or: whatName^ = "hello there", or whatever
```

To destroy an object:
```
delete (addrObject)
```

Typically, the object is a database entry; there is usually no need to delete a variable, since it will eventually go out of scope anyway. See Chapter 7, *The Scope of Variables and Handlers*. See also Chapter 11 for other uses of `delete()`.

To destroy a table's entries:
```
table.emptyTable (addrTable)
```

Information

To learn whether an object exists:
```
defined (objectReference)
```
To obtain the address of an object's containing table:
```
parentOf (objectReference)
```
To obtain a table entry's name:
```
nameOf (objectReference)
```

`defined()`, `parentOf()` and `nameOf()` are special forms; they take object references, not addresses, and they will never raise an error, even if *objectReference* doesn't exist or is somehow illegal. See Chapter 9.

`defined()` and `parentOf()` tell what will happen if you use a name. If `defined(`*objectReference*`)` is `true`, then it is legal to *get* the value of *objectReference*. If `boolean(parentOf(`*objectReference*`))` is `true`, then it is legal to *set* the value of *objectReference*.

Also, `defined()` and `parentOf()` together tell you whether you're dealing with a database entry or a local variable. If `defined(`*objectReference*`)` and `boolean(parentOf(`*objectReference*`))` are both `false`, then setting the value of *objectReference* will result in the creation of an implicitly declared local variable. This suggests the following utility script.

Example 17-1. willBeImplicitLocal()
```
on willBeImplicitLocal(addr)
    if not defined(addr^)
        if not parentOf(addr^)
            return true
    return false
```

To learn whether a defined object is a database entry, the simplest way is to find out, recursively, whether its ultimate parent is `"root"`.

Example 17-2. isDatabaseEntry()

```
on isDatabaseEntry(addr)
    if defined(addr^)
        if (addr == "root") or (isDatabaseEntry(parentOf(addr^)))
            return true
    return false
```

nameOf() is useful for obtaining just the last element of an address; for example, if addr is "suites.html.buildObject", we can use nameOf(addr^) to get just "buildObject". In this respect, parentOf() may be viewed as the complement of nameOf(), since we can use string(parentOf(addr^)) to get just "suites.html".

For the use of nameOf() with regard to array indexing, see Chapter 11.

To change an object's name:
```
table.rename (addrObject, nameString)
```

To obtain the address of the non-scalar currently selected in the frontmost window:
```
toys.getCursorAddress ()
```

To obtain the datatype of a value:
```
typeOf (value)
```

Moving and Copying

These verbs are all shortcuts, accomplishing nothing that could not be accomplished by direct assignment. For example, saying:

```
table.copy (@scratchpad.myString, @workspace)
```

has exactly the same effect as saying:

```
table.assign (@workspace.myString, scratchpad.myString)
```

As with table.assign(), there is no restriction on overwriting existing values. It might be wise to reread the warning near the start of this chapter. It is not an error to overwrite a value with itself.

To move or copy an object into a table:
```
table.copy (addrObject, addrTable)
table.move (addrObject, addrTable)
```

To move or copy all of a table's entries into a table:
```
table.copyContents (addrSourceTable, addrDestTable)
table.moveContents (addrSourceTable, addrDestTable)
```

To alter a table entry's address:
```
table.moveAndRename (addrSourceObject, addrDestObject)
```

There is no verb table.copyAndRename(); this is what table.assign() already does.

Age and Dirtiness

An object's *age* is the date when it was last modified. An object is *dirty* if it was modified more recently than the database was last saved.

Unfortunately, Frontier can report age only for a non-scalar; there is no way to find out when a scalar's value was last modified. The situation is somewhat alleviated by the fact that a table is considered modified if the value of one of its scalar entries is changed. However, changing the value of a non-scalar does not affect the age of its containing table; and many other things can cause a table to be considered modified, including simply scrolling its edit window.

Any time a non-scalar is modified, it and the database as a whole become dirty. That, at least, is how things *should* work. But, once again, scalars within tables can be troublesome; creating or setting a scalar in the database can dirty its containing table without dirtying the database. Also, Frontier might not share your idea of what modifies a non-scalar; for instance, changing the expansion state of subheads in an outline does not necessarily dirty the outline.

The situation is rather unsatisfactory, and somewhat diminishes the value of the following verbs.

To learn whether a given non-scalar is dirty:
```
window.isModified (addrObject)
```
To specify whether a given non-scalar should be considered dirty:
```
window.setModified (addrObject, dirty?)
```

Declaring the database as a whole to be non-dirty permits the database to be closed without saving and without the "Save?" dialog appearing. Otherwise, I find that `window.setModified()` with a *dirty?* value of `true` often doesn't work, and that the best way to ensure the dirtiness of an object is to alter its text.

To learn when a non-scalar was created or last modified:
```
timeCreated (addrObject)
timeModified (addrObject)
```

You can learn the `timeCreated()` and `timeModified()` of an object manually by selecting it within its containing table and choosing **Get Info** from the **Table** menu.

18

Non-Scalars

Non-scalars are datatypes that are editable in an internal Frontier edit window: tables, wptexts, outlines, scripts, and menubars. This chapter discusses the User-Talk verbs that edit non-scalars.

Even in UserTalk, a non-scalar is edited by way of its edit window. Only one window can be edited at any given moment; the object whose edit window is being edited is called the *target*. So the notion of the target is discussed first. Then the various verbs that operate within the target's edit window are enumerated.

Generally speaking, what you can do manually to a non-scalar in its edit window, you can do programmatically. But this isn't completely true; for instance, you can't perform multiple selections in an outline window programmatically. On the other hand, there are some valuable UserTalk edit window operations that no keyboard shortcut or menu item accesses (though of course you can always add one that does).

The object being edited may be either a database entry or a local variable. Beginners often don't realize this, because to open a database entry's edit window manually and edit it is common, whereas one never does the same to a local variable. Nevertheless, a local variable can be a non-scalar, and while it is in scope, its edit window can be opened and manipulated through UserTalk.

On editing non-scalars manually, see Chapter 2, *Edit Windows*. On edit windows as windows, see Chapter 24, *Windows*. On programmatic editing of the menu aspect of menubars (as opposed to the outline aspect covered in this chapter), see also Chapter 26, *Menus and Suites*. Pictures have an internal edit window too; see Chapter 28, *Pictures*. Database tables are a special case because their entries can also be accessed directly as objects; see Chapter 17, *Objects*.

The Target

UserTalk verbs that edit non-scalars operate in an edit window; but they don't take a parameter specifying what edit window to operate in. Instead, a non-scalar object must already be the target, and any verb that edits a non-scalar will automatically operate in the target's edit window. At any given moment, exactly one object is the target (or there might be no target).* An object is the target because it is either of the following:

- The object whose edit window is the frontmost window, meaning the window returned by `window.frontmost()`. This is the default situation. Only a non-scalar can be the target, so, in the default situation, if the frontmost window is not a non-scalar edit window, there is no target.

  During execution of a script set in motion by pushing a window's **Run** or **Debug** button, that window is ignored for purposes of determining `window.frontmost()` (and, in Debug mode, so is any window that comes to the front because the execution path runs through it). See Chapter 24.

- The explicitly declared target. We shall see in a moment how to declare a target explicitly. Since it is impossible to work within an edit window that isn't open, explicitly declaring a target opens its edit window if it is not already open.

  The explicitly declared target remains the target until either another object is explicitly declared the target, or until the default situation comes back into force, which happens when either (a) the default situation is explicitly declared, or (b) script execution comes to an end.

To learn the address of the target:
```
target.get ()
```
To declare an object explicitly the target, making its edit window visible and frontmost:
```
edit (addrObject)
```
To declare an object's parent table explicitly the target, making its edit window visible and frontmost:
```
table.gotoAddress (addrObject)
```
To declare an object explicitly the target, without making its edit window visible and frontmost:
```
target.set (addrObject)
```

The difference between `edit()` and `window.open()` is that `window.open()` doesn't change the target if there is already an explicitly declared target.

---

* That is, at any given moment in the execution of any one thread; any one thread can have at most one target at a time, but different threads running simultaneously can have different targets simultaneously. On threads, see Chapter 21, *Yielding, Pausing, Threads, and Semaphores.*

Both `edit()` and `table.gotoAddress()` accept either a non-scalar or a scalar as their object. A scalar cannot be the target, though, so when `edit()` receives the address of a scalar object, it makes the object's parent table the target, instead of the object itself. `table.gotoAddress()` makes the object's parent table the target no matter what kind of object it is.

One reason for calling `target.set()` instead of `edit()` is that `target.set()` doesn't affect the layering position and visibility of the object's edit window. Another is that if the object's edit window is not open already, `target.set()` opens it invisibly. The choice is not merely aesthetic; it is faster to operate on an invisible window than a visible one, because the window doesn't require any redrawing.

For greatest speed, call `op.setDisplay(false)` when an outline (or script, or menubar) is the target, or `wp.setDisplay(false)` when a wptext is the target. This prevents update events from being sent to the window at all, which speeds up operations considerably, regardless of the visibility or invisibility of the window. In one test I conducted, using a fairly slow computer, an operation that took 29 seconds on a visible window took 9 seconds on an invisible window, but only 3 seconds on a window whose display had been frozen. Be sure to unfreeze the window afterwards, with `op.setDisplay(true)` or `wp.setDisplay(true)`, or by closing the window; otherwise, it will be impossible to edit the window manually!*

To close a window, and return to the default situation if it was the target:
```
close (addrObject)
```
To return to the default situation, closing the target's edit window if invisible:
```
target.clear ()
```

The difference between `close()` and `window.close()` is that `window.close()` has no effect upon what object is the target; in fact, if you close the target's edit window with `window.close()`, that window may open again invisibly to accommodate editing in the target window. The difference between `close()` and `target.clear()` is one of orientation; as their descriptions show, `close()` is about a window and only secondarily about the target, but `target.clear()` is about the target and only secondarily about a window.

To understand the use of `target.clear()`, it will help to know the mechanics of targets. During script execution, if the target is explicitly declared, a variable `_target_` is created at the top level of local variables. If a different target is explicitly declared, then if the edit window of the old `_target_` is invisible, that window is automatically closed before changing the value of `_target_`. Thus

---

* You can learn the "frozenness" of the target with `op.getDisplay()` and `wp.getDisplay()`.

there is no need to call `target.clear()` if you are about to declare a new target. On the other hand, when execution ends, the default situation is returned to, but there is no automatic closing of any window, so the last target's invisible window may be left open. Calling `target.clear()` at the last minute prevents this.

So, a major use of `target.clear()` is to clean up at the end of a script or handler in which `target.get()` or `edit()` was used. It's a good idea to make a habit of doing this. There is no penalty for calling `target.clear()` in any case; if there isn't an invisible target window, nothing is closed. And if you don't do it, problems can arise; for example, this script causes a runtime error:

```
on makeOutline()
    local (otl); new (outlineType, @otl)
    target.set (@otl)
    « do stuff
    « oops! should have called target.clear() here!
makeOutline()
edit(@workspace) « error!
```

The problem is that `edit()` tries to close the old target, but the old target no longer exists because it was local to the handler.

There is only one target, which, as it is maintained at the top level of variables, is virtually a global variable. Thus, if a script sets the target and then calls another script, and the second script also sets the target, then, when the second script returns, the target may have changed, and the first script may break. For this reason, a script that explicitly declares a target, if it is likely to be called by another script, should save the old target beforehand and restore it afterwards:

```
local (oldTarget = target.get())
target.set(@myNewTarget)
« do things to the target
target.set(oldTarget)
```

There are some scripts in the database that take advantage of the global nature of the target, in the following way: they set the target and then call a second script which expects the correct window to be the target already. This approach works, but is probably to be deprecated. It's better to tell the second script explicitly, in one of its parameters, what object it is to operate on.

Verb Types, Window Types, and Modes

The verbs enumerated in the rest of this chapter would seem, from their names, to fall into classes corresponding to distinct types of object. There are op verbs (for outlines—"op" is short for "outline processor"), wp verbs (for wptexts), script verbs (for scripts), and table verbs (for tables).

But life is not so simple. The verb classes do *not* correspond precisely to the division between object types. First, there are editMenu verbs which work on all five types of window. There are also editMenu verbs which work only in wptexts. Second, the script verbs operate also on outlines and menubars, and the op verbs operate also on menubars and scripts; that's because all three are simply varieties of outline. Third, when a script, menubar, outline, or table is in content mode, the region containing the selection becomes a kind of miniature wptext "document," and wp verbs apply to it. (On the distinction between selection mode and content mode, see Chapter 2.) In the case of a script, menubar, or outline, this miniature wptext is the entire current line; in the case of a table, it's the entire current cell.

Suppose, for instance, that the target is an outline. Then if the window is in content mode, saying `wp.setText("hello")`, which in a real wptext object would replace the entire contents of the object with `"hello"`, would replace the entire current line with `"hello"`. (If the window is in selection mode, using this verb will raise an error.)

To work with modes, a pair of wp verbs is provided which work in script, menubar, outline, or table windows but are pointless in a wptext.

To set selection mode or content mode, or to learn which mode is current:
```
wp.setTextMode (boolean)
wp.inTextMode ()
```
> Content mode is true; selection mode is false.

Because of a bug, if an outline (or script) edit window was closed while in content mode, the next time it is made the target, `wp.setTextMode(true)` may fail silently. The workaround is to call `wp.setTextMode(false)` before calling `wp.setTextMode(true)`.

Verbs Operating in Edit Windows

The edit window of a non-scalar object must be the target before calling one of these verbs. As part of every verb's description, imagine that the phrase "in the target's edit window" is present. The verbs are classified according to what kind of object the verb actually operates on.

Scripts, Menubars, Outlines, Tables, and Wptexts

The current line of a script, menubar, or outline, or the current cell of a table, becomes a miniature wptext "document" when in content mode; the wptext verbs in this section may then be applied.

Selection and navigation

To learn or set the current selection:
```
wp.getSelect (addrStart, addrEnd)
wp.setSelect (start, end)
```

To select the word, line, or paragraph containing the end of the selection:
```
wp.selectWord ()
wp.selectLine ()
wp.selectParagraph ()
```

To select the whole "document," or all lines of the outline at the current level:
```
editMenu.selectAll ()
```

To move the insertion point relative to its present position:
```
wp.go (dir, count)
```

Text

To acquire the currently selected text:
```
wp.getSelText ()
```

To insert text:
```
wp.insert (string)
```

The inserted material takes the place of the current selection. To delete the current selection, insert the empty string.

To learn or set the text of the entire "document":
```
wp.getText ()
wp.setText (string)
```

See, for example, the implementation of **toys.readFileIntoTextObject()**.

Formatting

To set the text font and size:
```
editMenu.setFont (nameString)
editMenu.setFontSize (sizeInteger)
```

To copy, cut, paste, and clear:
```
editMenu.copy ()
editMenu.cut ()
editMenu.paste ()
editMenu.clear ()
```

All of these editMenu verbs function the same as choosing the corresponding menu items from the **Edit** menu.

Moving and deleting material can often be accomplished more efficiently without editMenu verbs, which use the clipboard. For example, to copy a table entry from one place to another, **table.copy()** is usually a better choice than **edit-Menu.copy()** and **editMenu.paste()**; to remove a table entry, **delete()** is usually better than **editMenu.clear()**.

Nevertheless, copying and pasting is the only way to preserve formatting within a wptext; and copying a line of an outline in selection mode is often the best way to transport it and its subheads to another outline.

Tables Only

To learn the address of the currently selected entry:
```
table.getCursor ()
```

To navigate within the table, relative to the current selection:
```
table.go (dir, count)
```

To navigate within the table, by numeric or name index:
```
table.goto (n)
table.gotoName (nameString)
```

To sort the table:
```
table.sortBy (byWhat)
```

Scripts, Menubars, and Outlines

These commands do not require us to be in selection mode beforehand.

Create, delete, and edit lines

To learn or set the text of the entire current line:
```
op.getLineText ()
op.setLineText (string)
```

To create (and select) a new line adjacent to the current line:
```
op.insert (string, dir)
```

To create (and select) a new line as the last subhead of the current line:
```
op.insertAtEndOfList (string)
```

To move the current line (and all its subheads):
```
op.reorg (dir, count)
```

To delete the current line (and all its subheads):
```
op.deleteLine ()
```

To delete the subheads of the current line:
```
op.deleteSubs ()
```

To delete everything in an outline:
```
op.wipe ()
```

Navigate

To navigate to the first summit:
```
op.firstSummit ()
```

To navigate relative to the currently selected line:

```
op.go (dir, count)
```

dir can be up, down, left, right, flatup, flatdown, or nodirection. flatup and flatdown navigate without regard to hierarchical levels. Thus, for instance, it is impossible to move up if we are at the top of a bundle, but it is always possible to move flatup unless we are at the first summit.

op.go() not only navigates, but also returns a boolean that tells whether any navigation was performed. Thus it is a powerful basis for a looping structure: it both acts and tests simultaneously. Consider, for example, the common problem of visiting each line of an outline. In Example 4-7, the following structure was used to loop over every line.

Example 18-1. Flatdown visit method

```
while op.go (flatdown, 1)
    « do something here
```

An even more elegant way is to combine:

```
if op.go (right, 1)
```

with recursion.

Example 18-2. Recursive visit method

```
on traverse()
    loop
        « do something here
        if op.go(right, 1)
            traverse()
            op.go(left, 1)
        if not op.go(down, 1)
            break
    return
traverse()
```

This structure visits each line in flatdown order, yet it is limited to the current bundle (often a desideratum) and respects the notion of depth; the result is often more efficient and flexible than the flatdown method. We shall illustrate this in Example 18-4.

The following example, based on suites.samples.basicStuff.buildTable-Outline(), illustrates another standard outline technique—recursively constructing an outline that mirrors the hierarchical structure of some other set of entities. We build an outline at addrOutline reflecting the hierarchical structure and the names of the entries in the table at addrTable. The traverse() recursion device already discussed in connection with files (see Example 12-2) is adapted for tables; a variable global to the recursion tracks where each inserted line goes.

Example 18-3. Outlining a table structure

```
on buildTableOutline (addrTable, addrOutline)
    new (outlineType, addrOutline); target.set (addrOutline)
    local (dir = right) « first line is to be inserted to the right
    on traverse (addr)
        local (addrSubitem, i)
        for i = 1 to sizeof (addr^) « do all the entries at this level
            addrSubitem = @addr^ [i]
            op.insert (nameOf (addrSubitem^), dir) « dir tells where to put it
            if typeOf (addrSubitem^) == tableType
                dir = right « prepare to insert deeper
                traverse (addrSubitem) « recurse, do all items deeper
                if dir ≠ right « we're back; did we in fact go deeper?
                    op.go (left, 1) « yes, so pop back out
            dir = down « prepare for next entry at same level
    traverse (addrTable)
    return (true)
```

To "bookmark" a line position:

```
op.getCursor ()
op.setCursor (bookmarkNum)
```

See `op.visit()` for an example: the loop works even if the proc deletes the line currently being visited.

Other formal rearrangements

To collapse or expand the subheads of the current line:

```
op.collapse ()
op.expand (depth)
```

To collapse or expand the entire outline fully:

```
op.fullCollapse ()
op.fullExpand ()
```

To promote or demote the lines below the current line:

```
op.promote ()
op.demote ()
```

To "zoom" the outline's focus in or out:

```
op.hoist ()
op.dehoist ()
op.dehoistAll ()
```

Hoisting is just a way of looking at the outline; the outline's actual contents are not changed. This feature is tremendously convenient in editing, and yet no menu items provide access to it, so it is helpful to create some which do.

To sort the lines of the current bundle:

```
op.sort ()
```

Information

See also `op.go()` and `op.reorg()`, which both act and inform.

To learn the level (depth) of the current line:

```
op.level ()
```

To learn whether the current line is expanded or collapsed:

```
op.subsExpanded ()
```

To learn how many summit lines the outline has:

```
op.countSummits ()
```

To learn how many subheads the current line has:

```
op.countSubs (depth)
```

Breakpoints and comments

To set, clear, and learn a line's breakpoint status:

```
script.setBreakpoint ()
script.clearBreakpoint ()
script.getBreakpoint ()
```

To set, clear, and learn a line's comment status:

```
script.makeComment ()
script.unComment ()
script.isComment ()
```

All these commands do work in menubars and outlines, but, except for the possibility of creating a comment in an outline, they are pretty meaningless outside a script.

A command line that is moved into subordination under a comment acquires a chevron for clarity, but is *not* in fact a comment. This is why such lines, when moved out from comment subordination, can regain their command status: they never really had comment status. Only a line that is explicitly created as or converted to a comment is truly a comment.

This means that there is a mistake in Example 4-7. The routine, you recall, was intended to create a text representation of an outline, suitable for use in this book; it uses `flatdown` looping to visit each line and to add its text to a string, preceded by an appropriate number of spaces and by a chevron if the line is a comment:

```
on obtainLineText()
    if script.isComment()
        return "« " + op.getLineText() + cr
    else
        return op.getLineText() + cr
op.firstsummit()
op.fullexpand()
local (s)
s = obtainLineText()
```

```
    while op.go (flatdown, 1)
        local (t = op.level ())
        while --t {s = s + "    "}
        s = s + obtainLineText()
    clipboard.putvalue (s)
```

But the test `script.isComment()` will fail for lines subordinate to comments, so these won't get a chevron even though they need one. An elegant solution is to rewrite the routine around the recursive visit framework of Example 18-2.

Example 18-4. Improved script-copying routine

```
    on traverse(inComment)
        local (c = inComment)
        loop
            if not inComment {c = script.isComment()}
            s = s + string.filledstring (' ', level*4)
            if c {s = s + "« "}
            s = s + op.getLineText() + cr
            if op.go(right, 1)
                level++
                traverse(c)
                op.go(left, 1)
                level--
            if not inComment {c = false}
            if not op.go(down, 1)
                break
        return
    op.firstsummit()
    op.fullexpand()
    local (s, level = 0)
    traverse(false)
    clipboard.putvalue (s)
```

The parameter handed to `traverse()` lets the next level down know whether it is inside a comment; if so, we don't test `script.isComment()` at all, but simply leave the local comment flag set to `true` throughout this level. Another refinement that the recursive structure enables is that we can track the level ourselves instead of calling `op.level()`, thus saving considerable time.

Supplementary routines

It is useful to create utility scripts (and menu items which call them) to split and join lines of an outline. Examples 18-5 and 18-6 show possible implementations.

Example 18-5. opSplit()

```
    if wp.intextmode()
        local (selStart, selEnd, s)
        wp.getselect(@selStart, @selEnd)
        if selStart == selEnd « error check, split at insertion pt only
            wp.setselect(selStart, infinity)
```

Example 18-5. opSplit() (continued)

```
s = wp.getseltext()
wp.insert("")
if op.countsubs(1)
    op.insert(s, right)
else
    op.insert(s, down)
```

Example 18-6. opJoin()

```
if !op.countsubs(1) « mustn't have subheads
    if op.go(up, 1) « we will join with what's directly above
        local (x = op.getcursor())
        op.go(down,1)
        local (s = op.getlinetext())
        op.deleteline()
        op.setcursor(x)
        local (s2 = op.getlinetext(), n = sizeOf(s2))
        op.setlinetext(s2 + s)
        wp.setTextMode(true)
        wp.setSelect(n,n) « leave insertion point in nice place
```

Two supplied utility verbs, `toys.outlineToList()` and `toys.listToOut-line()`, convert, respectively, an outline to a list of strings and a list of strings to an outline. The hierarchical structure of the outline is mirrored by lists within the list. These verbs start their actions at the current cursor location in the outline: `toys.outlineToList()` gathers the current bundle starting down from the selection point; `toys.listToOutline()` inserts starting down from the selection point.

So for example, if the outline **workspace.notepad** looks like this:

if the line "summit" is currently selected:

```
toys.outlineToList (@workspace.notepad)
    « {"summit", {"level 1", "level 1 again"}, "summit again"}
```

but if the line "level 1" is currently selected:

```
toys.outlineToList (@workspace.notepad)
    « {"level 1", "level 1 again"}
```

and if the line "level 1 again" is currently selected:

```
toys.outlineToList (@workspace.notepad)
    « {"level 1 again"}
```

Here is an example of these verbs in action. `op.sort()` is UserTalk's only sort function (except for sorting a table), and it is fast; so one way to sort something that isn't an outline is to convert it temporarily to an outline. So, the following script sorts a list of strings.

Example 18-7. sortList()

```
on sortList(theList)
    local (o); new (outlineType, @o); target.set(@o)
    toys.listToOutline(theList, @o)
    op.firstsummit(); op.deleteLine() « remove original blank line
    op.sort(); op.firstSummit()
    theList = toys.outlineToList(@o)
    target.clear()
    return theList
msg(sortList({"one", "two", "three", "four"}))
```

These verbs are useful also as a way of transferring information between outlines: read part of an outline out into a list, then read the list into a different outline. In many instances this is simpler than copying and pasting with editMenu verbs.

Outlines and Scripts

To work with a line's "secret" value:
```
op.setRefCon (scalarValue)
op.getRefCon ()
```

Each line of an outline has a "secret" scalar value associated with it (its "refCon"[*]), providing a further level of information storage. How to take advantage of this facility is entirely up to your imagination. Menubars use it to maintain the menu item's script, which is why these commands are not available in menubars. A convenient user interface for working with refCons is demonstrated in the `outliner` suite.

Wptexts Only

UserTalk support for the character and paragraph formatting features unique to wptexts is rather disappointing. There is no way, with UserTalk, to learn the current tab, justification, line spacing, or style of the currently selected text, nor its font and size.

---

[*] "RefCon" stands for "reference constant"; the name comes from a field in Mac OS window records where programmers can store extra information.

Character formatting

To set the text style of the selection or insertion point:
```
editMenu.plainText ()
editMenu.setBold (boolean)
editMenu.setItalic (boolean)
editMenu.setUnderline (boolean)
editMenu.setOutline (boolean)
editMenu.setShadow (boolean)
```

Paragraph formatting

To get and set margins:
```
wp.getLeftMargin ()
wp.setLeftMargin (pixels)
wp.getRightMargin ()
wp.setRightMargin (pixels)
```

To get and set first-line indents:
```
wp.getIndent ()
wp.setIndent (pixels)
```

To set a tab or clear all tabs:
```
wp.setTab (position, justification, leaderChar)
wp.clearTabs ()
```

To set paragraph justification:
```
wp.setJustification (justification)
```

To set line spacing:
```
wp.setSpacing (verticalPixels)
```

Ruler

To learn or set whether the ruler is visible:
```
wp.getRuler ()
wp.setRuler (visible?)
```

To learn the length of the ruler, in pixels:
```
wp.rulerLength ()
```

Ruler length is a function of the system's current printer page (as determined in the Page Setup dialog). The right margin is measured from the right end of the ruler.

19

Datafiles

Frontier can read data from and write data to a file on disk. The most common use of this facility is to read or write text, but the data can be of any type. The value that you receive from Frontier when you perform a read, or that you give Frontier when you perform a write, is a binary. If data read from a file is to be processed as text, it must be coerced to a string. (On binaries, see Chapter 10, *Datatypes.*)

The ability to read and write datafiles lies at the heart of some of Frontier's most powerful applications. In Frontier's Web site management facilities, for example, "rendering" a Web page consists of constructing the HTML for that page internally and then writing the result out to disk as a textfile. The Web site management facilities can also store GIF and other image files in the database and write them out as files to disk as part of the site.

This chapter discusses the UserTalk verbs and techniques for working with datafiles. On file pathnames, and for other verbs dealing with files and folders, see Chapter 31, *Driving the System.* For verbs dealing with a file's resource fork, see Chapter 20, *Resources.*

Utilities

Most often, what you'll want to do is read or write an entire file at once. Utility scripts are provided for this.

To read all of a file's data from disk into a Frontier object:

```
toys.readWholeFile (pathname)
```

> The result is a binary of the file's entire contents.

So, for instance, here is a script to let the user choose and read a textfile into a Frontier wptext.

Example 19-1. Textfile to wptext

```
local (thePath)
if file.getfiledialog("Choose a textfile, please.", @thePath, 'TEXT')
    if not defined(workspace.temptext)
        new(wptexttype, @workspace.temptext)
    edit (@workspace.temptext)
    wp.settext(string(toys.readWholeFile(thePath)))
```

The `string()` coercion is unnecessary here, because `wp.settext()` will cause implicit coercion of its parameter to a string; it is included in the example just to remind you that the result of `toys.readWholeFile()` is a binary, not a string. Most of Example 19-1 is unnecessary; reading a textfile into a wptext object is so common that a utility script is provided for that, too.

To read all of a file's data from disk into a wptext object:

```
toys.readFileIntoTextObject (pathname, addrDestination)
```

> The object is created if it doesn't exist, and replaced if it does.

So we can rewrite Example 19-1 as follows.

Example 19-2. Textfile to wptext again

```
local (thePath)
if file.getfiledialog("Choose a textfile, please.", @thePath, 'TEXT')
    toys.readFileIntoTextObject (thePath, @workspace.temptext)
```

To write a Frontier object out to disk as a file:

```
toys.writeWholeFile (pathname, data, type, creator, date)
```

Here, for example, is a script that makes a Microsoft Word textfile listing every file in a folder that the user chooses.

Example 19-3. ListFolderToDisk()

```
local(folder)
if file.getFolderDialog("Folder to list?",@folder)
    local (count = 0, f, theString)
    local (path = file.getSystemDisk () + "textfile")
    if file.putFileDialog("Textfile to save to?", @path)
        fileloop (f in folder, 1)
            local (oneLine = file.fileFromPath(f))
            msg (oneLine); count++
            theString = theString + oneLine + cr
        toys.writeWholeFile(path, theString, 'TEXT', 'MSWD', clock.now())
        msg (count + " files listed.")
```

To see that there's nothing special about dealing with a file whose data is not text, examine `html.loadImageFile()` and `html.data.standardMacros.image-Ref()`. In `loadImageFile()`, notice these two lines:

```
adrtable^.[name] = toys.readWholeFile (f)
setBinaryType (@adrtable^.[name], type)
```

The first line reads the data from the whole file directly into a database entry, which is therefore automatically of binary type; nothing else needs to be done. The binary's internal type is set purely as a way of labelling it, so that `image-Ref()` can later consult that label to decide what extension and type to give the file when it writes the database entry back out to disk.

Basic Verbs

The verbs we have looked at so far are utility scripts. Now we come to the more basic verbs upon which those utilities depend. You can use these for finer control as you read or write file data.

Before you can read from or write to a file, the file must exist, and it must be explicitly opened for access; when reading or writing is over, the file must be explicitly closed.* Let's call the period between opening and closing a file a *session.* Then during any session, the first read begins at the beginning and each successive read starts right after where the previous read ended, but every write appends to the end of the file. There is no problem with mixing reads and writes in a session.

The operating system will complain if certain file access problems occur, and Frontier will raise an error. It is wise, therefore, to embed sessions in a **try**, because if something goes wrong during a session, it is crucial that you close the file.

To create an empty file (replacing an existing file):
    `file.new (`*pathname*`)`

To open a file for read/write access, beginning a session:
    `file.open (`*pathname*`)`

To close a file previously opened for read/write access, ending a session:
    `file.close (`*pathname*`)`

To learn the size of a file:
    `file.size (`*pathname*`)`

See also Chapter 31, *Driving the System*, where `file.size()` is discussed again. On Mac OS, this verb reports the combined size of the resource fork and the data fork; hence, it can be used to learn the number of bytes of data only if you know

---

* In theory, if you're only going to write to a file, you don't have to open and close it (because the write action opens and closes the file implicitly). But in actual fact it's always more efficient to do so.

that there is no resource fork. You do know this if Frontier created the file (and never wrote to the resource fork). When in doubt, read the whole file and examine the size of the result; this works because reading a file reads its data fork only.

To read from a previously opened file:
```
file.read (pathname, howManyCharacters)
```

An interesting problem is that although successive reads within a session move the "pointer" at which the next read will begin, there is no way to ask Frontier where the pointer actually is. But you might need to know this, because if *howManyCharacters* is too large, the read may fail, and you won't be able to do any more reading in this session. One solution is to determine the size of the data beforehand, and then keep track of the number of characters read.

Similarly, you cannot ask Frontier to set the pointer just anywhere; for so-called "random access" reading, you have to begin a new session and then read up to the point where you actually want to start reading.

It is often simpler just to read the whole file and then process the result. For instance, here is a way of learning the last character of a textfile:
```
local (s = toys.readWholeFile("HD:textfile"))
msg(char(s[sizeof(s)]))
```

To read from a previously opened file until an end-of-line (cr) is reached:
```
file.readLine (pathname)
```

To check whether this session's reading has reached the end of the file:
```
file.endOfFile (pathname)
```

To append to the end of a file:
```
file.write (pathname, whatToWrite)
```

To append a string plus cr to the end of a file:
```
file.writeLine (pathname, whatToWrite)
```

As with reading, so with writing you cannot set the pointer just anywhere; your only choice is to append to the end of a file. To insert material at the beginning or middle of a file, you need to overwrite the entire file. For an example, see `suites.samples.NewIn96.add2lines()`, which inserts two lines at the start of a textfile by reading in the file as a string, inserting the lines at the start of the string, and overwriting the file with the result.

Finding in Textfiles

A couple of verbs help look for text on disk without opening any files.

To learn whether a file contains a certain string:
```
file.findInFile (pathname, whatToLookFor)
```

To learn what file in a certain folder contains a certain string:

    file.findInFolder (*folderPathname, depth, whatToLookFor*)

`file.findInFolder()` is a utility script that calls `file.findInFile()` in a `fileloop()`. You learn nothing from `file.findInFile()` about where in the file the string appears (to learn that, you could read the file and then use `string.patternmatch()`); but it is quick and convenient.

To determine the number of lines in a file:

    file.countLines (*pathname*)

The result of `file.countLines()` is *not* necessarily the same as the number of calls to `file.readLine()` that can be successfully performed, because it just counts end-of-lines, and the file might not end with an end-of-line.

20

Resources

On Mac OS, a file consists of both a resource fork and a data fork, either of which may be empty. The data fork is a single stream of data; access to it through User-Talk is described Chapter 19, *Datafiles*. The resource fork is a collection of individual structured chunks of data, called *resources*. Each individual resource is identified by a type (a string4) and an index, which is an ID number or a name; every resource has an ID, and it may or may not also have a name.[*]

Many interface-related data structures, such as sounds, dialog layouts, menus, and so forth, are stored as resources of types that the system knows how to deal with; applications are also free to define their own resource types to use as they wish. Users can access a file's resources, and edit many of them conveniently, with a utility application such as ResEdit.

This chapter describes UserTalk's verbs for accessing resources. It is possible to damage a file or an application's functionality beyond repair by altering its resources, so use caution.

Inspecting Resources

To obtain the number of different resource types in a file:

    rez.countResTypes (pathname)

To obtain the string4 identifier of the nth resource type in a file:

    rez.getNthResType (pathname, n, addrDestination)

To obtain the number of resources of a given type in a file:

    rez.countResources (pathname, type)

---

[*] For more information about resources, see *http://devworld.apple.com/dev/techsupport/insidemac/ MoreToolbox/MoreToolbox-10.html.*

To obtain the ID and name of the nth resource of a given type in a file:

```
rez.getNthResInfo (pathname, type, n, addrID, addrName)
```

`rez.getNthResType()` and `rez.getNthResInfo()` store the requested information by way of address parameters, but they also return a boolean reporting
success, so you can loop through all resources without counting them first. For
example, this script makes an outline giving the ID (and name if there is one) of
every resource in a file chosen by the user.

Example 20-1. Outlining all resources

```
local (f)
on whatDir()
    if y == 1 {return right}
    return down
if file.getFileDialog ("File to examine:", @f, 0)
    new (outlineType, @workspace.rezOutline); edit (@workspace.rezOutline)
    loop (local (x = 1, y, id, name, t); rez.getNthResType (f,x,@t); x++)
        op.insert (t, down)
        loop (y = 1; rez.getNthResInfo (f, t, y, @id, @name); y++)
            op.insert (id, whatDir())
            if name
                op.insert (name, right)
                op.go (left, 1)
        op.go (left, 1)
```

For a slightly different implementation of this example, refer to **suites.sam-
ples.basicStuff.resourceMap()**.

Resource Data

Verbs that do things with a resource's data generally come in two flavors, to let
you refer to the resource either by name or by ID. In getting a resource's data,
there is a third option, to refer to the resource as the *n*th of its type.

To obtain the data of a resource referenced by ID:

```
rez.getResource (pathname, type, ID, addrDestination)
```

To obtain the data of a resource referenced by name:

```
rez.getNamedResource (pathname, type, name, addrDestination)
```

To obtain the data of a resource referenced as the nth of its type:

```
rez.getNthResource (pathname, type, n, addrName, addrDestination)
```

To set the data of a resource referenced by ID:

```
rez.putResource (pathname, type, ID, addrSource)
```

To set the data of a resource referenced by name:

```
rez.putNamedResource (pathname, type, name, addrSource)
```

To delete a resource referenced by ID:

```
rez.deleteResource (pathname, type, ID)
```

To delete a resource referenced by name:

```
rez.deleteNamedResource (pathname, type, name)
```

To check the existence of a resource referenced by ID:

```
rez.resourceExists (pathname, type, ID)
```

To check the existence of a resource referenced by name:

```
rez.namedResourceExists (pathname, type, name)
```

The data of a resource is always a binary. Dealing with this data—parsing it further to extract particular pieces of information, for example—is up to your script. For one type of resource, parsing utilities are supplied, namely resources of type 'STR ' (a Pascal-type string, which starts with a length byte).

To obtain the data, as a string, from a 'STR ' resource:

```
rez.getStringResource (pathname, ID, addrDestination)
```

To set the data of a 'STR ' resource:

```
rez.putStringResource (pathname, ID, theString)
```

Otherwise, you must know the structure of the resource in order to parse it. Since resources are designed to be parsed by a machine, this is usually simple enough. Consider, for example, a 'STR#', which is a string collection. It is structured as follows: two bytes stating how many strings there are; then that number of Pascal strings, where a Pascal string is a length byte followed by the characters of the string. Knowing this, it is trivial to parse the resource; here, for example, is a utility that reads a 'STR#' and converts it to a table of strings.

Example 20-2. strListToTable()

```
on strListToTable (path, id, addrT)
    local (str, pos=3, howMany, x)
    rez.getResource (path, 'STR#', id, @str)
    howMany = number("0x" + str[1] + str[2])
    new (tabletype, addrT)
    for x = 1 to howMany
        local (length = string.nthChar (str, pos))
        addrT^.["s" + toys.padWithZeros(x,3)] = string.mid(str, pos+1, length)
        pos = pos + length + 1
```

The reverse utility, that writes out a table of strings as a 'STR#' resource, is left as an exercise for the reader.

To illustrate the rez verbs further, here is an example script that aids in the importing of XCMDs into the database; it lets the user choose a file, looks for

XCMDs in it (using the technique from Example 20-1), lets the user choose from a list of these, and imports the chosen XCMD into the **workspace** table.

Example 20-3. importXCMD()

```
local (f, x, id, name, theOutline)
new (outlineType, @theOutline)
if file.getFileDialog ("File to look for XCMDs in:", @f, 0)
    target.set(@theOutline)
    loop (x = 1; rez.getNthResInfo (f, 'XCMD', x, @id, @name); x++)
        if name {op.insert(name, down); rollBeachBall()}
    loop (x = 1; rez.getNthResInfo (f, 'XFCN', x, @id, @name); x++)
        if name {op.insert(name, down); rollBeachBall()}
    op.firstsummit(); op.deleteLine(); op.sort(); op.firstsummit()
    name = listselect.getChoice \
        (toys.outlineToList(@theOutline), "XCMD to import:", 1, 1)
    if name
        if not rez.getNamedResource (f, 'XCMD', name, @workspace.[name])
            rez.getNamedResource (f, 'XFCN', name, @workspace.[name])
    target.clear()
```

An interesting utility verb showing the rez verbs in action is **suites.sam-ples.basicStuff.resourceStealer**(); you specify a folder, a resource type, and a file pathname, and all resources of that type are copied from all the files in that folder, to a depth of infinity, into the specified file (which is created if necessary), which is then opened by ResEdit.

The Resource Chain

The *resource chain* is the set of resources available to the system at any given moment. Adding a resource to the resource fork of an open file brings the resource into the resource chain immediately. When Frontier is running, there is a file known to be open and whose resource fork we can modify: the database. A resource can be added to the database's resource fork, on the fly, while the database is open and Frontier is running, in order to make that resource available to the system. It can then be removed from the database's resource fork when we are finished using it.

Here, for example, is a script that plays a **'snd '** resource stored as a binary in the database; it adds the binary to the database's resource fork, plays it, and removes it from the database's resource fork (the database entry remains).

Example 20-4. Loading a resource into the resource chain

```
rez.putNamedResource \
    (frontier.getFilePath(), 'snd ', "mySound", @workspace.mySound)
speaker.playNamedSound("mySound")
rez.deleteNamedResource \
    (frontier.getFilePath(), 'snd ', "mySound")
```

Resource Attributes

With every resource is associated a set of resource attributes, coded as a short integer each of whose bits represents an attribute. Two verbs let you get and set this short integer.

To obtain a resource's attributes:

```
rez.getResourceAttributes (pathname, type, ID)
```

To change a resource's attributes:

```
rez.setResourceAttributes (pathname, type, ID, theShort)
```

The bit verbs (see Chapter 15, *Math*) can be used to manipulate the individual bits of the short. The bit flag meanings are:

- 1: resource has been changed

- 2: resource is to be preloaded

- 3: resource is protected

- 4: resource is locked

- 5: resource is purgeable

- 6: resource is to be read into the system heap

These are very advanced verbs, requiring an intimate understanding of Mac OS resources, and the vast majority of users will never call them.

21

Yielding, Pausing, Threads, and Semaphores

The Mac OS can run more than one process at the same time (multitasking), and one process can run more than one thread at the same time (multithreading). The UserTalk verbs discussed in this chapter allow Frontier to behave as a good Mac OS citizen by yielding time to other processes, and to control its own threads to accomplish several things at once. We also explain semaphores, a simple but effective device for preventing collisions among such simultaneous activities.

Yielding and Pausing

During a looping operation, especially when polling until some condition is met, you may wish to yield time to other processes, including giving the system a chance to update the screen; this is simply what all good citizen processes do from time to time on computers that lack true multitasking. To do so, call `sys.systemTask()`.

To pause for a fixed interval, call `clock.waitSixtieths()` or `clock.waitSeconds()`, depending on whether you'd like to specify the interval in sixtieths of a second or in seconds. These verbs are just like calling `sys.systemTask()` repeatedly until a certain time has elapsed.

Threads

A *thread* is a miniature process running within Frontier. Frontier is multithreaded, which means in essence that it can do several things at once. This feature is particularly valuable in connection with Frontier's ability to drive and be driven by other applications, because it lets Frontier maintain multiple active lines of communication. For instance, it makes Frontier a natural choice as a CGI application; Frontier can handle a server request while already processing other such requests.

Choosing the **StatusMessage** item from the agent selection popup at the left end of the Main Window displays a message which provides (among other things) a constantly updated count of the number of running threads. The "1 thread" present when Frontier is idle is the agent thread, in which Frontier's background processes are running. (For more about agents, see Chapter 27, *Agents and Hooks.*) Over and above this, Frontier automatically creates a new thread for every script initiated either by a human being or by another application; when the script finishes executing, its thread is destroyed.

For example, type `clock.waitSeconds(10)` into the Quick Script window and click the **Run** button. For the next 10 seconds or so, the Main Window's count of "1 thread" will be increased to "2 threads." Now let `workspace.testing()` go like this:

```
clock.waitSeconds(10)
```

Click the **Run** button in `workspace.testing`'s window, quickly switch to the Quick Script window and click its **Run** button. The Main Window now reports "3 threads." The thread count returns to 1 when both commands have terminated separately.

Next, start up another application that can speak to Frontier in AppleScript, such as the Script Editor or HyperCard. Suppose it's HyperCard. Make a new stack with a new button, open the button's script, use the popup at the top of the script edit window to change the language to AppleScript, and make the script read as follows:

```
on mouseUp()
    tell application "UserLand Frontier™" to do script "clock.waitseconds(10)"
end mouseUp
```

Close and save the script. Press the button to start the script, and switch to Frontier; the thread count climbs to 2 while the script is running.*

To learn how many threads are running:
```
thread.getCount ()
```

Simply calling a script from within a script does *not* generate a new thread; the caller pauses until the called script returns, so execution is linear, within a single thread. To enable a script to call another script so that the latter runs in a thread of its own, a verb `thread.evaluate()` is provided. The parameter is a string; that string is evaluated as a UserTalk expression, just like `evaluate()`, but it is evaluated from within a new separate thread. For example, let `workspace.testing()` be this:

```
thread.evaluate("workspace.testing2()")
clock.waitseconds(10)
```

---

* When a shared menu item is chosen in another application, or when an OSA-savvy application runs a UserTalk script, things are more complicated. The action takes place in the other application's context, and Frontier's thread count will not appear to rise; nonetheless, if the running script asks for the number of running threads, the count will be one more than Frontier's count.

And let `workspace.testing2()` be this:

```
clock.waitseconds(20)
```

Press the **Compile** button in both windows, make sure the "1 thread" message is visible in the Main Window, and press the **Run** button in `workspace.testing`. The thread count goes up to 3 for 10 seconds, because both scripts are running as individual threads; then comes down to 2 for 10 seconds, after `work-space.testing()` has finished; and finally returns to 1.

Since the parameter of `thread.evaluate()` is most often a verb call, and since it can be tiresome to construct a string representing a verb call with parameters, there is a utility script, `toys.threadCall()`. It takes the name of a verb (as a string) and a list of the actual parameters to be fed to that verb; it constructs the verb call as a string for you, and feeds it to `thread.evaluate()`.

The result of a call to `thread.evaluate()` is not the result of evaluating its parameter as a UserTalk expression, because the whole meaning of spawning a new thread is that the calling script doesn't wait for the new thread to return a result; the two scripts part company, as it were, and proceed simultaneously. To capture the result from the new thread, call instead `thread.evaluateTo()`, which takes the address of a database entry in which to store the result when the thread ultimately finishes.*

What, then, *is* the result of a call to `thread.evaluate()` or `thread.evalu-ateTo()`? It is a number identifying the new thread. The calling script will need this only if it intends to manage threads manually—which is the topic of the next section.

Managing Threads

The easiest way to manage threads is not to do so. Frontier manages threads automatically; taking control of threads yourself is complicated, rather as if control of an unconscious, involuntary bodily function, such as your liver, were suddenly entrusted to your conscious self. Nevertheless, it is sometimes necessary to take charge of threads—typically, where one wishes to maintain efficiency while awaiting a response from elsewhere (the Internet, for instance).†

---

* It is illegal for the address to be that of a local variable, because the new thread would have to interrupt the calling script to set the variable's value (assuming that the variable is even still in scope!).

† For an ingenious and highly educational example, see *http://siolibrary.ucsd.edu/Preston/scripting/root/ suites/VerifyURL.html*. One obstacle to managing threads manually is that they are difficult to debug. The only thread that Frontier's debugger can step through is the one initiated by the pressing of the Debug button; threads spawned by that thread will simply run independently, beyond the debugger's control.

For most thread verbs, you'll need to supply the ID number of a particular thread. The script that creates a thread can learn its ID number by capturing the result of `thread.evaluate()` or `toys.threadCall()` or `thread.evaluateTo()`. For other situations, further verbs are provided.

To learn the ID of the thread within which one is:
```
thread.getCurrentID ()
```

To learn the ID of the nth thread:
```
thread.getNthID (n)
```

To learn whether any thread has a given ID:
```
thread.exists (ID)
```

Once you have the ID of a thread, there are two sorts of thing you can do with that thread. One is to abort it.

To abort a thread:
```
thread.kill (ID)
```

For example, let's spawn an endless thread and then kill it. Let **work-space.testing()** go like this:

```
loop
    local (x)
    for x = 0 to 9
        msg(x + " I am thread " + thread.getCurrentID() + ".")
        clock.waitSixtieths(10)
```

and click the **Run** button. In the Main Window, you can see the thread cycling, and also learn its ID. Open Quick Script, type `thread.kill(ID)` (putting in the actual number of the running thread's ID) and hit the **Run** button. The counting in the Main Window stops; the thread is dead.

The other thing to do to a thread is put it to sleep. A thread that is put to sleep pauses and waits to be woken before executing its next command.

To put a thread to sleep:
```
thread.sleep (ID)
```

To awaken a thread:
```
thread.wake (ID)
```

To put to sleep the thread in which one is, either until woken or until a certain number of seconds passes, whichever comes first:
```
thread.sleepFor (seconds)
```

To learn whether a given thread is sleeping:
```
thread.isSleeping (ID)
```

As an example, let's have `workspace.testing()` put itself to sleep. Modify `workspace.testing()` so it says:

```
loop
    local (x)
    for x = 0 to 9
        msg(x + " I am thread " + thread.getCurrentID() + ".")
        clock.waitSixtieths(10)
    thread.sleep(thread.getCurrentID())
```

Now press **Run**. In the Main Window, we count from **0** to **9** and stop. Using the ID information from the Main Window, type into QuickScript `thread.wake(ID)` and press Quick Script's **Run**. The counting in the Main Window happens again, and pauses again until you hit Quick Script's **Run** again. Play this game as long as you like; when you tire of it, use Quick Script to kill the thread.

Threads are tracked in `system.compiler.threads`. Since the thread verbs provide a programming interface, you should have no reason to touch `system.compiler.threads` directly; but it is interesting to open it and watch the table change while playing with threads.

Semaphores

A *semaphore* is a device to enable cooperation among threads so that one thread doesn't try to access something in a way that could damage another thread's use of that same thing.

Here is a trivial example, just to illustrate the principle. Suppose `workspace.testing()` looked like this:

```
local (x)
workspace.ct = 0
thread.evaluate("workspace.testing2()")
loop
    x = workspace.ct++
    msg(x + ": I am thread " + thread.getCurrentID() + ".")
    clock.waitSixtieths(30)
    if x > 100
        break
```

And suppose `workspace.testing2()` looked like this:

```
local (x)
loop
    x = workspace.ct++
    msg(x + ": I am thread " + thread.getCurrentID() + ".")
    clock.waitSixtieths(40)
    if x > 100
        break
```

If we now run `workspace.testing()`, it spawns off `workspace.testing2()` as a separate thread. Both routines are looping simultaneously, and both routines will be repeatedly incrementing the value of the same database entry, **workspace.ct**. This is a bad thing. What if some of these accesses occur simultaneously? Will the results be messed up? Will Frontier crash?

Semaphores solve this kind of problem neatly and simply. There are two semaphore verbs.

To lock a semaphore, first waiting until it is unlocked:
```
semaphores.lock (name, timeOutTick)
```

To unlock a semaphore:
```
semaphores.unlock (name)
```

To return to our earlier example, suppose `workspace.testing()` looks like:

```
local (x)
workspace.ct = 0
thread.evaluate("workspace.testing2()")
loop
    semaphores.lock("workspace.ct", 600)
    x = workspace.ct++
    semaphores.unlock("workspace.ct")
    msg(x + ": I am thread " + thread.getCurrentID() + ".")
    clock.waitSixtieths(30)
    if x > 100
        break
```

And `workspace.testing2()`:

```
loop
    semaphores.lock("workspace.ct", 600)
    x = workspace.ct++
    semaphores.unlock("workspace.ct")
    msg(x + ": I am thread " + thread.getCurrentID() + ".")
    clock.waitSixtieths(40)
    if x > 100
        break
```

Compile both scripts and then press **workspace.testing**'s **Run** button. In the Main Window, you will see that the count increments smoothly even though it is sometimes one thread, sometimes the other that increments it. The threads are making cooperative shared use of a single entity.

Any time you are using multiple threads in Frontier, you should consider whether you need a semaphore to keep them from trying to access the same entity simultaneously. Here, our example was a database entry. Another might be a file on disk. Another might be a line of communication to an application that isn't multithreaded.

Semaphores are purely a voluntary, cooperative device. There is nothing magical about `semaphores.lock()`; it does not in fact lock anything! When we say:

```
semaphores.lock("workspace.ct",600)
```

access to `workspace.ct` is unaffected. In fact, the string `"workspace.ct"` is chosen arbitrarily, just to give this semaphore a unique, consistent name; we could equally have called it `"redFlannelUnderwear"`. The use of semaphores is nothing more than an agreement to be polite and safe; there will be politeness and safety so long as every running thread makes the same agreement.

Here is how semaphores really work. There is a semaphore lock registry, which is the table at `system.compiler.semaphores`; locking or unlocking a semaphore consists merely of adding or deleting its name in this table. A call to `semaphores.lock()` looks in the lock registry to see if the name is already there; if so, it waits until it isn't.* Then, it adds the name to the registry. `Semaphores.unlock()` removes the name from the registry.

The danger with semaphores is failure to unlock one, which will cause any other thread that tries to lock the same semaphore to be unable to proceed. To guard against this, use `try` (and possibly `else`) in such a way that regardless of what happens the semaphore will be unlocked. The following structure is fairly typical:

```
semaphores.lock("mySem", 600)
try
    « … do interesting but risky stuff here …
else
    « oops, an error occurred! release the semaphore and rethrow the error
    semaphores.unlock("mySem")
    scriptError (tryError)
semaphores.unlock("mySem")
```

In case of trouble, the semaphores table can also be cleaned out manually, by quitting and restarting Frontier, or by choosing **Unlock Semaphores** from the **Web** menu.

One final note about the names of the verbs. Some scripts exhibit a certain confusion about this, calling `semaphore.lock()` and `semaphore.unlock()`, the singular instead of the plural. This is not an error, but it is wrong behavior; it accesses the kernel directly. The correct interface is `semaphores.lock()` and `semaphores.unlock()`.

---

* Or until the timeout expires, whichever comes first; the expiration of the timeout will raise a runtime error. Clearly there is no intention of raising an error; the timeout is simply a safety device so that we do not accidentally pause forever if something goes wrong.

22

Stacks

A stack is a last-in-first-out storage structure, like a pile of plates: one can put a new plate on top at any time, and whenever one reaches to retrieve a plate from the top of the pile it is always the one most recently added to the pile. UserTalk implements stacks with utility scripts; they are at `suites.stacks`, and are worth studying, as they might give you ideas for implementing similar storage structures, such as queues.

To make a new stack:

```
addrMyStack = stack.create ()
```

The variable on the left of the assignment can be anything; the point is to capture the value of **stack.create()**, because this is the address of your new stack, and is needed as a parameter to the other stack verbs.

To add a value to a stack:

```
stack.push (addrMyStack, theValue)
```

To remove a value from a stack:

```
stack.pop (addrMyStack)
```

> The most recently pushed value is removed from the stack, and returned as the value of the verb call. If the stack is empty, a runtime error is generated; you can handle this gracefully by embedding the call in a **try...else** structure.

To learn the number of elements on a stack, get its **top**. For instance, if we have captured the stack's address as **addrMyStack**, then **addrMyStack^.top** is the number of elements of the stack.

To examine the *n*th element on a stack, where the first element is the one first pushed onto the stack, say this:

```
addrMyStack.vals.["item" + n]
```

When you are done using a stack, you should dispose of it or it will remain in the database. Simply say:

```
stack.dispose (addrmyStack)
```

The verb `stack.visit()` is discussed Chapter 12, *Control Structures*.

23

Extending the Language

Despite the many virtues of UserTalk, situations arise where its speed and functionality are deemed insufficient, and one wishes to extend the language at the level of compiled code. If the compiled code is in the right format, UserTalk can call it. The UserTalk script which calls the code, known as "glue," can itself be called like an ordinary UserTalk verb. Thus, the UserTalk programmer doesn't need to be at all conscious of the fact that a verb is implemented in this way.

The compiled code can be in one of three formats:

1. An XCMD (or XFCN)* is a scripting extension in a format originally devised for use with HyperCard and now rather more widely adopted. Many XCMDs will work when called from UserTalk.

2. A UCMD is a scripting extension in a format specifically devised for UserTalk.

3. An OSAX is a scripting extension in a format originally devised for AppleScript.

Use of OSAXen from UserTalk is discussed in Chapter 33, *AppleScript*; this book says nothing about how to *write* an OSAX. The rest of this chapter is about XCMDs and UCMDs; further technical details on how to write these are provided in Chapter 43, *XCMDs and UCMDs*.

To be called from UserTalk, an XCMD or UCMD must be living in the database as a binary. The standard location for these extensions is `system.extensions`; there is a search path to this table. For example, the `trigCmd` verbs discussed under Chapter 15, *Math*, are implemented by calling a UCMD at `system.exten-sions.trigCmd.code`. To see an XCMD in action, call `listSelect.example()`; it works by calling an XCMD at `system.extensions.listSelect.code`.

---

* The only difference between an XCMD and an XFCN is that the latter is intended to return a value, the former is not. Henceforward I shall generally refer to both as XCMDs.

Examination of a glue script such as `trigCmd.sqrt()` shows that the UCMD is called by way of a rather nasty-looking UserTalk verb, `appleEvent()`. The idea of the glue script is to shield the user from this sort of thing. Calling `trigCmd.sqrt()` is easy; calling `appleEvent()` is not, so `trigCmd.sqrt()` does it for you. XCMDs and UCMDs are often accompanied by a ReadMe wptext that explains how to call the glue.

Why XCMDs and UCMDs Exist

There are two questions to be answered. First, why do XCMDs and UCMDs exist at all? One reason is so as to be able to perform functions that are impossible through UserTalk alone. This is true of both `trigCmd` and `listSelect`.

Another reason is for speed. Some UCMDs started out as UserTalk scripts and were rewritten in compiled form just to make them faster. The speed increase is roughly an order of magnitude. Even a UCMD, though, won't be as fast as code incorporated into the UserTalk kernel. This is doubtless because of the overhead involved in packaging and passing the information to and from the UCMD. For example, `string.parseHTTPargs()` was about ten times faster as a UCMD than it was as a script, but now is incorporated in the kernel and is around five times faster than that. However, it is not open to the user to modify the kernel, which is part of the Frontier application itself and can only be changed by UserLand—whereas any technically minded user can write an XCMD or UCMD, put it in the database, construct glue for it, and call it.

Second, why do both XCMDs and UCMDs exist? After all, a single way of extending UserTalk ought to suffice. The reason is largely historical: the XCMD format was well established when Frontier first appeared; UCMDs were developed later to take advantage of Apple event architecture. XCMD support has not been abandoned, largely because, owing to the long history of HyperCard, there are so many XCMDs in existence.* Thus the user can leverage a large existing base of code.

Also, it is easy to import an XCMD into the database, and to write glue code for it; so if you need to extend UserTalk and you know of an XCMD that does what you want, you can, in most cases, make that functionality available to UserTalk without doing any coding. XCMDs are also faster and smaller than UCMDs. On the other hand, not every existing XCMD can be successfully called by UserTalk (only experimentation will show which ones can be).

---

* They are mostly available for free download via the Internet. See, for example, *http://www.glasscat.com/hypercard/makeIND.cgi/XCMD_etc.*

Importing and Glue

Suppose you have a HyperCard stack containing an XCMD you'd like to be able to call from UserTalk. To import the XCMD into the database, all you have to do is this: in the Finder, drop the stack's icon onto Frontier's icon. Frontier will search the stack for XCMDs and will offer to import each one in turn into `system.extensions.`*XCMDName*`.code` as a binary. Another option is to use a resource editor such as ResEdit to copy the desired XCMD into its own resource file, and drop that resource file's icon onto Frontier's icon.

It is also possible to write a script to import the desired XCMD into the database; this is demonstrated in Chapter 20, *Resources* (Example 20-3).

Some XCMDs may expect the presence of other resources, such as dialogs. It is possible to import these into the database as well and make them available in the resource chain as part of the action of the glue script. This is also discussed in Chapter 20.

Having imported an XCMD, you should write glue for it. The UserTalk interface to an XCMD binary is through the verb `callXCMD()`. It's easy to use. Suppose we have at `system.extensions.sortFieldByItem.code` an XCMD which in HyperTalk would be called like this:

```
sortFieldByItem (bg fld "myFld", 7)
```

where the first parameter is some text and the second parameter is a number. Then the glue would look like this:

```
on sort (s, n)
    return callXCMD (@sortFieldByItem.code, s, string(n))
```

The coercion is needed because XCMDs take only string parameters. If the glue lives at `system.extensions.sortFieldByItem.sort`, the user can then simply call `sortFieldByItem.sort()`.

A UCMD is easier to import because it will typically be delivered as a packed object; just open it from the Finder, and Frontier will import it automatically (see Chapter 29, *Import/Export*). The packed object will include both the UCMD and the glue, so the UCMD will be usable immediately with no further effort. The format of UCMD glue is explained later; see Chapter 32, *Driving Other Applications*, and Chapter 43.

IV

Interface

This section provides a tour of the features of the Frontier program itself. In Part I, *Getting Acquainted*, Frontier's basic edit windows and the database were introduced; here, the discussion is extended to windows and dialogs in general, as well as menus, agents, and pictures.

There is also an explanation of how database objects can be exported as separate files on disk and imported into the same or a different database. And the "care and feeding" of the database is described, with such topics as backing up, upgrading, and so forth. Finally, I talk about how Frontier can open more than one database at the same time.

These matters are all aspects of Frontier with which any user will wish to be familiar. At the same time, many of them can also be controlled with UserTalk, so the relevant UserTalk verbs and techniques for doing so are discussed here as well; readers who are less interested in the details of UserTalk can skip those parts of each chapter if desired.

24

Windows

This chapter discusses manipulation of Frontier's windows, both manually and through UserTalk. This means manipulation of the window as a physical object—its size and shape, for instance. Manipulation of a window's contents is discussed in Chapter 18, *Non-Scalars*.

The windows studied in this chapter include the Main Window, the Quick Script window, modeless dialog windows, and edit windows. Dialogs are more particularly discussed in Chapter 25, *Dialogs*. Windows belonging to other programs fall under Chapter 32, *Driving Other Applications*.

On navigation of the database by way of edit windows, see Chapter 2, *Edit Windows*.

Manual Window Manipulation

As in most applications that have windows, Frontier's windows are layered on the screen, and bringing a window to the front does not change the layering order of the other windows. Normally, clicking in a window that is not frontmost brings it to the front, and that's all; if you want to do something that requires clicking in the window (press a button, set the insertion point, etc.), you'll have to click again. By holding down the Command key, it is possible to drag a window by its titlebar without bringing it to the front.

A Frontier window can be hidden. This is not the same as closing the window; the window is open, but invisible. There is, unfortunately, no manual way to hide a window; it must be done by way of UserTalk, as explained later in this chapter.

The Main Window represents its database as a whole; closing the Main Window closes its database, offering to save it if it has been changed. (On opening multiple databases, see Chapter 30, *Multiple Databases*.) The Main Window has certain nonstandard physical features; see Chapter 3, *The Database*.

Window Menu

The **Window** menu lists all open windows, both visible and invisible, by title. A window can be brought to the front (and, if it was invisible, made visible) by choosing its title from the **Window** menu. The menu is divided into eight categorical sections, any of which will be omitted if no windows in a category are open: Main Windows (there can be more than one, if multiple databases are open); nonmodal dialogs, including Quick Script; tables; menubars; scripts; outlines; wptexts; and pictures. Within each section, windows are listed in the order in which they were opened.

When it is opened, the title of a database object edit window is the full path of the object, omitting `"root"`, except that the table edit window representing the top level of the database *is* titled `"root"`. The Main Window's title is the filename of the database it represents (the default database is *Frontier.root*). Dialog titles are as displayed in their titlebars. Once a window is open, it is possible to change its title programmatically; this changes what appears in its titlebar, and its listing in the **Window** menu. A changed title is forgotten when the window is closed.

In the **Window** menu, a diamond-mark appears beside the listing of the current database's Main Window. A checkmark appears beside the listing of the frontmost window. An underlined window listing indicates that that window is "dirty": its contents have been changed somehow, though Frontier's criteria for dirtiness can sometimes be unexpected.* Invisible windows are listed in italic.

Closing Windows and Close Hooks

A window can be closed by choosing **Close** from the **File** menu, or by hitting Command-w, or by clicking the close box at its upper left. Adding the Option key to any of these actions causes all windows to be closed, except for the Main Window—unless the Main Window is frontmost at the time, in which case it is closed (and so is the database).

Closing a database object's edit window manually† triggers the script at `system.misc.closeWindow` before the window has a chance to close. This script calls, in turn, any scripts in the table at `user.hooks.closeWindow`, passing three parameters: the title of the window, a boolean indicating whether the

---

* For instance: scrolling a window can be sufficient to dirty it, but expanding subheads may not be. See Chapter 18.

† Not programmatically, not a dialog window, and not a menu item script window.

window was modified within the last 15 seconds, and the address of a variable in which the called script may store **true** to prevent the closing of the window.

A script of this sort, initiated through some cause other than because the user explicitly ran it or it was called from another script, is called a *hook*. Other hooks are discussed in Chapter 27, *Agents and Hooks*.

How to take advantage of this particular hook is up to your imagination. There is already a script within **user.hooks.closeWindow** that calls **html.addTo-ChangedPages()**; this is part of Frontier's Web site management, and is used to maintain a list of Web pages that have been altered. A comment in **system.misc.closeWindow()** suggests how you could catch the closing of the Main Window and save the database programmatically, thus preventing the "Save changes?" dialog from ever appearing. Here's an (annoying) script that simply shows a confirmation dialog every time the user tries to close a window; it would be located at **user.hooks.closeWindow.confirmClose**.

Example 24-1. confirmClose()

```
on confirmClose(title, changed, addrWhoa)
    if dialog.yesno("Really close " + title + "?")
        addrWhoa^ = false
    else
        addrWhoa^ = true
```

There is a bug in the hook mechanism; while **system.misc.closeWindow()** is running, the **try...else** mechanism is broken. It should therefore not be used in your hook script.*

Formatting Windows

The size and position of edit windows may be manually altered in the usual ways: drag the titlebar to set position, drag the resize box to set size, and so on. Such changes are remembered when the window is closed; it will open in the same state the next time.

A splendid feature of Frontier is its "intelligent zoom." Clicking a window's zoom box resizes and, if necessary, repositions the window so that the window is just large enough to show its whole contents, to the limits that the screen will allow.

---

* In fact, the **try...else** mechanism is used in **html.addToChangedPages()**, and this can cause problems. **user.hooks.closewindow.addToChangedPages()** should be rewritten to read like this:

```
on addToChangedPages (adrpage, flchanged, adrstopclose)
    if defined(adrpage^) « add this line
        html.addToChangedPages (adrpage)
```

This change prevents **html.addToChangedPages()** from choking silently if **typeOf()** doesn't work on the window (because it is a menu item script window, for example).

Zooming repeatedly alternates the window's new size and position with the old ones, but zooming after changing a window's size or position always performs the intelligent zoom.

Except for wptext windows, every window has just one text font and size. These can be set with the **Font** and **Size** submenus of the **Edit** menu, or with the **Common Styles** submenu of the **Main** menu. (The **Common Styles** submenu can be customized as desired; see Chapter 26, *Menus and Suites*.) This applies even to the Main Window and Quick Script window. Programmatic manipulation of a window's font and size are described in Chapter 18, *Non-Scalars*.

When you make a new database entry by choosing from a menu, a script generates the entry; if that entry is a non-scalar, the script can also set the initial formatting of the entry's edit window. Unfortunately, Frontier is not entirely consistent about taking advantage of this opportunity. For example, when you choose **New Script** from the **Table** menu, the menu item script uses the values at `user.preferences.scriptFont` and `user.preferences.scriptFontSize` to set the text font and size for the new script object's edit window (and also inserts the first line of the eponymous handler). But there are also `user.preferences.outlineFont`, `user.preferences.outlineFontSize`, `user.preferences.tableFont`, and `user.preferences.tableFontSize`; yet these are not used when you choose **New Outline** or **New Table** from the **Table** menu.[*]

A script at `samples.basicStuff.windowFormatter`, called when you choose **Format Windows** from the **Table** menu, does some batch neatening of database object edit windows: it sets each window's text size and font, centers and sizes the window, and, if it's an outline, provides some "intelligent" expansion of subheads. It calls `table.visit()`, so it recurses infinitely deep in the original table. Running it on the whole database may cause an out-of-memory error, so it is best to run it on tables representing fairly restricted subsections of the database.[†]

Programmatic Window Manipulation

To be worked on programmatically, a window must be open (though it need not be visible). The only window verb that will have any effect upon a non-open window is `window.open()`. There is no penalty for calling a window verb on a window which is not open; the verb returns `false` without raising an error, and a script can test for this and respond as desired.

---

[*] A solution is to modify these menu item scripts; this is not recommended, as they "belong" more to UserLand than to the user, but I must admit that in my copy of the database I've done it anyway.

[†] This script pays no attention to the `user.preferences` settings. In my copy of the database, I have modified the script so that it does.

Most programmatic window manipulations require a reference to the window. The form of this reference, which we may call a *window reference*, is a little complicated to describe.

For the Main Window or a dialog window (including Quick Script), a window reference is simple enough: it is a string consisting of the window's title. However, for edit windows, Frontier will also accept the address of the object whose edit window this is—or, if you supply a string, Frontier will implicitly coerce it to an address. The complication is that since either a title or an address can be expressed as a string, Frontier doesn't necessarily know, to start with, which of the two you intend.

So when you give a window reference, Frontier begins by assuming that it is the address of a non-scalar object. If dereferencing the window reference resolves to a non-scalar, the resolution process is over. If either the resolution or the window operation fails, though, then if the window reference is a string, Frontier tries to understand it as a window title.

All of this is made trickier by the fact that object references are not case-sensitive, but string matching is. (See Chapter 45, *Punctuation.*) For example, on my computer, the Main Window must be referred to as `"Frontier.root"`; calling it `"frontier.root"` won't work. But the `workspace` table's edit window can be referred to as `"Workspace"` or `"workspace"` (or `@workspace` or `@Workspace`). And then, to top it all off, a window's title can be changed programmatically, so a database object's edit window might have a title different from its object reference.

Suppose, for instance, the `examples` table is open, and that its title has been changed to `"yohoho"`. Then all of the following yield `true`:

```
window.isOpen("examples")
window.isOpen("Examples")
window.isOpen("root.Examples")
window.isOpen(@examples)
window.isOpen("yohoho")
workspace.x = "Examples"; window.isOpen(workspace.x)
workspace.y = "yohoho"; window.isOpen(workspace.y)
workspace.z = @examples; window.isOpen(workspace.z)
```

But the following yields `false`:

```
window.isOpen("yoHoho")
```

because the capitalization is wrong (so there is no match on the window's title).

Now suppose we create a script in the `people.[user.initials]` table called YOHOHO and open it. Now if we say:

```
window.bringToFront("yohoho")
```

then, despite the capitalization, it is `people.[user.initials].YOHOHO` whose window comes to the front, because the resolution of `"yohoho"` as an object reference succeeds. Then if we close `people.[user.initials].YOHOHO` and say this:

```
window.isOpen("yohoho")
```

the result is **true**, because `"yohoho"` resolves to a database entry, that entry's edit window is not open, the operation fails, and Frontier tries again with `"yohoho"` interpreted as a title, which works—the **examples** window is open, and it has this title with this capitalization.

Opening and Closing; Showing and Hiding

To open, and make visible and frontmost, a non-scalar object's edit window:
```
window.open (addrObject)
```

It is also possible to open an edit window using `edit()`, but this additionally makes the window the target, whereas `window.open()` does not. See Chapter 18.

To close a window:
```
window.close (windowReference)
filemenu.close ()
filemenu.closeAll ()
```
To make an open window invisible:
```
window.hide (windowReference)
```

UserTalk is the only way to hide a window, so it is useful to make a menu item whose script includes:

```
window.hide(window.frontmost())
```

To open a non-scalar object's edit window invisibly (if it is not already open):
```
target.set (addrObject)
```
> Also makes the window the target; see Chapter 18.

Observe that there is no verb which opens a window invisibly without also making the window the target, as `window.open()` opens a window visibly without making it the target. It is aesthetically pleasing to be able to open a window invisibly, set its size, location, title as desired, and then show it, so that these changes do not happen before the eyes of the user. We illustrate this in Example 24-3.

To make an open window visible:
```
window.show (windowReference)
```
> Has no effect upon the window's position in the layering order.

To learn whether a window is open:
```
window.isOpen (windowReference)
```

To learn whether a window is open and visible:
```
window.isVisible (windowReference)
window.isHidden (windowReference)
```

A window that is not open counts as hidden. This yields some counterintuitive results: for example, `window.isHidden(@glubglub)` returns `true` even though there is no such window and no such object! There really is no need for `window.isHidden()` anyway, as its results are identical to saying `!window.isVisible()`.

Relocating and Resizing

To learn or set an open window's position:
```
window.getPosition (windowReference, addrX, addrY)
window.setPosition (windowReference, X, Y)
```

Global coordinates on Mac OS are measured in pixels from the upper left-hand corner of the main screen, with positive X to the right and positive Y downwards. A window's position is measured at the upper left-hand corner of its content region (*not* its titlebar!). The menubar is usually 20 pixels thick, and a titlebar is usually 18 pixels thick, so 38 is a good minimum Y value.[*] The Main Window has no titlebar, so its position is measured at its upper left corner.

When you use `window.setPosition()`, Frontier will curtail X and Y to prevent any of the window from ending up offscreen. To learn whether this has occurred, use `window.getPosition()`. See, for instance, Example 12-4. It's good that a script cannot put a window off the screen (where a user would be unable to access it), but it is odd that you can move a window manually to positions where you cannot move it programmatically.

To learn or set an open window's size:
```
window.getSize (windowReference, addrWidth, addrHeight)
window.setSize (windowReference, width, height)
```

You cannot use `window.setSize()` on the Main Window even though you can resize the Main Window manually.

A window's size is the width and height of its content region. When you use `window.setSize()`, Frontier will curtail *width* and *height* to prevent either from becoming smaller than a certain reasonable minimum, or to prevent the whole window (not just its content area) from becoming larger in either dimension than the size of the screen. To learn whether this has occurred, use `window.getSize()`.

---

[*] Under Mac OS 8, this will need to be larger—perhaps about 42.

The behavior of `window.setSize()` suggests the following trick for learning the dimensions of the screen.

Example 24-2. screenSize()

```
on screenSize()
    local (x, y)
    target.set(@readme) « or any rarely used window
    window.setSize(@readme, 5000, 5000)
    window.getSize(@readme, @x, @y)
    target.clear()
    return {x, y + 38}
```

To zoom a window:

```
window.zoom (windowReference)
```

> Equivalent to clicking the zoom box.

Here, we open a window invisibly, set its text font, its text size, its size, and its position, and then show it.

Example 24-3. "Pretty" open

```
target.set(@workspace)
editmenu.setFont("Geneva"); editmenu.setFontSize("10")
window.setSize(@workspace, 10, 10) « force intelligent zoom
window.zoom(@workspace)
window.setPosition(@workspace, 10, 45)
window.show(@workspace)
target.clear()
```

Layering Order

To obtain a reference to the frontmost visible window:

```
window.frontmost ()
```

If a script is initiated by pressing its edit window's **Run** or **Debug** button, then `window.frontmost()` will not return that script's edit window; it will return the first visible window behind it. Similarly, in Debug mode, windows that come to the front because the execution path runs through them are ignored. This is so that the script can pretend not to be open, for testing purposes.

To obtain a reference to the window behind a given window:

```
window.next (windowReference)
```

The string returned by `window.frontmost()` or `window.next()` is suitable for immediate use as a window reference.

Together, `window.frontmost()` and `window.next()` provide a way to cycle through all windows. Here, for instance, is a script that constructs a list of all open windows.* It takes elegant advantage of the fact that `window.next()`

returns the empty string when its parameter is a reference to the rearmost window.

Example 24-4. listWindows()

```
on listWindows ()
    local (theList = {})
    loop (local (w = window.frontmost()); w; w = window.next(w))
        theList[0] = w
    return theList
```

To move a window to the front or back of the layering order:
```
window.bringToFront (windowReference)
window.sendToBack (windowReference)
```

If a window is invisible, `window.bringToFront()`, but not `window.sendTo-Back()`, also makes it visible. This makes sense given that an invisible window cannot be `window.frontmost()`.

To assist me in navigating Frontier's open windows, I like to have on hand three menu items that do the following: send the frontmost window to the back, bring the second window to the front, and bring the rearmost window to the front. Given Example 24-4, the last of these is easy to write.

Example 24-5. cycleWindowsBackwards()

```
on cycleWindowsBackwards ()
    local (theList = listWindows())
    window.bringToFront(theList[sizeOf(theList)])
```

To learn whether a given window is frontmost:
```
window.isFront (windowReference)
```

Miscellaneous

To learn or set a window's title:
```
window.getTitle (windowReference)
window.setTitle (windowReference, newTitle)
```

To scroll the target window:
```
window.scroll (direction, count)
```

On the target, see Chapter 18. `window.scroll()` equates to pressing the window's scroll arrow the specified number of times, except that it's intuitively backwards: `window.scroll(up,1)` presses the window's *down*-pointing scroll arrow (because the window's content moves up). There is no penalty for supplying too large a value for *count*; so, for example, to scroll a window to its top, say `window.scroll(down,infinity)`.

---

* But if an invisible window is frontmost, it will not be included in the list.

To show and hide elements of the Main Window:

```
mainWindow.hideButtons ()
mainWindow.showButtons ()
mainWindow.hideFlag ()
mainWindow.showFlag ()
mainWindow.hidePopup ()
mainWindow.showPopup ()
```

The "buttons" commands do the same thing as clicking on the "flag" icon in the Main Window, but the other commands do things to the Main Window that the user cannot countermand, except programmatically; so be careful.

To force a window to update:

```
window.update (windowReference)
```

I'm not sure when you'd need `window.update()`; I've never seen it used except in `suites.samples.basicStuff.windowFormatter()`, which seems to work just as well without it.

An important problem is how to learn, in a script, what kind of window a given window is. The first question to ask is whether it is an edit window, and the way to find out is by obtaining from Frontier (with `window.frontmost()` or `window.next()`) a reference to the window and then testing:

```
defined(windowReference^)
```

The reason this works is that if the window is an edit window, the reference to it returned by Frontier is a string that can be dereferenced to the name of a defined object. At this point you can also find out what type of edit window it is by asking what type that object is:

```
typeOf(windowReference^)
```

Menu item scripts are a special case, because they are certainly script edit windows, and yet they do not represent any defined object (see Chapter 26). So if a window fails the edit window test, you still need to test whether it is a menu item script window. A special verb exists for this purpose.

To learn whether a given window is a menu item script:

```
window.oisMenuScript (windowReference)
```

If the window fails this test, too, it is the Main Window, the Quick Script window, or a non-modal dialog. Since the reference to such a window is its title, you should now be able to identify it.

25

Dialogs

This chapter describes the ways in which UserTalk lets your script communicate with the user through status messages and dialogs. Frontier comes with a number of modal dialogs ready for scripts to display. It is also possible to make your own custom dialog, modal or non-modal, which can interact with the user. A custom dialog may be constructed either as a resource or in a graphical dialog-editing environment (MacBird).

Dialog Types

There are three standard dialog window types. A *modal* dialog brings all action to a halt, and makes it impossible to click away to another window until the user dismisses it. *Movable modal* dialogs behave like modal dialogs, but they can be dragged by their titlebar, and they permit the user to switch to another application. *Modeless* dialogs are like ordinary windows.

A *windoid* is an ordinary window with a slightly peculiar appearance; for instance, its titlebar is unusually thin. The Quick Script window is a windoid.

On ordinary windows, including modeless dialogs, see Chapter 24, *Windows*.

Status Messages

A status message is brief informational text, not requiring any response, that a script posts to an existing window while continuing its activity; the text remains until explicitly removed or replaced by a new message. The Main Window is very commonly used as a destination for status messages. But there is also a status message area at the bottom of every edit window.

On the Main Window, see Chapter 24; and, on agent script messages in the Main Window, see Chapter 27, *Agents and Hooks*. For another way of providing feedback to the user, see on `rollBeachBall()` in Chapter 46, *Verbs*.

To display a status message in the Main Window:
```
msg (string)
```

To display a status message in the frontmost visible window:
```
window.msg (string)
```

`msg()` does not bring the Main Window to the front, because a status message shouldn't make the wrong window frontmost or obscure the user's view of other things. Still, it might be desirable to maximize the chances of the user seeing status messages. One possible approach is this: as a script starts, place the Main Window in the upper corner of the screen and behind whatever window is `window.frontmost()`, as in the following example.

Example 25-1. readyMainWindow()
```
if window.isFront("Frontier.root") {return}
window.sendtoback("Frontier.root")
window.setPosition("Frontier.root", 0, 20)
local (nextW, w = window.next(window.frontmost()))
if w
    loop
        if w == "Frontier.root" {break}
        nextW = window.next(w)
        window.sendtoback(w)
        w = nextW
```

But even this does not guarantee that the Main Window is unobscured.

Another possibility is to display the status message with `window.msg()`, which writes to the status message area of the window that is *actually* frontmost—not necessarily the one returned by `window.frontmost()`. Thus the message will almost certainly be visible. A window's status message turns invisible when the window is not frontmost, but reappears when it is made frontmost again.

Preconfigured Modal Dialogs

A variety of ready-to-roll modal dialogs are available in UserTalk.

To display an alert modal dialog:
```
dialog.alert (text)
```

To display an informational modal dialog:
```
dialog.notify (text)
```

To display a confirmation modal dialog:
    dialog.confirm (*text*)

To display a yes-or-no modal dialog:
    dialog.yesNo (*text*)

To display a two-button modal dialog where you determine the button names:
    dialog.twoWay (*text*, *OKbutton*, *cancelButton*)

To display a yes-no-or-cancel modal dialog:
    dialog.yesNoCancel (*text*)

To display a three-button modal dialog with custom button names:
    dialog.threeWay (*text*, *yesButton*, *noButton*, *cancelButton*)

To display a modal input dialog:
    dialog.ask (*text*, *addrResult*)

To display a modal integer input dialog:
    dialog.getInt (*text*, *addrResult*)

To display a modal password dialog:
    passwordDialog.run (*addrName*, *addrPassword*)

To display the database-entry creation modal dialog:
    table.promptNewItem (*newWhat*, *dataType*)

File Dialogs

Four modal dialogs let the user choose from files, folders, or volumes. Nothing happens to any files directly as a result of these dialogs; the dialogs just return a pathname, which the script may use as desired. Except for **file.getDisk-Dialog()**, these are familiar standard Mac OS dialogs. For more about file pathnames, see Chapter 31, *Driving the System*; for string verbs that manipulate pathnames, see Chapter 14, *Strings and Chars*.

To let the user choose a file:
    file.getFileDialog (*text*, *addrResult*, *fileTypes*)

To let the user choose a folder:
    file.getFolderDialog (*text*, *addrResult*)

To let the user choose a volume:
    file.getDiskDialog (*text*, *addrResult*)

To let the user specify a folder and enter a filename:
    file.putFileDialog (*text*, *addrResult*)

Another standard Mac OS modal dialog, the PPCBrowser, lets the user choose from applications running on the local (AppleTalk) network. There is no need to be logged into any of the other computers, and choosing an application does not actually link to it; there is no need to supply any passwords. But only processes on computers with program linking turned on will be visible.

To let the user choose a running process:

```
sys.browseNetwork (text, creatorCode, addrResult)
```

The value returned by **sys.browseNetwork()** at *addrResult* can be used subsequently to address communications to the chosen process; see Chapter 32, *Driving Other Applications.*

A purely informational dialog tells things like the size, creation and modification dates, and type and creator of a file; basically, it's a more informative version of the Finder's Get Info dialog. It's hard to imagine this being used *in* a script; rather, it's the sort of thing one might use while writing a script.

To display the file info dialog:

```
dialog.fileInfo (pathname)
```

Resource-Based Dialogs

You can create your own dialog layout as a `'DLOG'`/`'DITL'` resource pair with a resource editor such as ResEdit.* Such resources, if present in the resource chain, can be displayed as a dialog with a call to `dialog.run()` (to display a modal dialog) or `dialog.runModeless()` (to display a modeless dialog). On resources, see Chapter 20, *Resources.*

One of the parameters in the call to `dialog.run()` or `dialog.runModeless()` is the address of a handler. This handler (which I will call the "proc") is called when the dialog is about to be displayed, and is called again every time the user "hits" an item in the dialog (pushes a button, for example). Each time it is called, the proc can query or make changes to elements of the dialog, as well as performing any other desired tasks; it then returns **true** to continue displaying the dialog, or **false** to close it. Thus, the proc is in complete runtime control of the user's interaction with the dialog.

The proc must accept a single parameter. Each time the proc is called, the value of this parameter tells it what's going on, as follows:

• When the dialog is first about to run, but before it is actually displayed, the proc is called with a parameter of −1. This gives it a chance to initialize the dialog if desired.

---

* For more information about using ResEdit, see *http://devworld.apple.com/dev/techsupport/insidemac/ resedit/resedit-2.html.* For more information about elements of dialogs, see *http://devworld.apple.com/dev/ techsupport/insidemac/Toolbox/Toolbox-297.html* and *http://devworld.apple.com/dev/techsupport/inside- mac/Toolbox/Toolbox-370.html.* In our examples, a `'DLOG'` and `'DITL'` are the only needed resources; if your dialog uses icons or controls or other resources that aren't in the resource chain already, you will have to get them into the resource chain just as we do here with the `'DLOG'` and `'DITL'`.

- Thereafter, each time the proc is called, the parameter is a number specifying the item that the user has "hit."

- If the dialog is modeless, a parameter of –2 indicates that the user has clicked in the dialog's close box; the proc can perform last-minute tasks, including getting information from the dialog, but cannot prevent its closing. (On the other hand, the close hook mechanism is called, and can be used to prevent the dialog from closing; see Chapter 24.)

- If the dialog is modeless, the proc can "put it to sleep" by calling `clock.sleepFor()`. This is a way of getting Frontier to act as a timer: if the user doesn't "hit" any items in the dialog, the proc will be called with a parameter of –3 when the specified interval has expired. If the user does "hit" an item in the dialog during the sleep interval, the sleep interval ends immediately and the proc is called with that item's number as parameter.

All this may sound rather daunting, but in fact writing a proc is very easy. There are several examples in the database, and more are presented later in this chapter. Before getting to that, here are the relevant verbs for operating on the dialog. One of the first two will be used to display the dialog; the others are valid only from within a proc.

To display a resource-based modal or modeless dialog:
```
dialog.run (resourceID, defaultItemNum, addrProc)
dialog.runModeless (resourceID, defaultItemNum, addrProc)
```

To change the visibility of a dialog item:
```
dialog.showItem (itemNum)
dialog.hideItem (itemNum)
```

To change the enabled status of a dialog item:
```
dialog.setItemEnable (itemNum, enabled?)
```

To obtain or change the value of a dialog item:
```
dialog.getValue (itemNum)
dialog.setValue (itemNum, newValue)
```

Modal Example

For our first example, we make a dialog like Frontier's `dialog.confirm()`, except that it has the stop-sign icon (already present in the resource chain as the `'ICON'` whose ID is 0).

We will store the resources as database entries, bringing them into the resource chain just before showing the dialog. A major advantage of this strategy is that we can package in a subtable both the resources and the dialog "glue" script which the caller calls to display the dialog; this is aesthetically neat, and makes it easy to distribute the dialog to other users. We'll call the subtable `stopDialog`.

We start by drawing the dialog in ResEdit; in Figure 25-1, it is shown with its items numbered. Item 4, the static text of the dialog, will be set according to the parameter provided by the caller. Item 1, the **OK** button, will be the default.

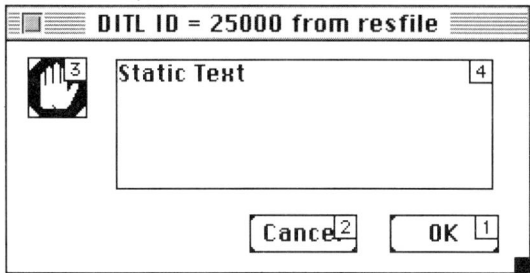

Figure 25-1. Modal dialog in ResEdit

We assign both the 'DLOG' and the 'DITL' ID numbers of 25000, because Frontier reserves this number, by convention, for temporarily loaded resources. Make sure that the 'DLOG' is set to use 'DITL' 25000, and that **Initially Visible** is unchecked in the 'DLOG' edit window so that we can set item 4 before the dialog appears. Now save the dialog resources. Let's say they are saved in a file called *ResFile*.

We will need a utility script to import the resources. Thus, we have an edit cycle where we edit the dialog resources in ResEdit, save them, import them with our utility script, and show the dialog in Frontier, until we are satisfied with the results. The utility script might look like this:

```
if not defined (people.[user.initials].stopDialog)
    new(tableType, @people.[user.initials].stopDialog)
local (dummy)
rez.getNthResource ("HD:resfile", 'DLOG', 1, @dummy, @stopDialog.stopDLOG)
rez.getNthResource ("HD:resfile", 'DITL', 1, @dummy, @stopDialog.stopDITL)
```

Our dialog glue script should obviously work like `dialog.confirm()`: it will take one parameter, the prompt text, and it will return **false** or **true** as the user clicks **Cancel** or **OK**. With this in mind, we can write the proc.

When first called, with a parameter of −1, the proc should set item 4 to the prompt, which we call **s**, and permit the dialog to be shown. Thereafter, if item 2 or item 1 is hit, it must close the dialog. To tell the script as a whole which item was hit, we use a variable global to the proc, which we call **userHitOK**. The proc is now complete!

Thus far, our glue script looks like this:

```
on run (s)
    local (userHitOK = false)
    on theProc(whatItem)
        case whatItem
            -1
                dialog.setValue (4, s) « s is the prompt
                return true « show the dialog
            1; 2
                userHitOK = (whatItem == 1)
                return false « close the dialog
```

We've defined the proc, but the glue script doesn't yet do anything. What it should do is load the resources into the resource chain, and show the dialog with **dialog.run()**. This will hand control over to the proc. When **dialog.run()** returns, the dialog has been dismissed, and the proc has set **userHitOK** according to whether **OK** was hit. So all we have to do is return **userHitOK**. There is no need to remove the resources from the resource chain; we don't care if a script overwrites them later on. So now we have:

```
on run (s)
    local (userHitOK = false)
    on theProc(whatItem)
        case whatItem
            -1
                dialog.setValue (4, s)
                return true
            1; 2
                userHitOK = (whatItem == 1)
                return false
    rez.putResource (frontier.getFilePath(), 'DITL', 25000, \
        @stopDialog.stopDITL)
    rez.putResource (frontier.getFilePath(), 'DLOG', 25000, \
        @stopDialog.stopDLOG)
    dialog.run (25000, 1, @theProc)
    return userHitOK
if run("Do you really want to erase your entire hard disk?")
    dialog.notify("You're lucky I was just kidding.")
```

I have added a two-line stub so we can test the glue script. Pressing its **Run** button brings up the test version of the dialog, as shown in Figure 25-2.

Modeless Example

Next, we illustrate a modeless dialog, with radio buttons for good measure. The purpose of the dialog will be to create a new non-scalar database entry. Again, we will package the glue script and the dialog resources in a subtable, which we will call **entryMaker**.

Do you really want to erase
your entire hard disk?
Cancel
OK

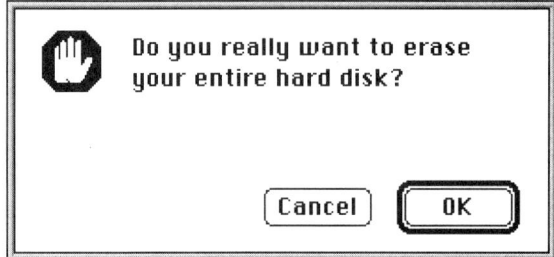

Figure 25-2. Modal dialog in action

Figure 25-3 shows the dialog as drawn in ResEdit. Make sure to give the `'DLOG'` window a title, so that it can be referred to in UserTalk with a window reference (in this case we call it "Make New Entry").

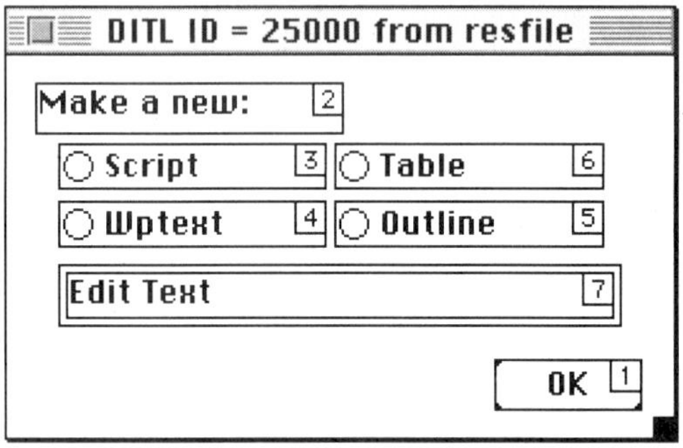

DITL ID = 25000 from resfile
Make a new:
2
Script
3
Table
6
Wptext
4
Outline
5
Edit Text
7
OK
1

Figure 25-3. Modeless dialog in ResEdit

Figure 25-4 shows how the dialog will look when displayed in Frontier.

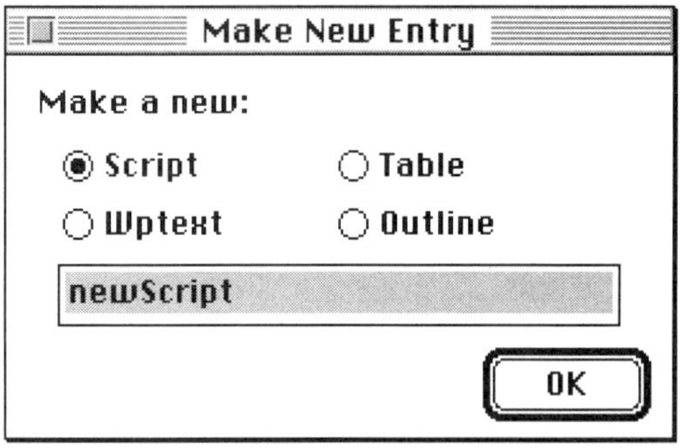

Make New Entry
Make a new:
Script
Table
Wptext
Outline
newScript
OK

Figure 25-4. Modeless dialog in action

The proc has rather more to do than in our previous example, though there is nothing particularly hard about it. Management of the radio buttons is up to us, and we must also give the dialog its actual functionality, so that it really does create a new table entry.

The proc is initially called with a parameter of **-1**, whereupon it initializes the radio buttons and edit text. Subsequently, if one of the radio buttons is hit, we set it, remember what datatype it signifies, change the edit text to match, and clear the other radio buttons. When the **OK** button is hit, we check whether the window behind the dialog is a table, and, if it contains an entry with the name the user has provided, we make sure it's okay to replace it. If all is well, we create the new entry as specified.

Notice the use of `dialog.setValue()` and `dialog.getValue()` throughout, the variables global to the proc to maintain state from one call to the next, and the fact that we always return **true** (because closing the window is entirely up to the user).

```
on run ()
    local (whatType = scriptType, whatName = "newScript")
    local (typeList = {scriptType, wptextType, outlineType, tableType})
    local (nameList = {"newScript", "newWptext", "newOutline", "newTable"})
    on theProc(whatItem)
        case whatItem
            -1
                dialog.setValue(3, true)
                dialog.setValue(7, "newScript")
            3; 4; 5; 6
                local (x)
                for x = 3 to 6
                    dialog.setValue (x, whatItem == x)
                dialog.setValue (7, nameList [whatItem - 2])
                whatType = typeList [whatItem - 2]
            1
                local (theTable = window.next("Make New Entry"))
                if typeOf(theTable^) ≠ tableType
                    dialog.notify (theTable + " isn't a table.")
                else
                    whatName = dialog.getValue (7)
                    if defined (theTable^.[whatName])
                        if !(dialog.yesNo("Replace existing "+whatName+"?"))
                            return true
                    new (whatType, @theTable^.[whatName])
        return true
    rez.putResource (frontier.getFilePath(), 'DITL', 25000, @entryMaker.ditl)
    rez.putResource (frontier.getFilePath(), 'DLOG', 25000, @entryMaker.dlog)
    dialog.runModeless (25000, 1, @theProc)
run()
```

File-Based Resources

In the previous examples, the resources live in the database and are loaded into the resource fork of the database file just before showing the dialog. This makes the edit cycle a two-step process, because the resources are edited in ResEdit and must then be imported into the database.

An alternative approach is to leave the resources in the ResEdit file, and then, when the time comes to display the dialog, to load the resources directly into the resource chain by calling `dialog.loadFromFile()`. This verb returns the ID of the `'DLOG'`, so you can hand the result to `dialog.run()` or `dialog.runModeless()`:

```
dialog.run (dialog.loadFromFile ("HD:resfile"), 1, @theProc)
```

In fact, there is another utility script, `dialog.runFromFile()`, consisting essentially of this very line.

This approach shortens the edit cycle: edit the resources in ResEdit, save the resource file, and display the dialog in Frontier immediately. The disadvantage is that now you have an extra file without which the dialog won't work; keeping the resources in the database is neater and makes it much easier to distribute your dialog to other users.

Card-Based Dialogs

Frontier comes with MacBird, a separate application for drawing and displaying dialogs; and MacBird Runtime, which displays dialogs, is built directly into Frontier. This cuts out the need for ResEdit (because the dialog is drawn with MacBird), the need to move resources into the database (because MacBird edits database entries), and the need to move the resources into the resource chain (because MacBird Runtime displays the dialog). Furthermore, instead of using a proc, programmatic management of items in the dialog is object-oriented: scripts are attached directly to the dialog items themselves, to dictate their behavior when they are "hit."

The dialog that MacBird draws is called a *card*; it is stored in the database as a binary of type `'CARD'`, referred to as a *card object*. That's why these are called card-based dialogs.[*]

As of this writing, MacBird Runtime,[†] which is responsible for the display of card-based dialogs from within Frontier, appears to be quite reliable. MacBird as a dialog construction tool, though, is unfinished and a bit shaky (certain actions consistently crash my machine). Still, it's easy and fun to use; just be sure that as you edit a dialog you save your work often.

---

[*] Possibly the "card" designation is homage to HyperCard. As for MacBird's name, some of us go back far enough into the '60s to remember what *MacBird* was, and the Texan boot used as an icon is a dead giveaway; but the name is actually said to be derived from a similar project (at another company) that was code-named "Blackbird."

[†] Also sometimes referred to, for historical reasons, as "Iowa Runtime."

You create a new card object in the database by choosing **New Card** from the **Table** menu. You edit the card graphically in MacBird by selecting the card object and choosing **Edit with App** from the **Main** menu; later, in MacBird, when you save the card, it is saved back into the database, updating the card object.* You display a card as a dialog from within Frontier by calling `card.run()`; or, for testing purposes, select the card object and choose **Run Selection** from the **Main** menu.

Editing a MacBird card graphically is very intuitive.† The meanings of the icons in the editor are (top to bottom, by row): selection tool and text editing tool, non-editable and editable text field, checkbox and radio button, button, popup menu, icon, picture, rectangle, color-picker button, labelled rectangle, and scrollbar. This tool set can be extended; see the Frontier SDK for instructions.

The radio button tool needs a bit of explanation. As you add radio buttons, they are linked with the already present radio buttons; to start a new set of radio buttons, "group" the existing radio buttons so that MacBird won't see them.

You can cancel out of a running card with Command-period (useful if things go wrong, or if you want to test a dialog before the card has scripts which close it).

Anatomy of a MacBird Card

When you work with a card from within MacBird, it contains three primary sorts of things. First, there are the items of the dialog, the buttons and text fields and such, usually referred to as *objects*, which you draw and edit as graphics.

Second, there are scripts associated with each object in the card. These are of two kinds. A button can have an *action script*; this is accessed by choosing **Button Info** from the **Object** menu. The action script runs when the button is pressed. And any dialog item can have a *recalculation script*; this is accessed by selecting the item and choosing **Recalculation** from the **Object** menu. A recalculation script runs in response to an external event: when first starting up the card, or when another object in the card is changed, or at preset intervals of time; the object's value is set to the result of the script.

Both recalculation scripts and action scripts are dynamically scoped: they are considered, while running, to exist at the point where `card.run()` was called. This means that they have access to variables and handlers that are in scope at that point of the calling script, making it possible to share information between the calling script and the card scripts.

---

* The mechanism of this symbiotic relationship between MacBird and the database is explained in Chapter 35, *External Editors*.

† There is a tutorial at *http://www.wrldpwr.com/frontier/macbird/index.html*.

Third, there is the card's *table* (which I'll call the "card-table"). This is a Frontier table embedded in the card's data; to access it, you choose **Edit Table** from the **Edit** menu in MacBird, which switches to Frontier to display the table.* The recalculation scripts and action scripts have an implicit `with` to this table; so they can access scripts and other objects that are stored there. Indeed, it is common to make recalculation scripts and action scripts mere one-liners that call card-table scripts. Scripts in the card table also have an implicit `with` to other objects in the card-table; but they do not have access to variables and handlers in scope in the caller at the point where `card.run()` was called.

If the card-table contains a menubar object called `menubar`, and if the dialog is modeless, the menubar will appear at the top of the screen whenever the dialog is frontmost. If a script in the card-table, or a handler in scope in the caller, has the name `startCard`, it is called automatically before the dialog becomes visible.

Since objects in the card-table are visible to action scripts, to recalculation scripts, and to card-table scripts, the card-table can be used as a kind of global storage while the card is running. But this is not persistent storage. When `card.run()` is called, a copy of the card-table within the card object is created temporarily in the database; this copy is not saved back into the card object, so any changes made in the card-table are lost when the card closes. Also, because it is impossible to form an object reference to the card-table, it is impossible for a script to create a new object in it.

Card Verbs

The card verbs are not well documented. Fortunately, most of them are identical to the corresponding cardEditor verbs, which are documented (in DocServer). And it is easy to find out what the card verbs are, by looking in `system.verbs.app.card`. For example, to learn about `card.getCardGrid()`, read about `cardEditor.getCardGrid()` in DocServer.

The vast majority of card verbs are seldom used (and several are not meant to be used directly at all). Here, only the commonest are discussed. Only `card.run()` requires that the card object be specified; the others are valid only from within card scripts, and apply automatically to the card in which they live or from which they are called.

To display a card-based dialog:
```
card.run (addrCardObject)
```

To close a displayed card-based dialog:
```
card.close ()
window.close (windowRef)
```

---

* So you're now using Frontier to display a table which lives inside a card being displayed by MacBird—a card that itself lives in the Frontier database. Got that?

`window.close()` is the way to close modeless dialogs, which are ordinary windows.

To learn whether a card is modal:
```
card.isModal ()
```

To learn or set the enabled status of a dialog item:
```
card.getObjectEnabled (objectName)
card.setObjectEnabled (objectName, enabled?)
```

To learn or set the visibility of a dialog item:
```
card.getObjectVisible (objectName)
card.setObjectVisible (objectName, visible?)
```

To learn or set the boolean status of a dialog item:
```
card.getObjectFlag (objectName)
card.setObjectFlag (objectName, boolean)
```

To learn or set the text of a dialog item:
```
card.getObjectText (objectName)
card.setObjectText (objectName, string)
```

To learn or set the menu item list of a popup menu:
```
card.popup.getMenu (objectName)
card.popup.setMenu (objectName, string)
```

To learn or set the visibility of a popup menu's title:
```
card.popup.getHasLabel (objectName)
card.popup.setHasLabel (objectName, visible?)
```

To learn or set which menu item of a popup menu is chosen:
```
card.popup.getCheckedItem (objectName)
card.popup.setCheckedItem (objectName, itemNumber)
card.popup.getSelectedText (objectName)
card.popup.setSelectedText (objectName, itemText)
```

Modal Example

To illustrate, we rewrite our earlier resource-based examples as MacBird cards. We start with our stop-sign dialog. We'll need a glue script and a card object. We'll call the glue script `run()` and the card object `stopCard`, and we'll keep them in a subtable called `stopDialog2`.

First, recall how we're going to want to call the dialog; the glue script needs to work like `dialog.confirm()`, taking as parameter a string which is the dialog's prompt, and returning `true` or `false` as the user clicked **OK** or **Cancel**. So here is the glue script:

```
on run (s)
    local (result)
    card.run (@stopDialog2.stopCard)
    return result
if run("Do you really want to erase your entire hard disk?")
    dialog.notify("Lucky for you I was just kidding.")
```

Now s and `result`, our local variables, will be available to the card's recalculation and item scripts while it runs. Keeping this mind, Figure 25-5 shows our stop-sign dialog drawn in MacBird.

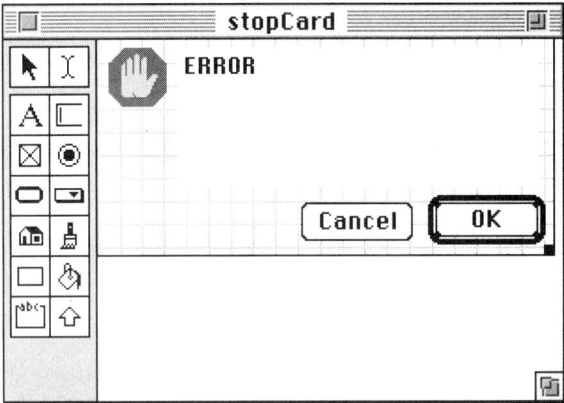

Figure 25-5. Modal dialog in MacBird

The dialog contains four items: an icon, a non-editable text, the **Cancel** button, and the **OK** button. The non-editable text has a recalculation script which runs on card startup and says simply:

```
s
```

The **Cancel** button has an action script, which says:

```
card.close(); result = false
```

The **OK** button has an action script, which says:

```
card.close(); result = true
```

That's all! The non-editable text says **ERROR** when we edit the card within MacBird, because there is no context in which to understand s; but if we click `stopDialog2.run()`'s **Run** button to test the dialog, we see the dialog shown in Figure 25-6.

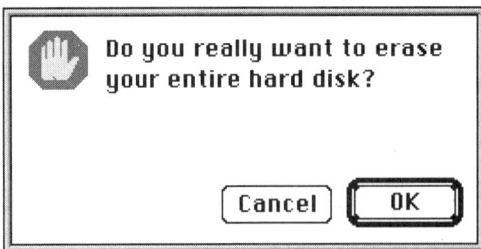

Figure 25-6. Modal card-based dialog in action

Modeless Example

The "Make New Entry" dialog is much like our resource-based version, except that the pieces of the action, instead of being distributed over callbacks to a proc, are distributed over various recalculation and action scripts, and instead of communicating with one another through variables global to the proc, they communicate with one another through entries in the card-table. One major difference is that in the card-based version, we don't have to manage the highlighting of the radio buttons: that's taken care of for us.

Let's name the card object `entryMaker2`. It needs to be designated as modeless. It is clear how to draw it in MacBird; Figure 25-7 shows the dialog in action, when `card.run(@entryMaker2)` has been called.

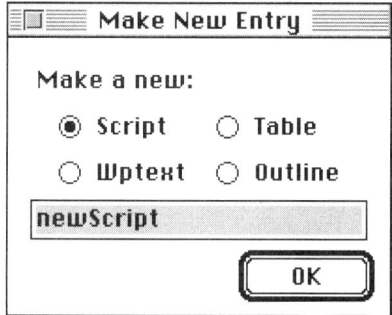

Figure 25-7. Modeless card-based dialog in action

For simplicity, the radio buttons have item names identical to their titles: "Script", "Wptext", "Table", and "Outline". The editable text field is named "editText", and has a recalculation script which is triggered when another item changes; the recalculation script calls `setEditText()`, which is in the card-table. The **OK** button has an action script which calls `makeNewEntry()`, also in the card-table.

The card-table contains five items. We have just met two of them: the scripts `setEditText()` and `makeNewEntry()`. There is also a boolean, `okPressed`, and a string4, `whatType`. These will allow the scripts to communicate with one another. Finally, there is a `startCard()` script, which says:

```
okPressed = false
```

The reason for this is that a recalculation script triggered a change in another item has no way of knowing what item was changed. We want the editable text to change when a radio button is pressed, but not when the **OK** button is pressed. So we use `okPressed` as a flag, to tell the recalculation script when to change and when not.

Here is `setEditText()`: remember, this is called by the editable text field's recalculation script every time the user clicks a button, and its result becomes the text of the text field:

```
local (x, buttonNames = {"Script", "Table", "Wptext", "Outline"})
local (typeList = {scriptType, tableType, wptextType, outlineType})
if okPressed
    okPressed = false
    return card.getObjectText ("editText")
for x = 1 to 4
    if card.getObjectFlag (buttonNames[x])
        whatType = typeList[x]
        return "new" + (buttonNames[x])
```

When the user presses the **OK** button, `okPressed` is set to true by the **OK** button's action script; so in order to keep the editable text field from changing its text, `setEditText()` returns its current text! If the user presses a radio button, that button highlights and the other radio buttons are unhighlighted, automatically; now we check each radio button to find which one is highlighted, and use the result to set `whatType` and the text of the editable text field.

Finally, here is `makeNewEntry()`, called when the user presses the **OK** button:

```
local (theTable = window.next ("Make New Entry"))
if typeOf(theTable^) ≠ tableType
    dialog.notify (theTable + " isn't a table.")
else
    local (whatName = card.getObjectText ("editText"))
    if (!defined (theTable^.[whatName])) \
    or (dialog.yesNo ("Replace existing " + whatName + "?"))
        new (whatType, @theTable^.[whatName])
okPressed = true
```

This is almost identical to the proc of the resource-based version; the chief difference is a slight rearrangement so that the very last thing we do, regardless of the execution path, is set `okPressed` to `true`, so that `setEditText()` won't change the value of the editable text field.

Dialogs in Other Applications

One dialog, `frontier.requestToFront()`, is specifically designed to appear in another application, as a way of telling the user that Frontier requires attention.

Other Frontier dialogs, except those called with `dialog.runModeless()`, *can* appear in other applications, provided they are called in threads initiated in the context of other applications; see Chapter 26, *Menus and Suites*, on shared menus. But when execution is initiated from within Frontier, if it is desired to bring another application to the front and then display a Frontier dialog in that application, there seems no general way to do it.

An alternative, if you have AppleScript on your machine, is to have it display one of its dialogs (using, for instance, the **display dialog** command) from within another application; see Chapter 33, *AppleScript*. This works even if the other application is not scriptable. For example, given the utility in Example 33-1, one can say:

```
sys.bringAppToFront("SimpleText")
doAsAppleScript("tell application \"SimpleText\" to display dialog " + \
    "\"Hello from Frontier!\"")
```

26

Menus and Suites

This chapter describes Frontier's menus and explains how to edit them, manually and programmatically. The chapter also discusses the implementation of *suites*, miniature applications written in UserTalk and included in the database.

Among Frontier's menus, the **File** menu, the **Edit** menu, and the **Window** menu are *intrinsic*. They are hard-coded into the Frontier application; they cannot be altered. This chapter is not about them.

All other menus are *custom* menus. The remarkable thing about custom menus is that they are implemented in UserTalk, within the database. The menus them-selves, as they appear at the top of the screen, are simply reflections of editable database objects called *menubars*. A menubar has an edit window, like a script or an outline. You can edit a menubar by hand; a menubar can also be changed through UserTalk. Edit a menubar, and you've changed the actual menu that appears at the top of the screen. Every item in a menubar has an attached script, which runs when the actual menu item is chosen at the top of the screen; edit the script, and you've changed what the menu item does.

NOTE   Because the term "menubar" refers in Frontier to an object datatype, I speak in this chapter of the *computer's* menubar (the place where menu names appear and from which menus pop down) as "the top of the screen." I speak of the menus and menu items that appear at the top of the screen as "actual" menus and menu items.

Menu Categories

Frontier manages its custom menus in some unusual ways. It will help to divide Frontier custom menus into four categories:

Stable menus

These menus are always present. These have been rearranged somewhat over the history of Frontier; at present they consist of the **Main** menu, the **Open** menu, the **Web** menu, the **=user.initials** menu (its actual name matches the user.initials database entry), and the **Suites** menu.

Modal menus

One of these appears automatically, depending on what kind of window is frontmost: the **Table** menu, the **Script** menu, the **Outline** menu, the **WP** menu, and the **Menubar** menu.

Suite menus

A suite is a miniature application implemented in UserTalk, and each suite can include a menu, to provide a menu interface to its functionality. The available suite menus are listed in the **Suites** menu, and choosing from that menu will cause the named suite menu to appear.

Shared menus

These appear in certain other applications (such as BBEdit, Eudora, Internet Explorer, and others), even though they belong to Frontier. Frontier's functionality can be conveniently accessed from them without leaving the other application.

Modifying Menus

Being able to make a menu item at any time, with any desired script attached to it, means that frequently needed utility scripts can be run by choosing from a menu. It is convenient and easy to make a new menu item. This is one of Frontier's nicest features.

As with the database, though, so with menubars—you should be careful not to make wanton or accidental changes to menus that belong more to UserLand than to you. Clearly you ought not delete the **Open** menu, or alter a **Main** menu item's script so that it breaks or causes harm.

The **=user.initials** menu is the one that belongs completely to you; it is the chief place where you are permitted and expected to add menu items as desired. It is also reasonable to customize a shared menu by adding your own items to it, but future Frontier upgrades may overwrite your changes. If you write a suite, its suite menu is yours to define. Other menus should probably be considered off limits, unless you are very sure of what you are doing.

As with the rest of the database, you are encouraged to explore the menubars and the scripts of the menu items to learn how they work.

Where the Menubars Are

Menubars for the four types of custom menu reside at different locations in the database:

- The stable menus are in a single object, `system.misc.menubar`.

- The modal menus are in the table at `suites.modes.menus`.

- The suite menus live each in its own suite in an object called `menu`; for instance, the **Droplets** menu, which appears when you choose **Droplet Developer** from the **Suites** menu, is at `suites.droplet.menu`.

- The shared menus are in the table at `system.menubars`.

Menubar Edit Windows

A menubar is a non-scalar: it has an edit window, like a script or an outline. You can open that edit window as you would that of any other non-scalar—for example, by double-clicking the table entry for that object, or by selecting the entry and hitting Enter.

You can also open a menubar from the actual menu at the top of the screen. To do so, choose a menu item while holding the Option key; the edit window of the menubar opens, with that item selected. Some menus (suite menus, mostly) also have an **Edit Menu** item which, when chosen, opens the menubar for editing.

You can open `system.misc.menubar` by hitting the **Menu Bar** button in the Main Window, or by choosing **Menu Editor** from the **Open** menu.

Menubars as Outlines

A menubar edit window is a species of outline. Each line of an indented bundle represents a menu item; the line that the indented bundle is subordinate to represents the menu containing those items. Lines at summit level in a menubar represent actual menus in the computer's menubar at the top of the screen; other lines that have subheads represent hierarchical submenus.

NOTE   Although menubars are outlines, the lines they contain represent menus and menu items, so I shall refer to the lines of menubars as "items."

Since a menubar is an outline, everything you already know about working with outlines applies. You can collapse and expand items, navigate, create, delete, reorganize, and edit items just as you do lines in an outline. The **Menubar** menu, which appears at the top of the screen when a menubar edit window is frontmost, is almost entirely a subset of the **Outline** menu.

Expanding and collapsing items in a menubar edit window has no effect on the look of the actual menu; it's just for convenience in editing. Similarly, you can change the text font and size of the menubar edit window, but this doesn't affect the actual menu.

Menubars and Actual Menus

While a menubar edit window is frontmost, the actual menu that it represents appears at the top of the screen (if it wasn't already showing). Changes that you make in the menubar edit window are reflected immediately in the actual menu, which you can drop down to see how it looks, or to choose an item and see how its script runs.

There are certain special coded correspondences between an item of a menubar and the actual menu item it represents:

- To make an actual menu item be a separator line, have its item in the menubar consist of just a hyphen. (When you navigate out of the line, it will take on the look of a separator line even in the edit window.)

- To make an actual menu item be checkmarked, have its item in the menubar start with an exclamation mark (!).

- To make an actual menu item be disabled, have its item in the menubar start with a left parenthesis (( ).

- To make the name of an actual menu item be calculated at the moment the actual menu is dropped down, have its item in the menubar start with an equal sign (=) and consist of a UserTalk expression.* (Try making an item of a menubar read =clock.now() and then drop down the actual menu a few times!) If the name, checkmark, or enabled status of a menu item needs to change depending on present conditions, make the item a call to a script, and have the script return the correct name. To see an example, look at how the **Enable Agents** item of the **Main** menu works; it calls system.misc.agents-MenuItem() for its name.

---

* Don't do this in the Finder's shared menu; there's a bug that can cause a crash.

A calculated menu name, as opposed to a menu item name, is calculated just once, when the actual menu is installed at the top of the screen; that's how the name of the **=user.initials** menu is generated.

Keyboard Shortcuts

Keyboard shortcuts let the user choose a menu item without dropping the menu down with the mouse. To assign a keyboard shortcut to a menu item, select its item in the menubar and choose **Set Command Key** from the popup menu at the edit window's lower left. When the dialog appears, type the key—not the actual shortcut. (So, to set a shortcut as Command-2, type **2**.)

Only simple Command key shortcuts are available; you can't make a shortcut be Control-a or Command-Shift-T. No visible distinction is made between regular keyboard keys and keypad keys, but they are different. (So, for instance, the keyboard shortcut for **Full Expand** in the **Outline** menu is Command-keypad-*.) The popup menu shows a list of shortcuts already set for this menubar; choosing from it navigates to the chosen item.

With so many menus and such a limited range of shortcuts available, conflicts are inevitable. When the user hits a keyboard shortcut within Frontier, Frontier searches each menu at the top of the screen starting with the rightmost; it is the first matching menu item thus encountered that will be chosen. Conflicts caused by menus that appear only temporarily (especially suite menus) are not regarded as highly problematic.

In a shared menu, the host application searches the shared menu after its own menus, so a keyboard shortcut in a shared menu that matches the shortcut for one of the application's own menu items causes no problems for the application— but the shared menu item can't be chosen with a keyboard shortcut, either.

Menu Item Scripts

Every menu item has a script attached to it, which is executed when the actual menu item is chosen by the user. This script is accessible for editing only when the menubar is open for editing. In the menubar, select the item and hit the **Script** button (or Command-Enter). A script edit window opens. This window does not, itself, represent a database object; it is a feature of the menubar object. When a menubar edit window is closed, any open menu item script windows associated with it are closed as well.

You can associate a script with an item that is a separator line or a menu (as opposed to a menu item); but such a script cannot be run by way of the menu interface, because the actual menu item cannot be chosen by the user.

Menu item scripts have no eponymous handler; the script is associated with no script object, and has no name for a handler to be eponymous to. When an actual menu item is chosen, its script runs just as when the script edit window's **Run** button is pushed: top-level commands are executed.

Menu item script windows have no **Compile** button; they are compiled afresh whenever the script runs. (This poses no danger, because an agent or other "involuntary" mechanism is not going to run a menu item script while it is open for editing. See Chapter 13, *Running and Debugging Scripts.*) Menu item scripts can be tested "live": when you make a change, choosing the actual menu item runs the changed script without your having to close the script edit window.

Because they are not precompiled, menu item scripts should be fairly short, so as to compile and start executing rapidly when the actual menu item is chosen. Often, a menu item script will consist of a single line, which calls some script object in the database. This makes editing and maintenance easier, too, since a database script object is more convenient to access and debug than a menu item script.

Debugging a menu item script can be a little tricky. You can't just choose the actual menu item; you have to open the menu item's script and press the **Debug** button. But now conditions on the screen may not match those under which the script normally runs. Suppose, for instance, that choosing the menu item is supposed to operate upon the window that is frontmost. Let's say this is expected to be an outline edit window. If an outline was frontmost before you started debugging, it isn't frontmost now: there's a menubar window in front of it, and a menu item script edit window in front of that!

Now, the menu item script edit window itself doesn't really count; as we've already seen (Chapter 24, *Windows*), when you press **Run** or **Debug** in a script edit window, calls to a verb like `window.frontmost()` return a reference to the window behind it. However, we don't want the window behind it to be the menubar edit window. Closing the menubar edit window won't help; that will cause the menu item script edit window to close too! What I do is hide menubar edit windows just before debugging a menu item script, using the following utility.

Example 26-1. hideMbars()

```
loop (local (w = window.frontmost()); w ≠ ""; w = window.next(w))
    if defined(w^)
        if typeOf(w^) == menuBarType
            window.hide(w)
```

Shared Menus

Shared menus are menus belonging to Frontier that appear in another application. That other application (the *host* application) cannot be simply any application; it has to have been deliberately coded in accordance with Frontier's menu-sharing protocol.* But even applications that don't support menu sharing can be made to do so with an extension, OSA Menu, by Leonard Rosenthol.† Frontier must be running for its shared menus to appear in other applications.

The shared menus mechanism works by way of the table at `system.menubars`. Each entry in this table is a menubar whose name is the creator code of that menubar's host application. So, for instance, `system.menubars.["R*ch"]` causes a **Site** menu to appear in BBEdit; `system.menubars.CSOm` causes a **Scripts** menu to appear in Eudora.

When the user chooses a shared menu item, its script runs, just as with any Frontier custom menu item. Typically, the purpose of the script will be to add functionality to the host application. This involves having Frontier drive the application; see Chapter 32, *Driving Other Applications*.

For example, in Eudora, choosing **Uppercase** from the **Selected Text** hierarchical submenu of the **Scripts** menu runs this script:

```
Eudora.replaceSelectedText (string.upper (Eudora.getSelectedText ()))
```

This makes up for the fact that Eudora has no facility for turning text to uppercase; Frontier does have such a facility. In Netscape Navigator, choosing **Edit Local File** from the **Scripts** menu drives a word processor to open the file one is viewing in the browser, so you can modify its HTML. This makes up for the fact that Netscape has no facility for telling an application to open a file; Frontier does have such a facility. The BBEdit **Site** menu provides tight integration with Frontier's Web site management facilities. See Chapter 41, *Web Site Management*.

The Mac OS Finder shared menu is implemented through OSA Menu, and provides a wide miscellany of utilities, including making an alias of the selected file, stuffing the selected file and attaching it to a new email message, backing up, reconciling two folders, changing a file's type and creator, sending a file to a specified application to open, and many others.

A special situation arises with regard to the Netscape Navigator and Microsoft Internet Explorer shared menus (`MOSS` and `MSIE`). UserLand can't know which browser you're going to use, but it wants their shared menus to have identical

---

* This is described in the Frontier SDK, which includes header and code files for inclusion in a project, plus a model application illustrating when and how to make function calls so as to support menu sharing.

† For the latest version of OSA Menu, see *http://www.lazerware.com/~leonardr/*.

functionality. Therefore, an agent script (see Chapter 27, *Agents and Hooks*), `system.agents.webBrowserAgent()`, watches to see if the shared menus differ, and reconciles them if they do. This can have some surprising effects if you're unaware of it; if you find this troublesome, you can comment out the lines of the agent that do this.

Within a host application, choosing a shared menu item while holding the Option key brings Frontier to the front and opens the menubar for editing. This makes it easy to explore or modify the scripts behind the existing shared menus.

Special Considerations

Although Frontier runs the script when a shared menu item is chosen, the host application is hosting the process, so that scripts which run perfectly when initiated from within Frontier may behave differently when initiated from a shared menu. The following sections describe the chief differences.

Threads

When a shared menu item is chosen, the host application must contact Frontier. If Frontier then starts driving the host application, wires can get crossed: the host application may be waiting for Frontier's response to what it told Frontier to do, so it isn't listening to what Frontier is telling it to do. A symptom is that everything seems to freeze up for a while. That, for instance, is what happened with a shared menu item script I wrote for Netscape Navigator. The script said simply:

```
people.man.driveNavigator()
```

But the script at `people.man.driveNavigator` was starting out with a `webBrowser.openURL()` command, and Navigator wasn't ready to respond to this, so nothing was happening.

The solution is to spawn a thread. My problem went away when I changed the shared menu item script to that the following.

Example 26-2. Spawning a thread

```
evaluateThread("people.man.driveNavigator()")
```

The verb `evaluateThread()` is an old, undocumented verb that should do exactly the same thing as `thread.evaluate()`. There seems to be a mysterious bug involving `thread.evaluate()` in shared menus, so I don't try to call it there.

Sometimes, you just can't get a direct threading call to work. In such cases, there is a trick which I call "stepping out to Apple events." The idea is to build a script and send it to Frontier as a `'dosc'` Apple event (see Chapter 34, *Driving Frontier*

From Outside). You just have to be careful of quoted strings inside quoted strings. Here is one way to modify Example 26-2 so that it uses Apple events.

Example 26-3. Stepping out to Apple events

```
local (s)
s = "thread.evaluate(\"people.man.driveNavigator()\")"
misc.doScript('LAND', s)
```

If you don't want to use Apple events, you can use another method, which I call "agent polling."* The shared menu item script puts a UserTalk expression into a preordained location in the database; that's all. Meanwhile, an agent script in Frontier is checking this location every second; if it finds anything there, it removes it (so it won't find it next time) and then spawns a thread to evaluate it. In my implementation, the agent script is at `system.agents.watchDoThis`, and is shown in Example 26-4.

Example 26-4. Agent polling

```
if people.man.doThis ≠ ""
    local (s = people.man.doThis)
    people.man.doThis = ""
    thread.evaluate(s)
clock.sleepFor (1)
```

Then the shared menu script would look like this:

```
people.man.doThis = "people.man.driveNavigator()"
```

It is possible to sophisticate the structure with semaphores, but you get the general idea.

Memory

Because the shared menu item script runs in the host application's context, there may not be enough memory to run it successfully. Sometimes you can solve the problem by increasing the host application's memory allocation, but the usual solution is to spawn a thread by one of the three methods just described. That way, the real work of the shared menu script is done in Frontier's context.

An example is at `FinderMenu.commands.reconciler()`, which is called by the Finder's shared menu item **Reconcile Folders**; the Finder hasn't enough memory to hold the operation, so the script calls `toys.threadCall()`, which in turn calls `evaluateThread()`, to do the real work.

---

* The idea for this comes from Cameron Smith.

Dialogs

If a script's execution was initiated by the user choosing from a shared menu, any Frontier dialogs put up during execution appear in the host application, not in Frontier. (Modeless dialogs run as modal, but at least they do appear.) Many shared menu scripts take advantage of this feature (see `FinderMenu.commands.reconciler()` again, for instance).

But now and then it isn't the desired behavior. Suppose, for instance, the script switches to Frontier, does something-or-other, and then puts up a dialog. That dialog will appear back in the host application; the user won't see it unless the script switches to the host application first.

Once again, a new thread can solve the problem, because it initiates a new context from within Frontier. If I call this script from a shared menu in Eudora:

```
frontier.bringtofront()
dialog.notify("howdy")
```

the dialog shows up in Eudora; but if I call this script instead:

```
frontier.bringtofront()
evaluateThread("dialog.notify(\"howdy\")")
```

the dialog shows up in Frontier.

WARNING   A dialog that is put up by a `dialog.runModeless()` call cannot be made to appear in a host application at all. The Find/Replace dialog is an example.

Giving up

Shared menus can be touchy creatures. Sometimes the threading tricks shown earlier can help, but sometimes menu sharing just doesn't work. For instance, on my computer, choosing any shared menu item in Microsoft Internet Explorer will crash the computer.

Fortunately, there's very little inconvenience in switching to Frontier and running the script from there instead. Or, I can usually save the script to disk (see Chapter 29, *Import/Export*) and run it from OSA Menu's script menu, or I can save it as a desktop script and run it from the Apple menu. So menu sharing is neat, but there are alternatives.

Modal Menus

The various modal menus appear depending on the nature of the frontmost window; for example, if a table edit window is frontmost, the **Table** menu appears. The mechanism behind this is an agent script at `system.agents.statusMessage`, which is called every second. The first thing this agent does is to call `suites.modes.monitor()`, which gets the datatype of the object represented by the frontmost window and looks up that datatype in `suites.modes.menus`, substituting the menubar there for the current modal menu, if there is one. This script is worth studying; it is simple and clever.

If agents are disabled, the modal menu system will stop operating. As explained in Chapter 27, you can disable agents by choosing **Disable Agents** from the **Main** menu. You might need to disable agents if you thought one of them was causing a problem—if, for instance, you edited an agent and compiled it, only to discover you'd made some serious mistake. Just be aware that you're also going to lose the modal menus and the functionality they contain.

Suites

A suite is a collection of scripts with related functionality—basically, a miniature application—stored in a subtable of the `suites` table, and accessed through a menu which appears when a certain item of the **Suites** menu is chosen.

Not every suite, though, meets this canonical description. The `suites` table has come to be a repository for all sorts of script collections. A suite need not have a menu specific to it, or if it does, it may not be a suite menu. For example, `suites.applescripts` contains the implementation for the three items of the **Applescripts** submenu of the **Main** menu; `suites.html` contains the implementation for the whole **Web** menu; `suites.regex` is the sort of thing you'd expect would live at `system.extensions`; `suites.samples` and `suites.toys` are just miscellaneous collections of unrelated scripts.

Accessing and Removing Suite Menus

A suite's menu doesn't appear until you summon it, which you do by choosing an item from the **Suites** menu, usually from its **UserLand Suites** submenu. For example, if you choose **Object Database Map**, the **Map** menu appears. If you choose **NetFrontier**, the **NetFrontier** menu appears. These are suite menus; the **Suites** menu is a menu for summoning suite menus.

Normally, only one suite menu is present at a time; if you summon a suite menu when a suite menu is already present, the old one goes away. To remove any suite menus, choose **Minimal Menus** from the **Suites** menu.

Writing a Suite

A set of scripts is a candidate for packaging as a suite if it forms a unified functionality that one needs only occasionally. But the usual reason for making a suite has to do with distribution. A suite can be packaged as a single exported file and distributed to other Frontier users; when a user imports a suite, the menu item in the **Suites** menu that summons its suite menu is automatically created. So a suite is a way of letting other users access your miniature Frontier-based application.

If you have a collection of scripts that you think would make a good suite, choose **Make New Suite** from the **Suites** menu. This calls `suites.samples.basic-Stuff.createNewSuite()`, a brilliant script which is worthy of study. You should try making a new suite even if you don't intend to do anything with it, just to see what it's like.

When you choose **Make New Suite**, a dialog box requests a title for the suite. This title will be used for, among other things, the name of your `suites` subtable, and changing this later will break things, so choose the title carefully.

Let's say you call your suite `toot`. What you now have is a framework for your `toot` suite. A `suites.toot` subtable is created, containing four items: a wptext called `readMe`, into which to place instructions for the user; your suite's menubar, called `menu`; and two scripts, `installMenu` and `importSuite`. You will add scripts and other objects to `suites.toot` as desired, to implement the functionality of your suite.

Your **toot** suite menu, in `suites.toot.menu`, initially contains three items: **Read Me**, **Edit Menu**, and **Edit Table**. Feel free to change the menu name **toot**, and to add menu items (and even other menus) to the menubar as needed to access the functionality of the scripts you'll put into `suites.toot`.

A new menu item in the **Suites** menu, called **toot**, summons your suite menu. It calls `menu.addSuite()`, which in turn calls `suites.toot.installMenu()`. `toot.installMenu()` calls `menu.install()` to put your suite menu at the top of the screen; you can modify it so that it also performs other tasks. For example, the `tableMap` suite has an extra line in its `installMenu()` script that calls `tableMap.zoomer()` to open the `tableMap.list` outline.*

---

* Similarly, you can define a `closeSuite()` script that will be called when your suite menu is removed by choosing **Minimal Menus** (or calling `menu.noSuite()`); unfortunately, it is *not* called when your suite menu is replaced by another, so it's not as useful as it might be.

If you're going to distribute your suite, `toot.importSuite()` becomes important. It is called, once and once only, when a user imports your suite into the database. It puts into the user's **Suites** menu the **toot** menu item that summons your suite menu; this is done by calling `menu.importSuite()`. The second parameter in the call to `menu.importSuite()` is the name of the menu item that will go into the **Suites** menu, so if you don't like the name **toot** for that menu item, `toot.importSuite()` is the place to change it.

Your `importSuite()` script is also your chance to prepare the database (kindly and carefully!) as may be required by your suite. For an example, study `suites.classAds.importSuite()`. It calls an `init()` script to create and initialize a dozen or so objects in the database, and a folder on the disk. These are all things that the suite will need in order to do its work later on. Typically, if a suite needs private ancillary objects in the database, the **user** table is the proper place to create them.

Clearly the onus is on the programmer to plan and test before distributing any suite. One should write only to acceptable places in the database, and be certain to do no damage. In my own copy of the database, I have disabled the automatic call to `importSuite()` when a suite is imported; I don't like the idea of running a script from another user before I've had a chance to study it. See Chapter 29.

Programmatic Menu Manipulation

Menubar items are strings, so matches on them are case-sensitive.

To add an item to a menubar:

```
menu.addMenuCommand (addrMenubar, menuName, itemName, scriptString)
```

scriptString is a string, but what's to be made from it is a menu item script. The way this works is that *scriptString* becomes the first and only line of the script; if the script is to consist of more than one command, it can be notated using semicolons and curly braces, but the new menu item script will still consist of only one line. Here, for example, I add to my **=user.initials** menu a menu item that beeps 10 times slowly.

Example 26-5. Making a menu item

```
local (s)
s = "for x = 1 to 10 {"
s = s + "speaker.beep();"
s = s + "clock.waitseconds(1);}"
menu.addmenucommand (@system.misc.menubar, "=user.initials", "BeepALot", s)
```

It is possible to use a different technique and edit the menu item script as an outline; an example appears later in this section.

The new item goes at the end of its menu. If that isn't where you wanted to put it, the menubar can be edited as an outline using the op verbs to move the item after it is created. This too is illustrated later.

To insert a copy of an existing menubar into another menubar:

```
menu.addSubMenu (addrDestMenubar, menuName, addrSourceMenubar)
```

To delete a menu item:

```
menu.deleteMenuCommand (addrMenubar, menuName, itemName)
menu.deleteSubMenu (addrMenubar, menuName)
```

> The difference is that deleteMenuCommand() lets you specify the item more precisely, and won't delete an item with subheads.

There is a certain element of danger to these verbs: menu.addSubMenu() and menu.deleteSubmenu(), especially, can replace or remove, respectively, a menu containing many items.

In actuality, menu.addMenuCommand(), menu.addSubMenu(), and menu.delete-SubMenu() do not distinguish between menus and menu items in the menubar they are altering: they simply look, scanning from top to bottom, for any item whose text matches *menuName*.

To open a menu item script edit window:

```
menu.zoomscript ()
```

To copy between a menu item script and a script object:

```
menu.getScript (addrScriptObject)
menu.setScript (addrScriptObject)
```

The three verbs menu.zoomscript(), menu.getScript() and menu.set-Script() do not specify a menubar or item because the item must already be selected in a menubar edit window which is already the "target." For more about the target, see Chapter 18, *Non-Scalars*.

Also in Chapter 18 are described the verbs that let you edit a script or menubar as an outline (the op verbs). These are needed for more sophisticated actions upon menu item scripts and menubars than the menu verbs permit. As an illustration, we modify Example 26-5 so as to place **BeepALot** first in the **=user.initials** menu and so that its script consists of three lines.

Example 26-6. Making a menu item, moving it, and editing its script

```
local (s); new(scriptType, @s)
bundle «construct the script in a script object variable
    target.set(@s)
    op.setLinetext("for x = 1 to 10")
    op.insert("speaker.beep()", right)
    op.insert("clock.waitseconds(1)", down)
    target.clear()
```

Example 26-6. Making a menu item, moving it, and editing its script (continued)

```
menu.addmenucommand (@system.misc.menubar, "=user.initials", "BeepALot", "")
bundle «edit the menubar, move the new item up and attach the script
    target.set(@system.misc.menubar)
    while op.getLineText() ≠ "BeepALot"
        op.go(flatdown,1)
    op.reorg(up, infinity)
    menu.setscript(@s)
    target.clear()
```

There are also several verbs that control which custom menus appear at the top of the screen.

To remove all custom menus from the top of the screen:
```
menu.clearMenubar ()
```

To restore the default custom menus at the top of the screen, with no suite menus:
```
menu.noSuite ()
```

To add a menubar to the top of the screen:
```
menu.install (addrMenubar)
```

To remove a menubar from the top of the screen:
```
menu.remove (addrMenubar)
```

To learn whether a menubar is present at the top of the screen:
```
menu.isInstalled (addrMenubar)
```

To add or remove a menubar at the top of the screen:
```
menu.toggle (addrMenubar)
```

To remove all suite menus and call a suite's **installMenu ()***:*
```
menu.addSuite (addrSuiteTable)
```

These verbs are part of Frontier's internal menu management, and would probably be called by your scripts only if you were taking over menu management yourself. You might, for instance, want to turn Frontier into a specialized application in which only your custom menus appear.

27

Agents and Hooks

An *agent* is a script that Frontier calls automatically from time to time in the background. A *hook* is a script that Frontier calls automatically when a certain high-level event takes place. Together, they let your scripts respond to what's happening around them.

Agents

Agents run in a thread of their own; the agent thread number is maintained at `system.compiler.threads.agents`. The agent thread is present even if Frontier is otherwise idle. See Chapter 21, *Yielding, Pausing, Threads, and Semaphores*.

Agent scripts live at `system.agents`. The database, as shipped, already includes several of them, which you are free to examine. Agents have no eponymous handler; Frontier calls them without parameters.

An agent script is called by Frontier repeatedly, at intervals. The default interval is one second after the agent script finishes its previous execution; or the agent script itself may tell Frontier to wait longer before calling it again, by calling `clock.sleepFor()`, which takes as its parameter the number of seconds to wait before calling again.

The call to `clock.sleepFor()` does not have to occur in the agent script itself; but it must occur either in the agent script or in a script called by the agent script. For example, the script at `system.agents.schedulerMonitor` calls `scheduler.monitor()`, which calls `clock.sleepFor()`. The call to `clock.sleep-For()` need not come last in its script, and it does not immediately return control to Frontier.

Writing and Debugging Agents

To create an agent, you can just make a new script in `system.agents`, or choose **New Script** from the **Main** menu and set the dialog popup to **Agent**—this creates the script in `system.agents` for you and supplies a call to `clock.sleepFor()`.

Testing an agent script requires some caution. You probably don't want Frontier to call the agent script while you're working on it. If you press the **Compile** button (or close the script edit window), Frontier will call the agent script one second later—so don't press it! You can, however, press the **Run** button to get implicit compilation; this lets you check that the script compiles and behaves as expected, but does not cause Frontier to call the script automatically as an agent. (See Chapter 13, *Running and Debugging Scripts.*)

You can't test an agent script by pressing its edit window's **Run** button if the script contains a call to `clock.sleepFor()`. That's because the script will run in a new thread, which is not the agent thread; and `clock.sleepFor()` is illegal outside the agent thread. The usual strategy is to comment out the `clock.sleepFor()` line, test the script with the **Run** button, and then, when you're done testing, uncomment the `clock.sleepFor()` line, and press the **Compile** button. Your agent is now live.

It's a good idea to supply a fairly large `clock.sleepFor()` value at first, so that there is time for you to stop the agent between calls to it if something goes wrong. I once wrote an agent into which I had put a call to `dialog.notify()` for debugging purposes. I then made the mistake of pressing the **Compile** button, and discovered that it was not easy to stop the agent; the dialog would appear, I would click its **OK** button, and the dialog would immediately appear again! I had a devil of a time getting things to pause. I would have had time to stop the agent if I had said `clock.sleepFor(15)` within the script.

How do you stop an agent? One way is to have the `system.agents` table ready to hand; selecting and cutting your agent script stops it from running. Another option is to select all the lines of your agent's script, turn them to a comment, and press the **Compile** button. Finally, choosing **Disable Agents** from the **Main** menu stops all agents; its name then changes to **Enable Agents**, and you can choose it again to start the agents up once more. This menu item relies on a call to `frontier.enableAgents()`.

Agent Messages

A question arises as to how agents are to share the message area of the Main Window. If five agents are running and they all include calls to `msg()`, is the Main Window going to be bombarded with a constant stream of messages, none of which will remain long enough to be legible? And what about regular scripts that call `msg()`?

The answer is as follows. At the left end of the Main Window is a popup menu. Choosing from this popup does two things. First, it actually calls the chosen agent, whether it was time to call it automatically or not. Second, it turns control of the Main Window's message area over to the chosen agent. Only one agent at a time can write to the Main Window with `msg()`, and it is the one most recently chosen with this popup.[*]

When a non-agent script calls `msg()`, agents stop writing to the message area of the Main Window. If a non-agent script calls `msg("")` with the empty string as its parameter, or the user clicks the top half of the Main Window, the Main Window goes back to displaying messages from the agent last chosen in the popup menu.

Choosing from the Main Window popup while holding the Option key opens the agent script's edit window—a handy shortcut.

UserLand Agents

Here are the agents supplied by UserLand, and what they do.

StatusMessage

This is the agent whose messages I usually like to have showing; it tells the number of threads and the amount of Frontier's free memory. It also has another job, a very important one: it calls `modes.monitor()`, which is responsible for managing modal menus (see Chapter 26, *Menus and Suites*).

WebBrowserAgent

Watches to see whether Microsoft Internet Explorer or Netscape Navigator is active, so that one can drive the browser by calling `suites.webBrowser` verbs, regardless of which browser it is. It also reconciles the shared menus belonging the two browsers (see Chapter 26).

Scheduler Monitor

Installed by the `scheduler` suite to check once a minute whether any tasks are scheduled to be run; see Chapter 37, *Scheduler*.

MinutesSinceShip

Provides a harmless bit of trivia.

FrontierPath

Isn't really an agent at all, in a sense. When you choose it in the popup menu, it displays the pathname of Frontier on your disk, and then calls `clock.sleepFor(infinity)`, which is as good as saying it is never to be called automatically. This is a useful trick: you can list in the agents popup menu, just because it is a convenient location, a utility without agent-like functionality.

---

[*] There is, as far as I know, no way to script an action equivalent to choosing from the popup.

Hooks

A *hook* is an internal Frontier mechanism that calls a script in response to a specific high-level event. Hooks are managed by associating particular areas in the database with particular events: scripts located in one of these areas are called when the corresponding events occur. Unless otherwise specified, such scripts take no parameters.

Open the database

> As a database is opened,* all the scripts in its `system.startup` table are called. Startup scripts are triggered, each in its own thread, and before any agents are called.

> If a suite has private values that must be initialized every time Frontier starts up, it will typically install a startup script, using its `installSuite()` to do so (see Chapter 26). For example, `scheduler.installSuite()` installs a startup script, `system.startup.["Scheduler Startup"]`, which sets a global value, `scheduler.startingUp`, to `true`. This way, when the scheduler monitor agent is first called, it knows that it is being called for the first time, and can take certain actions accordingly. See Chapter 37.

Close the database

> As a database is closed,† all the scripts in its `system.shutdown` table are called. This happens after the `closeWindow` hook has run, and after the "Save changes?" dialog has appeared if it needs to. It is impossible now to stop the database from closing, but it is possible to save it again (using `file-Menu.save()`); you'll need to do so if your shutdown script makes changes in the database.

Switch away from Frontier

> If Frontier is frontmost, then when the user or a script brings another application to the front, all the scripts in the frontmost database's `system.suspend` table are called while Frontier is still frontmost.

> A suspend script cannot prevent switching away from Frontier, but it can be used to bring Frontier back to the front again immediately. The following script, if placed in `system.suspend`, would effectively prevent the user or a script from being able to make any application frontmost other than Frontier:

```
while sys.frontmostapp() == "UserLand Frontier™"
    clock.waitsixtieths(1)
sys.bringapptofront("UserLand Frontier™")
dialog.alert("Don't leave me!")
```

---

* From the **File** menu, or with `filemenu.open()`; opening a database "invisibly" with `db.open()` does not trigger this hook. See Chapter 30, *Multiple Databases.*

† By closing its Main Window, or as a consequence of quitting Frontier.

Switch to Frontier

> If Frontier is not frontmost, then when the user or a script brings Frontier to the front, all the scripts in the frontmost database's `system.resume` table are called once Frontier is frontmost.

Close a window

> For the `closeWindow` hook, see Chapter 24, *Windows*.

Modifier-double-click text

> When the user double-clicks text while holding the Control key, the Option key, or the Command key, the script at `system.misc.control2click`, `system.misc.option2click`, or `system.misc.cmd2click`, respectively, is called with a single parameter, the double-clicked text. Since more than one modifier key can be down simultaneously, an order of precedence is defined: Control key, Command key, Option key (that is, if the Control key is down, `system.misc.control2click()` is called regardless of what other modifiers are down, and so on).

There are already scripts at `system.misc.control2click` and `system.misc.cmd2click`. The former switches to the double-clicked verb's DocServer documentation (or, if the Option key is also down, inserts the parameter information from DocServer into the script or outline where the verb was double-clicked). The latter jumps to the database object whose name was double-clicked (additionally closing the frontmost window if the Option key is also down). See Chapter 13, *Running and Debugging Scripts*.

Hooks as Interrupts

A hook script may interrupt a script in progress; the script in progress cannot proceed until the hook script has finished executing. And you can't be exactly certain as to just when the interruption will take place.

For instance, if you have a script that says:

```
sys.bringapptofront("finder")
clock.waitseconds(5)
sys.bringapptofront("UserLand Frontier™")
```

and a hook script in `system.suspend` that says:

```
dialog.alert("Hold everything.")
```

then running the first script will switch away from Frontier but won't switch back; the first line has triggered the hook script, which is now stopped, waiting for a response to the dialog, and the first script has been interrupted, so it too is stopped.

Similarly, if you have a script that says:

```
sys.bringapptofront("finder")
clock.waitseconds(5)
sys.bringapptofront('LAND')
for x = 1 to 10
    msg(x)
```

and a hook script in **system.resume** that says:

```
dialog.alert("Welcome back.")
```

then if you run the first script, it's hard to know how far through its counting **x** will actually get before the counting is interrupted by the dialog.

User-Based Pseudo-Hooks

When a database is opened, a call is made to (if it exists) **people.[user.initials].customStartup()**: it isn't a hook, properly speaking; it is triggered by the "startup" hook, though, because the startup hook calls the scripts in **system.startup**, and one of these is **system.startup.startupScript()**, which calls **user.login()**, which calls **people.[user.initials].custom Startup()**.

When you choose **Backup** from the **Main** menu, then after the backup has been performed, a call is made to **people.[user.initials].customBackup()** if it exists. This isn't a hook at all; it's part of the functionality of the menu item script for **Backup**.

These are "pseudo-hooks" in the sense that they aren't the result of high-level events; they're just scripts called by other scripts in the usual way. They are worthy to be considered along with real hooks, though, because their purpose is similar: they allow custom automatic actions to be triggered. You could use the same technique to introduce your own pseudo-hooks.

Because these pseudo-hooks live in the **people.[user.initials]** table, they may be thought of as associated with a particular user rather than the database as a whole. Frontier presently has a simple system of differentiation between users—a call to **user.switchUser()** brings up a dialog to let the user enter a new user name, organization, and initials. A different **people.[user.initials]** table is then in force.

28

Pictures

A Frontier *picture* is a non-scalar object type, like a wptext or an outline, except that it has no internal editing facilities: opening a picture's edit window simply displays the picture as a graphic in a window, without giving the user any ability to edit that graphic. On the whole, a Frontier picture is like a PICT (a standard Mac OS format for graphics), and verbs are provided to convert between the two; but a Frontier picture has the additional feature that its text can be calculated by evaluating a UserTalk expression in real time.

This chapter describes UserTalk verbs and techniques for manipulating Frontier pictures.

PICTs and Frontier Pictures

Since a Frontier picture cannot be edited as a graphic within Frontier, a common strategy is to create or edit it as a PICT in some other application, and then import the PICT into the database. It will be imported as a binary (of type `'PICT'`); User-Talk provides verbs that convert between a PICT binary and a Frontier picture object.

To convert a PICT binary object to a Frontier picture object:
    pict.PICTtoPicture (*addrBinaryPICT*, *addrPicture*)

To convert a Frontier picture object to a PICT binary object:
    pict.pictureToPICT (*addPicture*, *addrBinaryPICT*)

One way to import the PICT is through the clipboard. As we shall see in Chapter 31, *Driving the System*, Frontier has verbs that move data between the clipboard and an object. So, in another application that works with PICT graphics, we can select and copy a PICT onto the clipboard. Then, in Frontier, the following script will store the clipboard contents into the database as a Frontier picture and display the picture.

Example 28-1. Displaying a copied PICT as a Frontier picture

```
local (tempPICT)
clipboard.get('PICT', @tempPICT)
pict.PICTtoPicture (@tempPICT, @workspace.myPicture)
edit (@workspace.myPicture); window.zoom(@workspace.myPicture)
```

If a file contains a PICT resource, it can be imported using a rez verb; see Chapter 20, *Resources.* For instance, one might change the second line of Example 28-1 to:

```
rez.getResource("desire:docs:myFile",'PICT', 1000, @tempPICT)
```

If an application saves its documents as a PICT in the document file's data fork, the data can be read into Frontier and converted to a Frontier picture; see `suites.samples.basicStuff.readPictureFile()`. For example, Aldus Super-Paint can save a file in this format.

The verbs `pict.PICTtoPicture()` and `pict.pictureToPICT()` are actually utility scripts. They do their work through two other verbs which also convert between a PICT and a Frontier picture, but do not take any parameter saying what Frontier picture to work on—the picture's edit window must already be the "target." (For the target, see Chapter 18, *Non-Scalars.*)

To convert a PICT binary object to the target Frontier picture object:
```
pict.setPicture (binaryPICTvalue)
```

To convert the target Frontier picture object to a PICT binary object:
```
pict.getPicture (addrBinaryPICT)
```

Since the target can be a visible window, `pict.setPicture()` can cause the graphic in a visible Frontier picture edit window to change. There is a flicker, so this would not be a good way to implement animation, but one could conceivably make a crude "slide show" presentation.

Text in Pictures

A PICT can contain text objects. If the text of such a text object begins with an equal sign (=), it can be replaced in the Frontier picture version by evaluating as a UserTalk expression whatever text follows the equal sign and coercing the result to a string. For example, if a PICT contains a text object whose text is `=clock.now()`, the Frontier picture made from this PICT can display an actual date-time value. Whether this evaluation is performed for any given picture is governed by a boolean associated with the picture.

Moreover, a Frontier picture has associated with it an update interval, which dictates how often its expressions should be automatically re-evaluated. So, if a Frontier picture were made from a PICT which contained a text object whose text is =clock.now(), and the update interval for that picture was one second, then while that picture's edit window was open, the window would contain a counting clock.

To set whether the target Frontier picture object should have its expressions evaluated:

    pict.expressions (*boolean*)

To set the interval at which the target Frontier picture object should have its expressions evaluated:

    pict.scheduleUpdate (*seconds*)

The default value for the update interval, if not explicitly set, is infinity (I believe).

Other Graphic Formats

Frontier itself cannot convert between Frontier picture objects and graphic formats other than PICTs; but there are applications which can convert between PICTs and other graphic formats, and some of these applications are scriptable, so Frontier can drive them (see Chapter 32, *Driving Other Applications*).

Suppose, for example, we have read a GIF image file into the database, where it becomes a binary. Then it is possible to write a utility script which drives Clip2GIF to convert such a binary to a PICT; the PICT may then be converted to a Frontier picture. I use such a utility in conjunction with Frontier's Web site management facilities; I can store a GIF in the database, and use the utility to display it within Frontier.

29

Import/Export

This chapter describes ways in which objects may be separated from the database and saved as independent files. One use for such files has to do with backing up the database, so this chapter also discusses the "care and feeding" of the database: how to back it up, maintain it, and upgrade it.

Types of Exported Object

Objects can be made into an independent file, or *exported*, in several forms. Each form has a different use. After an overview of these forms, they are each discussed at length.

Packed object

A packed object is a database entry copied to a separate file. When a packed object is opened, Frontier automatically imports it as a normal object into the database. Thus, a packed object is a good way to communicate an object to other users, or to a different database.* Packed objects are useful also for backing up parts of the database; and, as we shall see, a typical way to maintain or upgrade the database is to export customized parts of the database as packed objects, then import them into a clean root.

Desktop script

A desktop script is a script which has been exported as a packed object, but also has a special property: when it is opened, instead of being imported into the database, it is executed by Frontier. It isn't really an executable, since

---

* If two copies of Frontier can see each other over a network, there is another way to send objects between them, using NetFrontier, discussed in Chapter 39, *NetFrontier.*

Frontier must be running; but it *feels* like an executable, and is a convenient way to access Frontier functionality from outside Frontier.*

Droplet

A droplet is a very minimal application, a true executable, onto which icons of files or folders can be dropped; the application contains a script which is then called by Frontier, and which has access to a list of those files and folders, so that it can process them.

Packed Objects

To export a database entry to a packed object file, select the entry in a table edit window, and choose **Export** from the **Main** menu. The Export dialog appears. The editable text box in the dialog is for the object reference of the object to be exported; this should already be the entry you selected.

The first popup menu should be set to **Packed Object**. The second popup menu gives two options. One is simply to make the file. The other is to make the file and attach it to the current outgoing Eudora message.

In the first case, where you save the packed object is up to you. As the packed object is about to be created on disk, a StandardPutFile dialog appears. Feel free to alter the default filename, as it is without significance to Frontier: when the packed object is imported, the object reference of the originally exported object will be retrieved from inside the packed object.

In the second case, a temporary version of the file will be placed into the *Outgoing Frontier Objects* folder inside the *Eudora* folder in the *UserLand* folder in your system's *Preferences* folder; it is this temporary file that Eudora is instructed to attach to its frontmost message. The temporary file is not automatically deleted after the message is sent, so it is up to you to clean the *Outgoing Frontier Objects* folder periodically.†

Radio buttons in the Export dialog provide options to compress the packed object in StuffIt format, or to compress it and then BinHex-encode the compressed version. If you choose one of these options, the StuffIt Engine will be driven to carry out your wishes (or the StuffIt application, if you don't have the StuffIt Engine), and the uncompressed packed object file will be deleted.

---

* For example, I keep a number of frequently used Frontier desktop scripts in my Apple menu, where I can run them no matter what application is frontmost.

† An alternative is to modify `suites.export.card.doExport()` so that when it calls `Eudora.attachFiles()`, it "spools" the temporary file and deletes it.

You can avoid the Export dialog by holding the Shift key as you choose **Export**; the selected object will be exported according to the dialog settings from the last time you used the dialog.

Menus

A menubar object can be exported as an ordinary packed object, but frequently what is desired is to export a particular menu item (along with its script, and any menu items hierarchically dependent upon it). This is done by selecting the menu item in a menubar edit window, and choosing **Export SubMenu** from the **Menubars** menu.

Folders

You can export a table and its entries as individual packed objects, rather than exporting the whole table as one packed object. To do this, select the table as an entry in its parent table, and choose **Export** from the **Main** menu; when the Export dialog appears, set the first popup to **Folder**.

The table will be represented on disk by a folder (despite the Save dialog, which asks for a "file" name). All entries in the table are exported as individual packed objects, except for tables, which are represented as folders. Thus the end product is a hierarchy of folders and packed object files, mirroring the structure of the original table.

One advantage of this format is that the exported objects, though many, are relatively small; Frontier has an easier time importing and exporting smaller objects. On the other hand, a single packed object file is far more space-efficient than its contents broken out into individual packed object files.

Importing Packed Objects

A packed object may be opened from the Finder like any other file—by double-clicking it, or by selecting it and choosing **Open** from the **File** menu, and so on. It is also possible to open a packed object from within Frontier, using **Open** from the **File** menu; or the packed object's icon may be dropped onto Frontier's icon.

When a packed object is opened, Frontier comes to the front and presents a dialog offering a chance to cancel importing, or to determine the imported object's destination (its object reference) in the database. The default is the object's path from before it was exported. You can change this destination if you wish: you must supply a complete path (except for `"root"`), and it must be a valid object reference. If an object of the same name already exists, a second dialog asks whether you wish to overwrite it with the imported object; if you decline, the first dialog appears again.

Security concerns

After the object is imported, if it is a table, and if the table contains a script entry called `importSuite`, that script is called. This is in order that suites may initialize themselves as they are imported (see Chapter 26, *Menus and Suites*). But the notion that a script supplied by someone else is to be run before it can be examined seems dangerous, since Frontier scripts can do very powerful things, such as erasing the hard disk. In my database, I have commented out the line in `suites.export.importer` that calls `importSuite`, replacing it with a dialog just letting me know that the `importSuite` script exists, like this:

```
if typeOf (adr^) == tableType «see if it has an importSuite script
    if defined (adr^.importSuite)
        « adr^.importSuite (adr)
        « previous line removed, and next one added, for safety
        dialog.notify (string(adr) + " has an importSuite routine")
        return (true)
```

Alerted by the dialog, I can later study the `importSuite` script, and call it manually if I approve of it.

My approach to importing other objects is equally circumspect. If an imported object offers to replace an existing object, then I generally import it to a different location, study it at leisure, and, if it passes muster, manually paste it in place of the existing object.

Menu items

When importing a menu item, no dialogs appear; all destination information comes from the packed object, and there is no way of knowing what menu item in what menubar is about to be imported. If the menu item comes from a menubar that doesn't exist in the database, that menubar is created. If it comes from a menu that doesn't exist, that menu is created. If the menu item already exists, it is replaced; if it doesn't exist, it is added at the end its menu (no control over the menu item's location within its menu is provided). See Chapter 26 on the verb `menu.addSubMenu()`.

I find this situation unacceptably insecure. The script performing the import is `suites.export.importSubMenu`; in my database, I have modifed it to add dialogs and choices to let me control the import process, as with other types of object.

Folders

There are three ways to import a folder of exported packed objects. One is to call `suites.export.importFolder()`. There is no menu interface; you have to call it manually, or, if you are using folder format as a means of distribution, you might supply an "installer" desktop script for the recipient to run—the desktop script could then call `suites.export.importFolder()`.

`importFolder()` takes two parameters: the address of a table into which to import the folder contents, and the pathname of the folder. Because of the unfortunate way in which `importFolder()` is coded, things will go badly wrong if the address of the destination table doesn't match that of the table which was originally exported. This is because the entry names are absolute, not relative.

For example, if you did a folder export from `workspace`, and it contained an entry `workspace.myEntry`, then if you call `importFolder()` with a destination address `@workspace.newTable`, `workspace.newTable` will be created but `myEntry` will not go into it—it will go into `workspace`, because that's where it came from.

If the destination table exists already, everything in it is destroyed; so take appropriate precautions. Subtables are created automatically. No dialogs appear. No `importSuite` scripts are called.

The second way to import a folder is with the batch importer. The batch importer, which is better written than `importFolder()`, is discussed later in this chapter.

The third way to import a folder is to do it manually, one packed object at a time. This is time-consuming, but gives you the most control over the process.

Desktop Scripts

To create a script which is to be exported as a desktop script, choose **New Script** from the **Main** menu and set the popup to **Desktop Script**. This creates a new script object in `system.deskscripts`. Once the script is written, select the script object and export it, choosing **Desktop Script** in the first popup of the Export dialog.

Since any script can be exported as a desktop script, why create the script in `system.deskscripts`? Because when a desktop script file is opened, Frontier reads the script into its original location in the database in order to call it, wiping out whatever is already at that location; `system.deskscripts` is a safe place to read it into.

When the desktop script file is opened, in any of the same ways as a packed object, the sequence of events is as follows. First, the pathname of the desktop script file will be placed in `system.deskscripts.path`; the script might wish to use this, for instance, to obtain the file's containing folder, so that it can somehow process all files in the same folder as the desktop script file. Frontier will come to the front; the context is Frontier, and any Frontier dialogs will appear within Frontier.* The script will then be copied into the database and called without parameters (the script cannot contain an eponymous handler). After execution, the Finder comes to the front, unless at some point the script sets `frontier.finderToFront` to `false`.

After a desktop script is run, the copy of the script in the database which was actually called is usually deleted automatically. But if an object already existed at this location when the desktop script was opened, the imported copy is not deleted after running. This is in order to facilitate the development cycle for a desktop script: one can repeatedly edit the script in `system.deskscripts`, export it, and run it.

A script that lives in a desktop script file can be imported without running it: open the file and hold the Command key. This lets you edit the script, or inspect a desktop script you receive from someone else, so as not to run it imprudently.

Script Editor Format

If you're using OSA Menu, you can have it run an exported script for you. The chief advantage of this is that the script runs in the context of the frontmost application. Also, such scripts are not imported into Frontier (though of course Frontier must be running). The whole operation works just like choosing a shared menu item (see Chapter 26).

Here's an example. Suppose we are in Microsoft Word. From OSA Menu's special script menu towards the far right of the menubar, choose **Open "Microsoft Word Scripts" Folder**. In Frontier, make a script; as a test, it might say simply:

```
dialog.notify("Hello there.")
```

Export the script into the *Microsoft Word Scripts* folder (it's in a folder called *Scripts* in the System folder), choosing **Script Editor format** from the first popup menu in the Export dialog. Suppose we call the exported file *Hello*. Now go back into Microsoft Word. OSA Menu's script menu now contains an item **Hello**; choosing it brings up our dialog within Microsoft Word, just as if we had called the same script from a shared menu.

---

* That is, a desktop script is not like a shared menu script. But there is a way around this, discussed next.

Droplets

To create a droplet, choose **Droplet Developer** from the **Suites** menu, and then, from the **Droplets** menu that appears, choose **New Droplet**. Provide a single-element name for the droplet, not an entire object reference; the new droplet will be created as a table in `user.droplets.tables` (we will refer to the created table as the "droplet table").

There are two ways to edit a droplet for which you do not possess the original droplet table. One is to open it and, if it is not faceless, press the **Edit** button in the resulting dialog. The other is to choose **Import App** from the **Droplet** menu.

There are two approaches to making a droplet, script-based and dialog-based; you must decide upon one or the other (they cannot be combined). Either way, you make certain changes in the droplet table; then you create the application by bringing the droplet table edit window to the front and choosing **Export to App** from the **Droplets** menu.

Since the droplet is a genuine executable, the Finder's Get Info dialog can be used on it. The droplet table's `finderComment` and `version` entries allow you to set some of the information that will appear in this dialog.

The purpose of a droplet is to process the files and folders whose icons are dropped onto it. The pathnames of those files and folders will be supplied while the droplet is running, but Frontier does not automatically look inside folders (though of course the droplet can do so).

Script-Based Approach

In the script-based approach, you begin by deleting the `card` entry in the droplet table. You then edit the script entry called `script`.

The script will be treated rather like the proc of a visit verb: it will be called multiple times. On the first call only, `system.droplet.startup` will be `true`, to give the script a chance to perform any initializations; in that same call, the first file to be processed will be in `system.droplet.path`. Then in subsequent calls, successive files to be processed will be in `system.droplet.path`. After all files have been provided, there is one last call, where `system.droplet.closedown` will be `true`. To maintain globals between calls to your script, just store them in `system.droplet`.

When you create a droplet table, it contains a script object called `script` that provides an illustrative minimum framework:

```
if system.droplet.startup
    msg ("starting up")
```

```
if system.droplet.closedown
    msg ("closing down")
local (f = system.droplet.path)
```

Suppose we modify this to read as follows.

Example 29-1. Sample droplet script

```
if system.droplet.startup
    frontier.bringtofront()
    dialog.notify ("starting up")
if system.droplet.closedown
    dialog.notify ("closing down")
else
    local (f = system.droplet.path)
    dialog.notify(f)
```

The resulting droplet, when icons are dropped onto it, brings Frontier to the front and presents a succession of dialogs: the first dialog says **starting up**, then dialogs give the pathnames of the icons dropped onto the droplet, and finally, there is a dialog reading **closing down**. The script explicitly brings Frontier to the front first, because that's where the dialogs will appear.

An entry in the droplet table is named `faceless`. If its value is `false` when the application is built, then when icons are dropped onto the application, a "splash screen" dialog will appear in the Finder, displaying the text from the `helpText` entry of the droplet table and offering a choice of running the droplet, editing it, or quitting it. You are expected to modify `helpText` to make the splash screen informative about what the droplet does.

A non-faceless droplet also has its own menus at the top of the screen. By default, these consist of just a **File** menu with a **Quit** item. To add further menus, create a menubar called `menubar` in the droplet table; its menus will appear at the top of the screen in addition to the **File** menu. While the splash screen is showing, the user might choose from your menu items in order to initialize settings affecting how the droplet will behave.

Dialog-Based Approach

In the dialog-based approach, the `card` entry is not deleted, but is modified to be a MacBird dialog (see Chapter 25, *Dialogs*) that will appear within Frontier. The `script` entry is ignored. All of the pathnames of files and folders dropped on the droplet are available in a list of fileSpecs, `system.droplet.fileList`.

As an example, we'll make a dialog-based droplet which lists in a popup menu the files and folders dropped. The dialog might look like Figure 29-1 in MacBird.

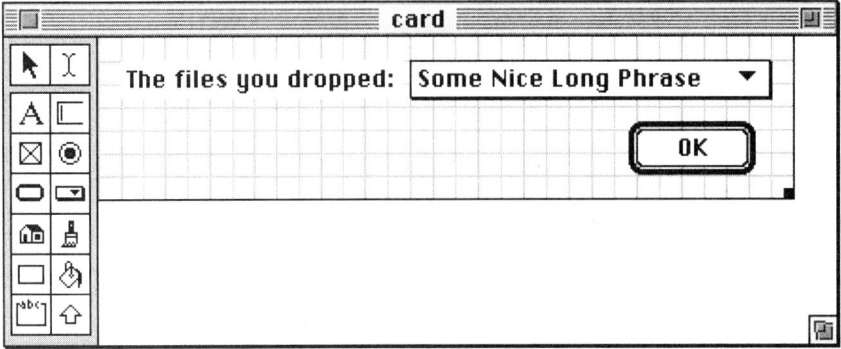

Figure 29-1. Dialog for a droplet

The popup menu is called theMenu, and an initial phrase has been entered into it just to space the layout nicely. The **OK** button has an action script, card.close(); and the card table contains a startCard script, as follows:

```
local (s, f)
for f in system.droplet.filelist
    s = s + string(f) + ";"
s = string.delete(s, sizeof(f), 1)
card.popup.setMenu("theMenu",s)
```

When files and folders are dropped onto the droplet, Frontier comes to the front automatically, unlike with script-based droplets.

Implementation of Exported Objects

The exporting of packed objects and desktop scripts is implemented by scripts in suites.export. The exporting of droplets is implemented by scripts in suites.droplet.

When a packed object or desktop script is opened, Frontier reacts by calling frontier.finder2click(). This passes the call along to **frontier.clickers.type2CLK()**, which imports the script contained in the object and calls it. In the case of a desktop script, this results simply in the execution of the exported script. In the case of a packed object, it results in the execution of a copy of export.callImporter() (or, in the case of an exported menu item, **export. callImportSubmenu()**), which was exported along with the actual object; this results in the calling of export.importer() (or export.importSubmenu()), which imports the object.

When icons are dropped onto a droplet, Frontier receives an Apple event of class 'OHIO', which is routed to the appropriate script within **system.verbs.**

`traps.OHIO`. For more about the Apple event mechanism, see Chapter 34, *Driving Frontier from Outside.*

Backing Up the Database

The condition of the database should be a matter of some concern, for the following three reasons:

The database can become inefficient.
> The database is an extremely space-efficient storage mechanism, but in the course of usage, the amount of unclaimed free space increases, wasting disk space and slowing down database access.

The database can become corrupted.
> Although Frontier's treatment of the database is robust, unexpected errors are possible in any computer application, and corruption of the database can result. UCMDs have access to deep memory and can accidentally harm the database's structure. And you yourself have the power to damage the database directly, as this book constantly points out.

The database can become outdated.
> From time to time, a new version of the database is released by UserLand. When this happens, you will want to use the new version, without losing your customized work from your current copy of the database.

In the rest of this chapter, we discuss strategies for avoiding database trouble, and for being ready if it occurs. We also suggest some work habits that will make protecting and upgrading the database easier.

Backing Up

You can create a backup copy of the database by choosing **Backup** from the **Main** menu. This calls `suites.backups.backuproot()`, which simply makes a Finder copy of the database. The new copy goes into a folder called *backups* in the same folder as Frontier; backups get consecutively numbered names, and old backups are not deleted, so it is up to you clean out the *backups* folder occasionally.

Afterwards, the pathname of the file just created is deposited at `backups.lastfile`. Then, your `people.[user.initials]` table is consulted to see if it contains a script called `customBackup`; if so, it is called with no parameters. How you use this mechanism (a pseudo-hook) is up to you; see Chapter 27, *Agents and Hooks.*

Saving a Copy

Frontier provides a mechanism for saving a compacted version of the database. There are two reasons why this is useful.

First, although making frequent backups provides a measure of security, it is not a panacea. To be sure, the backup preserves entries in case they are overwritten or deleted in the working copy of the database. But if the database has become structurally corrupted in some way, backing up does not eliminate the corruption: whatever problems lurk in the database will lurk in a copy.

On the other hand, saving a compacted copy does help assure the database's integrity, because it causes Frontier to "rethink" the whole structure of the database. This affords a considerably greater measure of security.

Some sense of the database's structural status can be gained by calling `table.validate()`; this checks the internal consistency of a table and all its subtables to infinite depth. However, I'm told that it is possible for certain kinds of lurking problem to pass this test undetected. I'm also told that you can get a false positive if a table contains an entry whose value has not yet been set.

Second, as the database is used, free space opens up in it, and pointers to the free blocks are added to a list called the "avail list," which must be traversed each time Frontier searches the database. This strategy makes saving and accessing the database rapid and robust in the short term, but over time its accumulated effects reduce the database's efficiency.

To examine this aspect of the database, call `window.dbStats()`; the third line of the windoid that appears tells the number of nodes in the avail list, and the fourth line shows the amount of free (meaning wasted) space in the database.

To compact the database, choose **Save a Copy** from the **File** menu. (You can drive this command programmatically with `filemenu.saveCopy()`.) The process may take considerable time, but afterwards both the third and fourth lines of the dbStats window will be 0. Frontier never performs this operation automatically; it is incumbent upon the user to do it from time to time, and it is important to do it, as a way of keeping the database in good condition.

The dialog that appears when you choose **Save a Copy** will offer to overwrite the existing database with the new copy, but this is probably not a good idea; instead, back up the database just in case, save a copy under a different name, such as *Frontier.rootNew*, and then quit Frontier, throw away the old *Frontier.root*, and rename the compacted copy *Frontier.root*.

Batch Export

The most reliable way to assure the integrity of your database material is to export it as a number of packed objects, thus altogether isolating it both from the database and from its structure. This is called a *batch export*. A batch export is also a very good way to transport data between roots, or to migrate into a clean root (discussed later in this chapter).

For this purpose, the `batchExporter` suite is provided. To access it, choose **Batch Exporter** from the **Suites** menu. You then choose **Export** from the **Batch** menu that appears. This causes `suites.batchExporter.batchExport()` to generate packed object files.

The process works like this. A single-level outline at `suites.batchExporter.list` says what tables should be exported. Frontier tries to export the items of each of these tables as a single packed object; if any of the items proves to be a table too large to permit this, Frontier instead performs a folder-style export, representing the table as a folder and trying to export each of its entries as a single packed object—and so on. The result is sort of limited folder-style export. The results of the operation are logged at `user.batchExporter.log`, which you can check afterwards to determine that all went well.

`suites.batchExporter.list` is of interest because it tells you what parts of the database UserLand thinks a user might have modified—implying, *mutatis mutandis*, that the rest of the database should be considered off limits.

people
> You might have put anything at all into `people.[user.initials]`, and you might have imported third-party scripts into other tables.

suites
> You might have written your own suite, or imported a third-party suite.

user
> Any suites and commands maintain important data here, and some suites expect you to store your own customized settings, scripts, and other objects here.

system.agents
> You might have written an agent, or a suite might have installed one.

system.extensions
> You might have written a UCMD, installed a third-party UCMD, or imported an XCMD.

system.menubars
> You might have customized or created a shared menu.

`system.misc`

In `system.misc.menubars`, you might have customized your **=user.initials** menu, added bookmarks, installed suites that changed the suites menu, and so forth; I'm not sure there's any other part of `system.misc` that you would really be free to personalize.

`system.startup`, `system.suspend`, `system.resume`, `system.shutdown`

You or a suite might have added a hook script.

`system.verbs.apps`

You might have built or imported glue for driving an application.

`system.verbs.traps`

You might have added a trap so that Frontier can be driven from outside.

`workspace`

You might have put anything at all here.

Notice that `scratchpad` is not exported, because by definition its contents are expendable.

It is possible that, outside these areas of the database, you might "illegally" have reached deeper into Frontier's functionality to customize it. For example, I don't like the way the **Find & Replace** dialog "forgets" where I put it, so I have modified `search.dialog()` to remember the dialog's location between calls. But, being a built-in verb, `search.dialog()` belongs to UserLand, not to me, and is not included in a batch export. In cases like this, I maintain a copy of the customized script in my `people.[user.initials]` table. That way, my customized version is preserved during a batch export.

Import from Batch Export

To import material exported with batch export, you have a choice of two strategies:

* You could perform a batch import of a whole folder. This is done by choosing **Batch Import** from the **Batch** menu, which asks what folder to import and then calls `suites.batchExporter.batchImport()`. This routine performs a much better folder import than `suites.export.importFolder()`: it doesn't empty out the table into which it imports, and it creates subtables intelligently. It will, however, replace database entries without notifying the user.

* You could import packed objects individually, by opening them in the Finder. This lets you pick and choose what objects to import, and gives you a chance to think twice before deleting an object already present in the database. However, your importing of packed objects can only be as "individual" as the export was; the batch export might have exported as a single packed object a

table of which you want to import only one particular entry. What I like to do, therefore, is this: if an object exists already, I import the packed object into the `workspace` table temporarily, where I can inspect its contents. Then I can copy and paste just the desired entries into their proper place in the database.

Importing from a batch export, no matter how you do it, can be a little tricky. Experience has suggested the following cautions:

- If you do a batch import of the *Suites* folder, when Frontier comes to the packed object file for `suites.batchExporter`, it will use the imported object to overwrite the currently running script. That's very bad. Move that file so it won't be batch-imported.

- Don't open a packed object representing `suites.export` and replace the existing `suites.export`: that, too, overwrites the currently running script.

- Don't do a batch import that includes `system.misc.paths`. In fact, don't ever touch `system.misc.paths`.

- Don't do a batch import that includes `system.misc.menubar`. You can, however, import `system.misc.menubar` by opening it in the Finder and then replacing the existing object.

Clean Root

The database as distributed by UserLand is called a *clean root*. To move all your customized material from your present database to a clean root is called *migrating* to a clean root.

There are two reasons why you might want to migrate to a clean root. First, from time to time UserLand releases a new version of the clean root. At such moments, UserLand usually does also provide an "upgrade" installation to let you upgrade your database in place. But I don't recommend this approach, because it adds to your root without cleaning it out (so that some bug fixes may not "take"), and it sometimes includes extra components that need installing but are not installed automatically. Instead, I always bite the bullet and download the entire clean root, and migrate into it.

Second, you might want to migrate to a clean root because you suspect or want to ward off corruption in your present database. In fact, I keep a copy of the current clean root and migrate into it periodically just as a matter of course.

How to Migrate

The usual first step in migrating to a clean root is to do a batch export from your present database. Then you throw out your present database, make the clean root your *Frontier.root*, and import into it your customized material from the batch-exported files.

This import process can be a little tricky. The clean root may contain improvements from UserLand, so you don't want to replace an object if the imported version would be less up to date than what's already there. For example, you wouldn't want to import your exported `suites.Eudora` if the new clean root comes with better Eudora "glue" than the old one did. On the other hand, perhaps you customized the old Eudora glue somehow, so you'd like to use the new glue but keep your customization.

The problem, in my view, is one of *organization*. You can use the usual techniques to import carefully and conservatively; but if you don't remember what changes you've made to the database, you're afraid of losing those changes if you import too conservatively.

The solution is to have good work habits. Don't make changes to parts of the database that won't be exported in a batch export; if you do, make copies in your `people` table. Try to keep running documentation, in the `people` table, of changes that you make to the database. As you change or create a script, mark it using **Insert Timestamp** in the **Scripts** menu; later, before doing a batch export, you can search the database for your initials to find all such places.

A Different Approach

If you're extremely organized, and have kept track of every database entry you've customized, then instead of using batch export and subsequent import, you could take advantage of the fact that it is possible to have two databases open at once, to have a script copy your customized material from the old database to the new one. See Chapter 30, *Multiple Databases*.

30

Multiple Databases

The Frontier database is a remarkable thing: a hierarchical name–value storage system, with hash-indexed access that stays rapid as tables get large, and with efficient memory management both in RAM and on disk. (In fact, the database is such cool technology that UserLand distributes it separately as a cross-platform library for C and Java applications.*)

As with other things, so with the database: if it's good to have one of them, you might, not unreasonably, want two. *Frontier.root* is the "working" database, constantly being accessed for verbs and other pieces of Frontier's "system." The storage for a sizeable storage task might preferably be separated from the working database, out of considerations of convenience, speed, safety, and size.

For this reason, it is possible to open a second database simultaneously with the first, and communicate with it manually or in UserTalk. This chapter explains the process.

Another option for storing data outside the database is to drive a serious, dedicated database application such as FileMaker or FoxPro, or even a small flat-file database application such as uBASE which is included with the Frontier distribution. This involves inter-application communication, which might mean a speed hit, but such an application might have capacities that a Frontier database does not provide. For example, uBASE provides a flatfile structure of records with identically named fields. A serious database application might provide indexing and rapid search of fields, ability to work with very large numbers of records, relational structure, and so forth. On such matters, see Chapter 32, *Driving Other Applications.*

---

* For more information, and to obtain the database SDK, see *http://www.scripting.com/Gimme5/odb/.*

Manual Interface

You can open a second database manually. There is little point to doing this except to examine or alter data by typing, copying and pasting, and so forth. That's because the second database is very minimal: it has a restricted set of verbs (just the kernel, in `system.compiler`) and no custom menus (just the built-in **File**, **Edit**, and **Window** menus).

You could add scripts and an interface to the second database, but this would be rather against the spirit of the thing. The second database isn't intended as a place to work. It contains sufficient functionality to enable the manipulation of data in an edit window, and no more.*

To create and open a second database manually, choose **New** from the **File** menu; to open an existing second database manually, choose **Open** from the **File** menu. You can perform these actions programmatically with `filemenu.new()` and `filemenu.open()`.

When a second database is opened in this way, there is a second Main Window representing it. This Main Window's popup menu gives access to an agent which shows the pathname of the database; this is useful so that you know which Main Window goes with which database. For instance, you might wish to close the second database by clicking in the go-away box of its Main Window; you'll want to be sure you're closing the right database.

Frontier keeps track of what database owns any window you bring to the front; this determines which database you are "in," and the menus at the top of the screen change accordingly, as does the repertory of verbs available to you.

Programmatic Interface

To work with a second database programmatically, the db verbs are provided. These verbs do not cause any windows to open on the second database, and they communicate with the second database through a completely different mechanism than if the second database were opened manually. (For example, a database that is opened with **Open** from the **File** menu is not necessarily open in the `db.open()` sense.)

References to entries in the second database must be strings, because if they were addresses their validity would be checked against the main database, which might

---

* If for some reason you need a reduced working database, make a copy of your *Frontier.root* and selectively delete entries from it.

not work. These strings must be full paths: they cannot be partial paths, and they cannot be abbreviated by means of `with` constructions. Hence it makes sense to create tables at or close to root level.

To create a database on disk:

    db.new (*pathname*)

To open an existing database:

    db.open (*pathname, readOnly?*)

The advantage of opening a database read-only is that it remains possible for another thread, or another copy of Frontier, also to open it read-only.

To make a new table:

    db.newTable (*pathname, tablePathString*)

To set a database entry value, possibly creating the entry:

    db.setValue (*pathname, entryPathString, value*)

To learn a database entry value:

    db.getValue (*pathname, entryPathString*)

To learn whether a database entry exists:

    db.defined (*pathname, entryPathString*)

To learn whether an existing database entry is a table:

    db.isTable (*pathname, entryPathString*)

To learn the number of entries in a table:

    db.countItems (*pathname, entryPathString*)

To obtain the value of the nth item of a table:

    db.getNthItem (*pathname, entryPathString, n*)

`db.getNthItem()` makes up for the fact that, in the absence of direct references, array notation is impossible.

To remove a database entry:

    db.delete (*pathname, entryPathString*)

To save a database:

    db.save (*pathname*)

`db.save()` returns `true` or `false`, as the database was saved or not. It is well to check this value, just to be certain that Frontier realized the database was dirty. If you call `db.close()` without saving first, changes will be lost.

To close a database without saving:

    db.close (*pathname*)

In this example, we create a second database, make a table at root level called `myTable`, and copy into it all the entries in the main database's `workspace` table.

Example 30-1. Copying a non-scalar between databases

```
local (f = "HD:testing")
try
    db.new(f)
    db.open(f, false)
    db.setValue (f, "root.myTable", workspace)
    db.save(f)
    db.close(f)
else
    db.close(f)
    scripterror(tryerror)
```

The whole operation is embedded in a `try` so that, if something goes wrong, the second database is not left open; this approach is useful when working with second databases, as there is no way to learn the pathnames of open databases.

As mentioned in Chapter 29, *Import/Export*, one can take advantage of the ability to open multiple databases as a way of migrating to a clean root. Imagine that as you customize the database you note the address of each customized entry in an outline at, for example, `workspace.myMigrateList`.[*] So `workspace.myMigrateList` might look like this:

```
people.man.customcommands
system.verbs.apps.MORE
system.verbs.apps.Eudora.examples.man
people.man.notepad
user.hooks.closewindow
```

And so forth. Then, having renamed your old root and migrated to a clean *Frontier.root*, you could run a script like the following.

Example 30-2. migrate()

```
local (x, o, f = "HD:Frontier folder:Frontier.root.old")
on grab (s)
    x = db.getValue (f, s)
    s^ = x
db.open (f, true)
o = workspace.myMigrateList
target.set(@o); op.firstSummit()
grab (op.getLineText())
while op.go(flatdown, 1)
    grab (op.getLineText())
db.close (f)
```

---

[*] Granted, this assumes a degree of self-discipline and organization beyond most people's imagination. However, I owe this example to Doug Baron, who really *is* this organized and really *does* use this technique.

V

Border Crossings

One of the most important aspects of Frontier is its ability to reach out beyond itself into the wider world of your computer. It can get information from the system, and it can manipulate files in many powerful ways. What's more, it can make other programs do its bidding, and respond to other programs when they ask Frontier to do their bidding.

These are the powers of Frontier that make it a command center for your computer as a whole, and around which Frontier was originally designed. Thanks to them, UserTalk scripts can bring to bear the specialized abilities of other programs, building systems where programs perform automated tasks, operate in concert, and exchange data.

This section is exclusively about UserTalk verbs, but even readers less concerned with UserTalk may wish at least to skim these chapters for what they reveal about the sorts of thing that Frontier is capable of doing.

This section also talks about how to run AppleScript scripts inside Frontier, how to extend UserTalk with OSAX code fragments, and how to edit Frontier objects, such as wptexts, with other applications.

31

Driving the System

UserTalk can query the operating system to find out what's happening at system level. What time is it? Has the user clicked the mouse? What programs are running? It can make system-level events occur, such as noises from the computer's speaker. Some verbs obtain system-level information about Frontier itself, such as how much free memory it has. And UserTalk can drive the computer's file system, to get or set file information, or to create, copy, or delete files.

This chapter lists the verbs that do these things.

Frontier

To learn how much free memory Frontier's heap contains:[*]
```
memAvail ()
string.memAvailString ()
```

To learn the pathname of the Frontier application file:
```
frontier.getProgramPath ()
```

To learn the pathname of the frontmost database:
```
frontier.getFilePath ()
```

To learn the version number of the Frontier database:
```
frontier.version ()
```

To learn the version number of the Frontier application, as opposed to the database:
```
file.getVersion(frontier.getProgramPath())
```

---

[*] It also possible to put a strain on Frontier's stack size, quite without regard to the value of memAvail(), which measures the heap. This might occur, for instance, if you call subroutines to a great depth, or import a table with very deep table nesting. There is no verb for directly enlarging the stack size, but a desktop script, *SetFrontierStackSize*, lets you do it.

System

UserTalk's system-level powers are here categorized as follows: noticing when the user holds a modifier key or manipulates the mouse; getting and setting the contents of the clipboard; learning what processes are running and which is front-most; making sounds from the speaker and manipulating the cursor; and obtaining some miscellaneous information such as RAM memory and clock time.

User Events

To learn whether the user is holding down a modifier key:
```
kb.cmdKey ()
kb.controlKey ()
kb.optionKey ()
kb.shiftKey ()
```

The kb verbs are often used to give a menu item a slightly different functionality if a modifier key is held down when it is chosen.

To learn whether the user is holding down the mouse button:
```
mouse.button ()
```

To learn the present location of the cursor:
```
mouse.location ()
```

`mouse.location()` is supposed to measure from the top left corner of the front-most window, but because of a bug, it usually measures from the top left corner of the Main Window instead. There are two workarounds. One is to call `window.frontmost()` immediately before calling `mouse.location()`; this disables the bug temporarily. The other approach is to accept the bug and compensate by calculation. For example, the following script displays the mouse's position in global coordinates continually for 20 seconds:

```
local(start = clock.ticks(), x, y, xx, yy)
loop
    point.get(mouse.location(), @xx, @yy)
    window.getPosition("Frontier.root", @x, @y)
    msg((x+xx) + ", " + (y+yy))
    clock.waitsixtieths(1)
    if clock.ticks() - start > 60 * 20
        break
```

To learn how long the user has been "idle":
```
frontier.idleTime ()
```

Clipboard

The clipboard commands move data between an object and the clipboard. They are intended primarily for use before or after a context switch: for example, to get some data into a Microsoft Word document, we might put it on the clipboard, switch to Microsoft Word, and tell Word to paste it.

This clipboard is technically known as the *system scrap*. It is not quite the same as the place where data is copied and pasted within Frontier; Frontier maintains what is known as a *private scrap*.* The private scrap allows Frontier to maintain internally copied data in its own special formats. The system scrap's data, on the other hand, must be in a universally usable format—usually either TEXT or PICT.

When there is a context switch to or from Frontier, or when you ask for the contents of either scrap, Frontier reconciles the two scraps, so that they both contain whatever most recently went into either one of them. If the private scrap was more recently changed, Frontier reconciles the scraps by placing onto the system scrap a text version of the private scrap.

Unfortunately, there's a slight bug. If a clipboard verb changes the system scrap and there is then a context switch, Frontier sometimes fails to realize that the system scrap is more recent than the private scrap, and overwrites the system scrap with a text version of the private scrap. I have found that a reliable workaround is to read from the system scrap by saying `clipboard.getValue('TEXT')` just before changing it.†

To move an object's value onto the system scrap:

```
clipboard.put (MacOSFormat, addrObject)
```

Coercion from Frontier's internal formats is *not* performed automatically when you move data directly onto the system scrap; you must coerce explicitly beforehand. For example, this is *not* the way to put the text version of an outline onto the system scrap:

```
clipboard.put ('TEXT', @workspace.notepad) « this is not going to work!
```

If you say this, and then switch to Microsoft Word and paste, you'll see some extra "garbage." Instead, say this:

```
local (o = string(workspace.notepad)) « convert
clipboard.put ('TEXT', @o)
```

To move a value onto the system scrap, setting its type automatically:

```
clipboard.putValue (value)
```

---

* For more about the scrap and the difference between the system scrap and a private scrap, see *http://devworld.apple.com/dev/techsupport/insidemac/MoreToolbox/MoreToolbox-110.html.*

† To see the bug, select some text in a wptext, copy it, and paste it. Then, in Quick Script, execute `clipboard.putValue("haha")`. Switch to another application such as SimpleText, and paste. The result will be what you copied from the wptext, not `"haha"`.

`clipboard.putValue()` is a convenient utility, because *value* can be a literal or the result of a verb call—it doesn't have to be put into an object, as with `clipboard.put()`. So, we might rewrite the previous example:

```
clipboard.putValue (string(workspace.notepad))
```

To move the system scrap's contents into a binary object:
```
clipboard.get (MacOSFormat, addrObject)
```

To return the system scrap's contents:
```
clipboard.getValue (MacOSFormat)
```

The reason why both `clipboard.get()` and `clipboard.getValue()` require that *MacOSFormat* be supplied is that the system scrap can contain more than one piece of data, provided they are in different formats.* Thus it is necessary to specify which piece of data is desired. `clipboard.get()` returns a binary; if (as is usually the case) the desired scrap data is text, `clipboard.getValue()` is usually more convenient, because its result will be text. For an example of `clipboard.get()` in action, see Example 28-1.

Processes

These verbs have to do with applications that are actually running; verbs that deal with applications as files are discussed later in this chapter. Starting up and quitting processes is discussed later in this chapter, and also in Chapter 32, *Driving Other Applications*.

Process identifier parameters are very flexible about the form of the identifier: it can be the name of the process, the pathname of the application file, or the application's string4 creator code. The creator code is frequently the most convenient, especially since an application's name can contain surprises such as trademark symbols, version numbers, and so on; for example, it is easier to type `'LAND'` than `"UserLand Frontier™"`.

To learn whether a particular process is running:
```
sys.appIsRunning (processIdentifier)
```

To learn what process is frontmost:
```
sys.frontmostApp ()
```

To learn how many processes are running:
```
sys.countApps ()
```

`sys.countApps()` includes faceless background applications. To obtain a count of non-faceless applications (and excluding the Finder itself), ask the Finder for its count, like this:

---

* For example, copying text in SimpleText puts two pieces of data onto the clipboard, a `'TEXT'` and a `'styl'`.

```
with finder
    count (objspec(nil), process)
```

To learn the name of the nth process:
```
sys.getNthApp (n)
```

To bring a particular process to the front:
```
sys.bringAppToFront (processIdentifier)
```

Frontier can also be brought to the front with **frontier.bringToFront()**, and scriptable applications with "glue" have a **bringToFront()** verb of their own; these also start up the application if it isn't running (see Chapter 32).

To learn the pathname of the application file for a running process:
```
sys.getAppPath (processIdentifier)
```

Sound and Cursor

To play the system alert sound:
```
speaker.beep ()
```

To play a brief, high-pitched "blip" sound:
```
speaker.ouch ()
```

To play a customized square-wave sound:
```
speaker.sound (duration, amplitude, frequency)
```

For an elaborate use of **speaker.sound()**, see **suites.samples.basic-Stuff.rimshot()**. There seems to be some sort of logarithmic relationship between *frequency* and what one normally thinks of as frequency. The following two notes are an octave apart:

```
speaker.sound (10, 200, 1000); speaker.sound (10, 200, 10000)
```

and to express a note between them, you express it as a fraction of the distance between 1000 and 10000. So, for example, the following routine plays a major scale:

```
theList = {0, 2, 4, 5, 7, 9, 11, 12}
for x in theList
    speaker.sound(1, 200, 1000 + 9000 * x / 12)
    clock.waitsixtieths(20)
```

To play a sound resource:
```
speaker.playNamedSound (name)
```

When using **speaker.playNamedSound()**, the resource must be in the resource chain already; see Example 20-4.

To put up Frontier's "turning beachball" cursor:
```
rollBeachball ()
```

Miscellaneous

To learn (in bytes) how much free RAM memory there is:

```
sys.memAvail ()
```

To learn what version of the system is running:

```
sys.OSVersion ()
```

To learn or set the date and time on the system clock:

```
clock.now ()
clock.set (dateTime)
```

To learn how many "ticks" have elapsed since the computer was started up:

```
clock.ticks ()
```

`clock.ticks()` is useful for measuring the time elapsed between two moments of script execution:

```
local (startTime = clock.ticks())
« do some operation here
local (elapsedTime = clock.ticks() - startTime)
```

To get an accurate measurement, it is usual to repeat the operation some large number of times, since ticks are imprecise, and are a very gross unit in comparison to the speed of computer operations.

Another useful trick is to use `clock.ticks()` to generate identifiers that are guaranteed distinct. For example, to save several files under different names, one could name each of them `"temp"+clock.ticks()`.

To query the system using the Gestalt Manager:[*]

```
gestalt (selector)
```

File System

For verbs that read and write the resource and data forks of files, see Chapter 19, *Datafiles*, and Chapter 20, *Resources*.

The verbs described in this section drive the filing system at system level; but since the Finder is a scriptable application, and since the Finder, too, drives the filing system, Frontier can also do things with files by way of the Finder. See Chapter 32.

---

[*] For more information about the Gestalt Manager, see *http://devworld.apple.com/dev/techsupport/insidemac/OSUtilities/OSUtilities-9.html*. A number of selectors are listed there; however, the selector list is extensible, so for the most complete list, see *http://www.bio.vu.nl/home/rgaros/gestalt/*.

Pathnames of Files

The pathname of a file is a string identifying the file as a series of elements—the volume name, the names of nested folders, and the filename—delimited by a file separator character, which on Mac OS is a colon.

Strictly speaking, a volume or folder pathname should end in a colon; but in actual fact, Frontier is often quite forgiving if the final colon is omitted. This is because when you supply a string pathname, it is coerced to a fileSpec before use, and the coercion process sometimes supplies the missing final colon. For the same reason, file pathnames are not case-sensitive.

Frontier maintains a pathname prefix, which it will prefix to any one-element file pathname (not ending in a colon) in an attempt to resolve the pathname. When Frontier starts up, this pathname prefix is the folder containing the Frontier application. For example, on my computer it is possible to refer to the database file as `"frontier.root"`; the file is in the same folder as the Frontier application, so Frontier is able to resolve the partial pathname.

To learn or set the current pathname prefix:
```
file.getPath ()
file.setPath (pathString)
```

The fact that there is a pathname prefix, and the fact that Frontier is forgiving about the final colon, can combine to give unexpected results. On the whole it is best to be strict with oneself about the colon. To learn how Frontier will interpret a pathname string, coerce it to a fileSpec in Quick Script.

See Chapter 14, *Strings and Chars*, on how to extract elements from a pathname string. See Chapter 25, *Dialogs*, for dialogs that let the user choose a file.

Opening Things

There are various sorts of things the system can open, and various ways in which they can be opened. The launch verbs, described in this section, take advantage of this "open" metaphor; they cause the item to open in the sense applicable to that item. See also Chapter 32.

To open a Finder object:
```
launch.anything (pathname)
```

To open a control panel:
```
launch.controlPanel (name)
```

To open an Apple menu item:
```
launch.appleMenu (name)
```

Since, under System 7, control panels and Apple menu items are double-clickables, `launch.controlPanel()` and `launch.appleMenu()` do nothing you couldn't do yourself with `launch.anything()`.

To start up an application:
```
launch.application (pathname)
```

To start up an application with a particular document:
```
launch.appWithDocument (appPathname, docPathname)
```

To make certain that an application is running, `launch.application()` and `launch.appWithDoc()` are much better choices than `launch.anything()`. They are faster, they don't bring the application to the front, and they return `true` immediately if the application is already running. Also, they operate interactively with the launch process so as not to return until either the application has fully opened or an error comes back from the system. `launch.anything()`, on the other hand, returns immediately, with no ability to learn from the system how things went or when the launch process was over. These verbs should perhaps be embedded in a `try`, to make absolutely certain all is well.

To start up an application given its creator code:
```
launch.usingID (string4)
```

`launch.usingID()` is a utility script which is not recommended, because it calls `launch.anything()`. You're better off resolving the ID yourself:

```
try
    launch.application (file.findApplication (theID))
```

To run an FKEY or other code resource:
```
launch.resource (typeString4, IDnumber)
```

Volumes

To learn how many volumes are currently mounted:
```
file.countVolumes ()
```

To obtain a list of the names of currently mounted volumes, use `fileloop()`, as follows:

```
local (theList = {}, f)
fileloop (f in "")
    theList[0] = f
« now the list is in theList
```

To mount a remote server volume:
```
file.mountServerVolume (path, userID, password)
```

To learn whether a volume is ejectable:
```
file.isEjectable (volumePathname)
```

To eject an ejectable volume:

    `file.eject (`*`volumePathname`*`)`

To unmount a volume:

    `file.unmountVolume (`*`volumePathname`*`)`

To learn (in bytes) the size of a volume:

    `file.volumeSize (`*`volumePathname`*`)`

To learn (in bytes) how much of a volume is occupied:

    `file.bytesOnVolume (`*`volumePathname`*`)`

To learn (in bytes) how much of a volume is free:

    `file.freespaceOnVolume (`*`volumePathname`*`)`

As might be expected, the following is always **true**:

    `file.freespaceOnVolume() + file.bytesOnVolume() == file.volumeSize()`

To learn the block size of a volume:

    `file.volumeBlockSize (`*`volumePathname`*`)`

Under the Mac OS HFS filing system, the larger a volume, the larger the block size. The significance of block size is that a file takes up a whole number of blocks on disk, so the smallest amount of disk a file can occupy is `file.volumeBlockSize()` (twice `file.volumeBlockSize()` if the file has a resource fork). This wastes huge amounts of space on large disks containing many small files.[*]

To learn the number of files and folders on a volume:

    `file.filesOnVolume (`*`volumePathname`*`)`

To learn the number of folders on a volume:

    `file.foldersOnVolume (`*`volumePathname`*`)`

Folders

To learn the total number of bytes occupied by all files in a folder:

    `file.bytesInFolder (`*`folderPathname`*`)`

To learn the number of files and folders in a folder:

    `file.filesInFolder (`*`folderPathname`*`, `*`depth`*`)`

To learn the number of folders in a folder:

    `file.foldersInFolder (`*`folderPathname`*`, `*`depth`*`)`

These verbs are utility scripts dating back to a time before the Finder was scriptable; they use a time-consuming technique of recursively traversing every file

---

[*] As this book goes to press, Apple is preparing to rectify this situation with the release of System 8.1 and HFS+.

individually. It is now far faster to drive the Finder to get the same information (see Chapter 32). Here is the faster Finder equivalent of `file.bytesInFolder()`.

Example 31-1. folderSize()

```
on folderSize (path)
    with objectmodel, finder
        return (get (folder[path].size ))
```

The following is the faster Finder equivalent of `file.filesInFolder()`. In this version, folders are not counted, and there are only two choices of depth.

Example 31-2. filesInFolder()

```
on filesInFolder (path, infiniteDepth)
    with objectmodel, finder
        if not infiniteDepth
            return (count (folder[path], file ))
        return (count (folder[path].entirecontents, file ))
```

It is easy to modify this to count folders instead of files.

System Folder

To obtain the pathname of the startup volume:
```
file.getSystemDisk ()
```

To obtain the pathname of the active system folder:
```
file.getSystemFolderPath ()
```

To obtain the pathname of any system folder's standard subfolders:
```
file.getSpecialFolderPath (volumePathname, folder, create?)
```

File Information

Changes made to files with these verbs take place at system level. The Finder may take a while to hear about them, so if a Finder window containing the file is already open, it may not immediately reflect the changes; to force it to do so, call `finder.update()` with one parameter, the pathname of a file whose changes you want the Finder to reflect immediately.

To learn (in bytes) the size of a file:
```
file.size (pathname)
```

`file.size()` returns the so-called "logical size," the sum of the total bytes of data in the file's resource fork and data fork—not the "physical size," the amount of space it occupies on the disk. For the reason why these two numbers differ, see the earlier description of `file.volumeBlockSize()`. You can obtain a file's physical size by asking the Finder.

Example 31-3. physicalSize()

```
on physicalSize (pathname)
    with objectModel, finder
        return get (item[pathname].physicalsize)
```

To obtain the size of just the data fork of a file, one way is to say this:

```
sizeOf (toys.readWholeFile (pathname))
```

To learn or set the minimum or normal RAM partition for an application:

```
sys.getMinAppSize (pathname)
sys.setMinAppSize (pathname, bytes)
sys.getAppSize (pathname)
sys.setAppSize (pathname, bytes)
```

To learn or set a file's creator code:

```
file.creator (pathname)
file.setCreator (pathname, string4)
```

To learn or set a file's type code:

```
file.type (pathname)
file.setType (pathname, string4)
```

To obtain the pathname of the application associated with a given creator code:

```
file.findApplication (string4)
```

To learn or set a file or folder's created date:

```
file.created (pathname)
file.setCreated (pathname, date)
```

To learn or set a file or folder's modified date:

```
file.modified (pathname)
file.setModified (pathname, date)
```

Modifying a file does not change its containing folder's modified date. This may not be what you want. A utility script, `toys.touchPath()`, changes the modified date to `clock.now()` on a file and all its enclosing folders.

To learn or set a file or folder's comment:

```
file.getComment (pathname)
file.setComment (pathname, commentString)
```

To learn or set a file's version string:

```
file.getVersion (pathname)
file.setVersion (pathname, versionString)
```

To learn or set a file's full version string:

```
file.getFullVersion (pathname)
file.setFullVersion (pathname, versionString)
```

To learn or set a file or folder's label:
```
file.getLabel (pathname)
file.setLabel (pathname, labelString)
```

To learn or set the state of a file or folder's visibility:
```
file.isVisible (pathname)
file.setVisible (pathname, visible?)
```

To learn or set a file or folder's locked state:
```
file.isLocked (pathname)
file.lock (pathname)
file.unlock (pathname)
```

To learn or set the state of a file's bundle bit:
```
sys.hasBundle (pathname)
sys.setBundle (pathname, set?)
```

To learn or set the state of any of a file or folder's finder flags:
```
finderflags.get (pathname, whichBit)
finderflags.set (pathname, whichBit)
finderflags.clear (pathname, whichBit)
```

It is unlikely that one will want to manipulate Finder flags directly, and the most important ones are accessible through other commands.[*]

To learn or set the icon position of a file or folder:
```
file.getIconPos (pathname, addrHoriz, addrVert)
file.setIconPos (pathname, horiz, vert)
```

The icon position in these commands is measured with reference to some point I can't figure out. It is easier to drive the Finder to get and set icon positions; that way, the icon position is measured from the top left of the window's content area, as expected.

File Manipulation

Commands that can overwrite or delete a file cannot be undone and should be used with utmost care.

Moving, copying, and renaming

To copy a file or folder:
```
file.copy (source, dest)
```

To copy a file or folder selectively:
```
file.filteredCopy (source, dest, addrProc)
```

To move a file or folder within its volume:
```
file.move (source, destFolder)
```

To copy a file's data fork or resource fork to a new file:
```
file.copyDataFork (source, dest)
file.copyResourceFork (source, dest)
```

---

[*] The Finder flags are listed at this URL: *http://devworld.apple.com/dev/techsupport/insidemac/Toolbox/ Toolbox-464.html.*

To "drop" a copy of a file into the system folder:
```
file.copyToSystemFolder (source)
```

To rename a file or folder or volume, in place:
```
file.rename (pathname, nameString)
```

Deleting

To delete a file:
```
file.delete (pathname)
```

To delete a folder even if it has contents:
```
file.deleteFolder (pathname)
```

`file.deleteFolder()` is a utility script which calls `file.delete()` in a simple, elegant recursive structure worthy of study.

To delete all files and folders in a folder:
```
toys.emptyFolder (pathname)
```

Creation and existence

To create a new empty file:
```
file.new (pathname)
```

To create an alias to a file or folder or volume:
```
file.newAlias (original, dest)
```

To create a new empty folder:
```
file.newFolder (pathname)
```

To create a new empty folder without raising an error if the folder exists:
```
file.sureFolder (pathname)
```

To create all folders along a proposed file pathname:
```
toys.sureFilePath (pathname)
```

All other verbs that create files and folders require that all the folders along the destination pathname exist already. By first passing the pathname to `toys.sure-FilePath()`, you can proceed to create the file or folder, secure in the knowledge that all those folders now do exist.

To learn whether an existing item is a folder or volume:
```
file.isFolder (pathname)
file.isVolume (pathname)
```

To learn whether an existing item is an alias:
```
file.isAlias (pathname)
```

To test whether a file or folder or volume exists:
```
file.exists (pathname)
```

`file.exists()` is an important test, because many other file commands raise an error if the item doesn't exist.

Miscellaneous

To obtain the pathname of an alias's original:
```
file.followAlias (pathname)
```

Since Frontier can resolve aliases, they provide a way to maintain, in the Finder, a sort of physical analogue to a list. For an example, see `suites.samples.basic-Stuff.nightlyBackup()`, where aliases tell the script what files to back up and where to back them up.

To learn whether a file is "busy" (open for use by an application):
```
file.isBusy (pathname)
```

Some verbs in this chapter that make changes to a file cannot make them if the file is busy; the application that has it open for use locks out other users.

Testing whether a file is busy is also useful as a way of discovering whether an external process has come to an end. For example, I once used Frontier to drive a non-scriptable application to open and process several files in succession. The process on each file took about an hour. Frontier needed some way of knowing when the application was finished with each file, so that it could make the application go on to the next one; because the application was non-scriptable, it had no way of letting Frontier know this. The solution was a loop which checked once a minute to see if the file was still busy.

To learn whether the dataforks of two files are identical:
```
file.compare (pathname1, pathname2)
```

Nomenclature Woes

Nowhere is the lamentable hodge-podge inconsistency of Frontier's verb names more evident than in connection with the file verbs. Consider the following get/set pairs:

```
file.getLabel()                    file.setLabel()
file.modified()                    file.setModified()
file.isVisible()                   file.setVisible()
file.isLocked()                    file.lock(), file.unlock()
```

How simple it would have been to have named these verbs consistently:

```
file.getLabel()                    file.setLabel()
file.getModified()                 file.setModified()
file.getVisible()                  file.setVisible()
file.getLocked()                   file.setLocked()
```

Instead, one must deal with a different set of expressions for each function, and as a result the verb names are almost impossible to remember. Personally, I typically spend more time in the `file` table looking up the verb names than I do actually writing code that uses them.

32

Driving Other Applications

UserTalk can drive applications other than Frontier, sending them commands and asking them questions. This means that Frontier can make other applications do its bidding, and can take advantage of their specialized abilities to perform actions and manipulate data in ways that it cannot do by itself.

For example, Frontier is not a relational database program, but it can drive File-Maker Pro, so it can work with a relational database just as if it were. Frontier is not a spreadsheet program, but it can drive Excel, so it can work with a spread-sheet just as if it were. And so on.

This chapter explains how other applications are driven with UserTalk scripts. You might be wishing to drive another application yourself, or you might be trying to understand an existing script which drives another application. (Any verb that lives in a table within `system.verbs.apps` probably drives another application.)

Some readers may have experience driving other applications with AppleScript. Frontier scripts can be written in AppleScript instead of UserTalk; see Chapter 33, *AppleScript.*

How to Use This Chapter

How you use this chapter depends on just how much you want to learn. The minimum that you, the UserTalk programmer, will need to do in order to write scripts that drive other applications will be merely to put into your scripts some UserTalk commands such as those you are already used to. Just as you would call `dialog.notify()` or `file.new()`, now you call `Eudora.createMessage()` or `Filemaker.find()`. Such verbs are called *glue*. That's because they act as a link between you, the UserTalk programmer, and what Frontier is really doing, which is communicating with those other applications using structured system-level messages called *Apple events*.

You will also have to learn some new UserTalk syntax. That's because Apple events have their own way of specifying to an application what feature of that application's world it should operate on or report information about. Such a feature is called an *object*. For example, in FileMaker, an object might be the "Name" field of the second record currently being browsed. In Excel, it might be cell B2 in the frontmost spreadsheet. Such objects are specified using the *object model*, and UserTalk has special object model syntax that allows you write object specifiers to use as parameters to glue verbs.

This chapter starts out with an explanation of Apple events and how UserTalk sends them. This is the nitty-gritty level of communicating with other applications, and you do not have to know about it in order to use glue verbs. In fact, the whole idea of glue verbs is that you don't have to worry about Apple events. So, if you like, you can skip or skim the discussion of Apple events. You might wish to learn the details of Apple events later, so that you can modify glue verbs or write some of your own, or just because you want a deeper understanding of how glue verbs work.

Next, there is a brief discussion of four sets of UserTalk glue verbs which you will rarely, if ever, need to call directly, because their purpose is mostly to be called by other glue verbs. They are the required verbs, the core verbs, the misc verbs, and the app verbs. Every reader can skip or skim this section; it is provided only for completeness, and in any case the details are relegated to a chapter in the reference section of this book, Chapter 47, *Apple Event Suites*.

Then we come to the parts of the chapter that you should read, at a minimum, if you want to know how to drive other applications with UserTalk. Object model syntax is explained, and finally, there is discussion of glue verbs.

You may be surprised to find that this chapter doesn't explain in detail how to drive any particular application. For example, there's no full-scale discussion of driving FileMaker Pro or Excel. That's partly because to talk about each application completely would require a huge book in itself. There are lots and lots of scriptable applications, and the details are completely different for each one. What this chapter does explain is how to study and experiment with the glue verbs so that you can learn more about driving any particular application on your own.

Non-Scriptable Applications

Frontier cannot drive just any application. That's because not every application knows how to receive Apple events. An application that deliberately exposes its functionality with a defined repertoire of Apple events to which it is prepared to respond is called *scriptable*.

Otherwise, the application is not scriptable, and Frontier cannot drive it. But Frontier may be able to drive some other application which can. There exist various commercial "macro" programs which can simulate user actions in just about any application, doing such things as choosing from menus, typing phrases, and clicking buttons. This is often sufficient to accomplish what you need.

The best such "macro" program for use with Frontier is probably Player, from PreFab Software, Inc.* It is a faceless background application intended to be driven from Frontier, and both the glue and the extensive help are beautifully implemented in UserTalk.

Other scriptable macro programs include OneClick, KeyQuencer, and QuicKeys.†
If you own one of these, you can drive it with Frontier to help you script the unscriptable.

Apple Events

This section explains the basic structure of Apple events, the system-level messages whereby processes communicate with one another on Mac OS, and the UserTalk verbs that send them.

Anatomy of an Apple Event

We will learn about Apple events by analyzing one Apple event;‡ and we will obtain our sample Apple event by intercepting it as it flies through the system. One of my favorite ways to do this is with a control panel called Capture AE.§ When it is open, Capture AE watches the system for Apple events, and decodes them into a text window.

Let's get the Finder to send an Apple event. To do this, we'll turn on Apple event recording: this means that if we perform an action manually in the Finder, it will be "echoed" through the system, showing us the Apple event that could have been sent to make the Finder perform that same action. Our action will be: show the current window in Name view.

---

* See *http://www.prefab.com/player.html*.

† For OneClick, see *http://www.westcodesoft.com/OC-Overview.html*. For KeyQuencer, see *http://www.binarysoft.com/kqmac/kqmac.html*. For QuicKeys, see *http://www.cesoft.com/quickeys/qkhome.html*.

‡ For even more information about Apple events, see *http://devworld.apple.com/dev/techsupport/insidemac/IAC/IAC-2.html*.

§ Available, for example, at *ftp://ftp.westcodesoft.com/OneClick_Scripting_Tools/CaptureAE.sit.hqx*.

So, I open Capture AE. Then I switch to Frontier, open any script edit window and press the **Record** button; this turns on Apple event recording. Now I switch to the Finder, open a window, and, from the **View** menu, choose **Name** view. Now I go back to Frontier and press the **Stop** button, to turn off recording. Now I go to Capture AE, copy out the text it contains, and close Capture AE so it stops intercepting Apple events.

Now we can paste the text of the Apple event we've captured into a Frontier wptext, and study it. Here it is:

```
Process("Finder").SendAE "core,setd,'----':obj {want:type(prop), from:obj
{want:type(prop), from:obj {want:type(cfol), from:obj {want:type(cdis),
from:'null'(), form:name, seld:"book disk"}, form:name, seld:"book folder"},
form:prop, seld:type(cwin)}, form:prop, seld:type(pvew)}, data:pnam"
```

Pretty horrible! Its structure will be clearer, though, if we rewrite it in a sort of outline form, like this:

```
Process("Finder")
    core,setd
    '----':obj {
        want:type(prop)
        from:obj {
            want:type(prop)
            from:obj {
                want:type(cfol)
                from:obj {
                    want:type(cdis)
                    from:'null'()
                    form:name
                    seld:"book disk"}
                form:name
                seld:"book folder"}
            form:prop
            seld:type(cwin)}
        form:prop
        seld:type(pvew)}
    data:pnam
```

Think of our Apple event as a kind of verb. The first line just says what application the verb is directed to, like the address on an envelope. What follows consists of a verb name and the parameters that that verb expects. The second line is the verb name, which has two parts: it is the verb `'core'/'setd'`. Every Apple event is identified in this two-part way, somewhat as a Frontier verb is called `dialog.notify`.

Everything else is the parameters of the `'core'/'setd'` verb. These parameters are expressed like a UserTalk record: a series of name-value pairs, where all the names are string4s and the name is separated from the value by a colon. Some of the values are themselves records, so they have curly braces around them.

A 'core'/'setd' verb takes two parameters, whose names are '----' and 'data'. (The chief parameter of an Apple event is very often named '----'; this is known as the *direct object*.) The verb itself means: set the value of whatever is specified in the '----' parameter to whatever is in the 'data' parameter.

The value in the 'data' parameter is 'pnam', which is Apple event code for "name"—we set the window to Name view, remember?

But what is the value of the '----' parameter? It's the whole indented "bundle" in curly braces. Now, just before the left curly brace we see the term obj. This means that the value of this parameter is packaged up as an *object specifier*, which is a way of designating a particular feature of the application's world—in other words, it specifies one of its objects! So now we need to understand about object specifiers.

An object is specified by saying which object of its type it is, in relation to its container (which is usually some other object). An object specifier has four parts:

* 'want' tells what class of object it is.

* 'seld', the selector, tells which member of that class the object is, typically in relation to the object's container. *

* 'form' tells what datatype the 'seld' is; to a human being this seems obvious, but a computer has to be told explicitly so that it can read the 'seld'.

* 'from' tells what the container is; this is usually another object specifier, until we reach the top of the hierarchy, at which point the 'from' is null.

To give an analogy: if I ask you what is the fourth word of this sentence, you can answer precisely. That's because you know I'm asking for a *word* (the 'want'), that it's the *fourth* word (the 'seld'), that "fourth" is a *numerical* index (the 'form'), and that it's the fourth word *of* that particular sentence (the 'from').

So now we can see that the direct object is made up of a series of nested object specifiers, and we understand why: in each case, the 'from' is another object, climbing through the application's object hierarchy until we reach the top.

And now it's easy to read the direct object: it's the 'pvew' (view), of the 'cwin' (window), of the 'cfol' (folder) whose name is *book folder*, of the 'cdis' (disk) whose name is *book disk*. That's the window whose view I changed to Name view.

---

* Classes are actually of two types, and the 'seld' works in two different ways. An *element* is a class which the container might have many of, so the 'seld' specifies which element (e.g., the *fourth* word). A *property* is a class of which the container has only one, in which case the 'want' is always 'prop', and the 'seld' specifies what property it is (e.g., the word's *length*).

How UserTalk Constructs Apple Events

To make Frontier send an Apple event, UserTalk has the `appleEvent()` verb. This verb takes parameters which are laid out just like an Apple event, namely:

- An application ID, telling what application the Apple event is to be sent to

- The type of the Apple event (the first part of the name of the "verb")

- The subtype of the Apple event (the second part of the name of the "verb")

- Any number of name-value pairs, submitted as separate parameters: in other words, a parameter which is a string4 name, and then a parameter which is its value, and so on

So, imagine that we wanted to send the above Apple event, constructing the `appleEvent()` call ourselves. Here is a partial sketch of its form:

```
appleEvent( 'MACS', 'core', 'setd', '----', ????, 'data', 'pnam')
```

Our main problem is that we don't know how to denote an object specifier; so I've put *????* where it should go.

One way to make an object specifier is to use the verb `setObj()`, which takes four parameters corresponding to the `'want'`, the `'from'`, the `'form'`, and the `'seld'`. Knowing this, we can build the whole `appleEvent()` call directly from our model (`null` is coded here simply as 0):

```
appleEvent('MACS', \
    'core' ,'setd', \
    '----', setObj ( \
        'prop', \
        setObj ( \
            'prop', \
            setObj ( \
                'cfol', \
                setObj ( \
                    'cdis', \
                    0, \
                    'name', \
                    "book disk"), \
                'name', \
                "book folder"), \
            'prop', \
            'cwin'), \
        'prop', \
        'pvew'), \
    'data', 'pnam')
```

If we run this script, though, we get an error. That's because of the last line. In an `appleEvent()` call, values in name-value pairs are sent with the datatype they actually possess.* So we're sending `'pnam'` as a string4; but what the Finder actually expects is an object specifier, an `objSpecType`. So we coerce it, with `objSpec()`. Our script now looks like this:

```
appleEvent('MACS', \
    'core' ,'setd', \
    '----', setObj ( \
        'prop', \
        setObj ( \
            'prop', \
            setObj ( \
                'cfol', \
                setObj ( \
                    'cdis', \
                    0, \
                    'name', \
                    "book disk"), \
                'name', \
                "book folder"), \
            'prop', \
            'cwin'), \
        'prop', \
        'pvew'), \
    'data', objSpec('pnam'))
```

Guess what—it works! Executing this script really does cause the Finder to change that particular window to Name view.

That's the basic way to code an Apple event in UserTalk. But a nice thing about UserTalk is that you can express parameters as variables, which can make the Apple event much easier to read. To illustrate, we recode our Apple event, using variables to build the object specifier in stages:

```
theDisk = setObj ('cdis', 0, 'name', "book disk")
theFolder =  setObj ('cfol', theDisk, 'name', "book folder")
theWindow = setObj ('prop', theFolder, 'prop', 'cwin')
theView = setObj ('prop', theWindow, 'prop', 'pvew')
appleEvent('MACS', 'core' ,'setd', '----', theView, 'data', objSpec('pnam'))
```

Basic Object Model Syntax and Glue

In real life there is no need to use `setObj()` at all; UserTalk permits us to construct an object specifier directly, using *object model syntax*. In the case of elements, the object class is followed by the selector in square brackets, as in

---

* The UserTalk datatypes are defined to correspond to the Apple event datatypes. If a list is expected, a UserTalk list can be used. If a file specifier is expected, a UserTalk fileSpec can be used. And so on. If the type is `binaryType`, the value is sent with the binary's internal datatype.

array notation; containment is indicated as a container followed by a dot followed by what is contained, as in a database object reference. So, here's the schema:

```
class[selector].class[selector].class[selector]
```

A property is just the same, except that there's no selector. A schema might look like this:

```
class[selector].class[selector].property.property
```

There's one more thing to know: Frontier is expecting each class or property to be a name, not a literal, so a string4 literal needs to be enclosed in square brackets, to avoid a syntax error. Now we can rewrite our Apple event like this:

```
theView = ['cdis']["book disk"].['cfol']["book folder"].['cwin'].['pvew']
appleEvent('MACS', 'core' ,'setd', '----', theView, 'data', objSpec('pnam'))
```

This is good, but it gets better. Thanks to glue, there is no need to use any string4 literals! Instead, we can use English-like names. If you examine the Finder's glue (at `system.verbs.apps.finder`), you will see that it defines `disk`, `folder`, and `view`, respectively, as `'cdis'`, `'cfol'`, and `'pvew'`. Also, the Finder's application ID, `'MACS'`, appears as the `id` entry. There is a search path to `system.verb.apps`, so we can write:

```
with Finder
    theView = disk["book disk"].folder["book folder"].['cwin'].view
    appleEvent(id, 'core' ,'setd', '----', theView, 'data', objSpec('pnam'))
```

But what about `'cwin'` and `'pnam'`? These are universal object classes, whose English-like names are defined in `system.macintosh.objectModel`. There is a search path to `system.macintosh`, so we can write:

```
with objectModel, Finder
    theView = disk["book disk"].folder["book folder"].window.view
    appleEvent(id, 'core' ,'setd', '----', theView, 'data', objSpec(name))
```

Finally, the glue verb `finder.set()` constructs the `appleEvent()` call for us, and so what we'd actually say in real life is this:

```
with objectModel, Finder
    theView = disk["book disk"].folder["book folder"].window.view
    set(theView, objSpec(name))
```

Object model syntax and glue are more fully discussed later in this chapter.

Variant Forms of appleEvent()

In an `appleEvent()` call, any name-value pair or series of name-value pairs, instead of being expressed as individual name and value parameters, may be expressed as a single parameter, which is either a record or a table. This works because both records and tables are themselves collections of name-value pairs.

The first parameter (the application ID) in an `appleEvent()` call can be the address of a database object; the Apple event will then be sent to that object, which should be a code fragment that knows how to receive Apple events, such as a UCMD or an OSAX. Another option is that the first parameter can be 0; this sends the event directly to the system, permitting system-level OSAXen (those in the *Scripting Additions* folder) to be called. Alternatively, calling `systemEvent()` has the same effect.

The value returned from an application in response to an Apple event is usually a single value, in which case it simply becomes the value of the `appleEvent()` call. But it can instead consist of multiple values. In this case, use `complexEvent()` instead of `appleEvent()`; this lets you supply the address of an object where the result will be placed as a table of name-value pairs. For an example, see `passwordDialog.run()`.

A further variant is `tableEvent()`. This is just like `complexEvent()`, but it also lets you supply the address of a table of name-value pairs for the input parameters. This is merely a syntactical variant on `complexEvent()`. For an example, see `card.setCardAttributes()`. The script begins:

```
on setCardAttributes (tableAdr)
    local (result)
    tableEvent (tableAdr, @result, 0, card.id, 'sacd')
```

but exactly the same effect could have been achieved by saying:

```
on setCardAttributes (tableAdr)
    local (result)
    complexEvent (@result, 0, card.id, 'sacd', tableAdr^)
```

Local and Remote Processes

To communicate with a process running locally, supply as the application ID either the process's string4 creator code or its string name.

To communicate with a process running elsewhere on the network, supply as the application ID either a network address string of the form `"zone:machine-ID:processName"` (where *zone* can be * to signify the local zone) or the binary that is returned from `sys.browseNetwork()` (see Chapter 25, *Dialogs*). An example of communication with a remote process appears later in this chapter.

Asynchronous Events

Sometimes it is desired to send an Apple event asynchronously. This is analogous to spawning a new thread: we send the event and then go on with the script, without waiting for the recipient application to respond.

The way to do this is to substitute a `finderEvent()` call for the `appleEvent()` call. The reason for the odd name is that asynchronous calls were necessary when driving older versions of the Finder, because it never sent back any response.

Transactions

Some applications have the ability to conduct *transactions*. A transaction is a session during which only one interlocutor is permitted to communicate with the application; Apple events from other interlocutors are rejected.

For example, we might be about to send a series of Apple events to FileMaker Pro; if, between these events, someone else sent FileMaker Pro an Apple event which brought another database to the front, we might wind up fetching (or deleting) the wrong data. So we operate within a transaction, to lock out this "someone else" temporarily.

The mechanics of transactions are simple. When you ask to begin a transaction, the application provides you, in effect, with a ticket (a transaction ID). From then on, with every Apple event you send to that application, you must show your ticket. No one else knows your ticket number, so everyone else is locked out. When you're finished with the transaction, you tell the application to end the transaction, so that you don't monopolize it unnecessarily; after that, the ticket is no longer necessary.

The way to show your ticket when you send the application an Apple event during a transaction is to substitute for each `appleEvent()` call a `transactionEvent()` call, which lets you supply a transaction ID. But it is usually impractical for you to call `transactionEvent()` yourself, because you aren't calling `appleEvent()` yourself either—you're using glue. But the glue verbs make `appleEvent()` calls, not `transactionEvent()` calls.

To get around this, you call `setEventTransactionID()`, giving it your transaction ID. Now, every subsequent `appleEvent()`, `complexEvent()`, or `tableEvent()` call in the current thread is automatically converted to a `transactionEvent()` before it is sent to the system. After you've told the application to end the transaction, you call `setEventTransactionID(0)`, which turns off conversion of your `appleEvent()` calls to `transactionEvent()` calls.

In this example, I monopolize FileMaker Pro while obtaining information about my recorded music collection. The entire interchange with FileMaker involves glue verbs, including beginning and ending the transaction; but it is not important to understand these right now. The important thing to grasp is the overall form. Notice in particular the use of the `try...else` structure; without this, an error interrupting our script before we could end the transaction would leave FileMaker in transaction mode, unable to respond to *any* Apple events; since the transaction

ID was in a local variable, its value would be lost, and the only way to remedy the situation would be to quit FileMaker and restart it.

Example 32-1. Using transactions

```
on doTheFind()
    local (theDB = "disks", myFind, theList = {}, x, ticket)
    with objectmodel, filemaker
        try « monopolize FileMaker
            ticket = beginTransaction(); setEventTransactionID(ticket)
            goto (database[theDB].layout[1]) « establish window and layout
            delete (database[theDB].request [all]) « make a Find
            myFind = create (request)
            set (myFind.cell["last"], "dvorak")
            set (myFind.cell["date"], "1887")
            find (database[theDB]) « do the Find
            for x = 1 to count (window[theDB], record) « gather results
                theList[0] = \
                    get (record[x].cell["title"]) + ", op. " + \
                    get (record[x].cell["opus"])
        else « always clean up!
            try{endTransaction()}; setEventTransactionID(0)
            scripterror(tryerror)
        endTransaction(); setEventTransactionID(0)
    return theList
doTheFind()
    « {"quintet, op. 81", "piano quartet, op. 87"}
```

Notice the order of commands at the very end. You mustn't tell Frontier to `setEventTransactionID(0)` before telling FileMaker to `endTransaction()`, or the `appleEvent()` call in FileMaker's glue for `endTransaction()` will not be transmuted to a `transactionEvent()`, and FileMaker will reject it!

You might wonder whether, while `setEventTransactionID()` is in force, it is possible to speak to an application other than the one you're having the transaction with. Apparently, it isn't a problem; if an application is not in transaction mode, it is not an error to send it a transaction ID as part of an Apple event.

Timeouts

Every Apple event has an associated *timeout*, the length of time for which the sender declares a willingness to wait for a reply from the recipient application. If the time expires without a response from the recipient application, an error is raised by the system.

Frontier supplies a default timeout of `infinity`, but there are situations where you might wish to change this, to avoid your script hanging for a very long time if the recipient gets into some sort of trouble. To change the timeout value, call `setEventTimeout()`; the number of seconds you supply as parameter will remain in force for the remaining execution of the current thread, or until you restore the default with `setEventTimeout(-1)`.

User Interaction Levels

Some applications respond to a user interaction level setting which is allowed to accompany an Apple event. You alter this setting by calling `setEventInteraction()`; the parameter is a boolean saying whether user interaction is permitted. If you disallow user interaction and then ask the application to do something which normally requires user interaction, it may return an error.

I have not discovered much practical use for this setting, because most applications don't seem to respond to it. One that does is FileMaker Pro. If you say:

```
with fileMaker
    setEventInteraction(false)
    open("power:myDatabase")
```

an error results.

Apple Event Suites

A collection of standardized Apple events intended to be applicable to more than one application is called a *suite*.

Apple Computer, Inc., has promulgated a number of such suites under the title *Apple Event Registry*, and developers of scriptable applications have been encouraged to make their applications respond to as much of these standard suites as makes sense, while of course also defining Apple events for those facets of the application that are unique.

In this section, we describe how Apple events of the standard suites are implemented in UserTalk. This implementation involves both verbs and constants. The verbs reside in three tables of scripts in `system.macintosh` (to which there is a search path): `required`, `core`, and `misc`; these are known as the required, core, and misc verbs, and they implement, respectively, the verbs for the Required Suite, the Core Suite, and the Miscellaneous Standards. Constants for the Required and Core Suites are defined in `system.macintosh.constants` (to which there is a search path). Constants for the Text, Graphics, and Table Suites, as well the Miscellaneous Standards, are defined in `system.macintosh.objectModel`.

It isn't likely that you would need to call one of these verbs directly; but you might encounter one while studying glue verbs, and many glue verbs are in fact implemented as calls to these same Apple events even if they don't call these verbs. Hence, the details may be useful; they are relegated to a chapter of their own, Chapter 47, to make them easier to find.

The *appID* becomes the first parameter of the Apple event. For its possible forms, see "Local and Remote Processes" earlier in this chapter.

Required Events

All applications, even if they aren't scriptable, are required, under System 7, to respond to four Apple events usually sent by the Finder. These make up what is called the *required suite*, implemented in UserTalk in `system.macintosh.required`. They are:

```
required.openApplication (appID)
required.openDocument (appID, documentPathname)
required.printDocument (appID, documentPathname)
required.quitApplication (appID)
```

Core Events

All scriptable applications are strongly encouraged by Apple Computer, Inc., to support the entire *core suite*. It is composed of Apple events regarded as universal and essential to scriptable applications. These Apple events are implemented in UserTalk at `system.macintosh.core`.

In general, these verbs must be called inside a `with objectModel` bundle. This provides access to constants that may be used in specifying an object, as well as to verbs used in specifying locations (locations are explained later in this chapter). To refer to elements and properties unique to a particular application, the verb call must also be inside a `with` for that application's glue.

```
core.open (appID, whatObject)
core.close (appID, whatObject, saving?, savingIn)
core.save (appID, whatObject, filePathname, fileType)
core.saveAs (appID, whatObject, filePathname, fileType)
core.print (appID, whatObject)
core.quit (appID, saving?)
core.create (appID, whatClass, data, propertyList, where)
core.delete (appID, whatObject)
core.count (appID, container, whatToCount)
core.exists (appID, whatObject)
core.duplicate (appID, whatObject, toWhere)
core.move (appID, whatObject, toWhere)
core.dataSize (appID, whatObject, whatType)
core.get (appID, whatObject, whatType)
core.getAs (appID, whatObject, whatType)
core.set (appID, whatObject, toWhat)
```

Miscellaneous Standards

The body of Apple events called the Miscellaneous Standards is implemented at
`system.macintosh.misc`. Particular applications may or may not support any of
these. Some, for all I know, may be unsupported by any application.

The remarks about the core verbs apply also to the misc verbs, listed here:

```
misc.beginTransaction (appID)
misc.endTransaction (appID, transactionID)
misc.copy (appID)
misc.cut (appID)
misc.paste (appID)
misc.undo (appID)
misc.redo (appID)
misc.revert (appID, whatObject)
misc.select (appID, whatObject)
misc.doMenu (appID, whatMenuItem)
misc.show (appID, whatObject, whatWindow, whatPoint)
misc.isUniform (appID, whatObject, whatProperty)
misc.editGraphic (appID, graphicArea)
misc.imageGraphic (appID, graphic, format, antialiasing?, dithering?,
    rotationRec, scale, translationPoint, flipHorizontal?, flipVertical?,
    quality, structuredGraphic?)
misc.createPublisher (appID, whatObject, whatFile)
misc.doScript (appID, string)
```

`misc.doScript()` is for applications that have an internal scripting language (such
as HyperCard with HyperTalk, Microsoft Word with WordBasic, Nisus Writer with its
macro language). It tells the application to execute a script written in that language.

Applications with internal scripting languages can be just as scriptable from other
applications as they are internally, provided they support this one Apple event.
And even when such applications are also heavily scriptable through Apple
events, it is often easier to drive them with `misc.doScript()`, because it lets you
capitalize on your knowledge of the application's internal scripting language.

In this example, we suppose that we have placed text in the clipboard which we
wish to make the content of a new Microsoft Word document. We tell Microsoft
Word, in WordBasic, to make a new document, paste the clipboard's contents,
and put the insertion point at the start of the document.

```
local (s)
on add(t)
    s = s + t + cr
bundle « build the wordBasic script
    add("FileNewDefault")
    add("EditPaste")
    add("StartOfDocument")
if not sys.bringAppToFront('MSWD')
    dialog.notify("Couldn't bring Word to the front."); return
misc.doScript('MSWD', s)
```

Frontier's Private Apple Events

There is one more Apple event suite, but it is defined by UserLand, not by Apple Computer. It goes back to the days when Apple events were very new, and no other standardized suites had yet been proposed. UserLand developed some applications in accordance with this suite, some of which—DocServer, BarChart, and MacBird—are included with the Frontier distribution.

It is fairly unlikely that you would want to drive one of these applications yourself. They are ancillary to Frontier, which should be permitted to control them for you. The applications in question are not scriptable in a general sense: only Frontier can speak to them.*

The suite involves a set of `'app1'` Apple events. The glue for these Apple events is in `system.verbs.builtins.app`; there is a search path to `system.verbs.builtins`, so these are known simply as the app verbs. On the whole, the app verbs may be regarded as an interesting historical remnant, a sketch for a standard that was never widely adopted; they are not documented in this book. A couple of app verbs do remain important, though, for driving other applications; they are discussed later in this chapter. The following remarks are for those who wish to explore the app verbs independently.

Before speaking to an application with app verbs, we call `app.start()`, handing it as parameter the string name of the application's glue table in `system.verbs.apps`. This starts up the application if it isn't running.

It also sets `app.id`, so that all future app verb calls will be directed to that application. This is why the remaining app verbs do not take any parameter indicating what application they are directed to. The purpose of each of these verbs is fairly clear from their names: they do such things as manipulate windows, obtain selected text, change font and size, and so forth. Each application also has verbs unique to itself in its glue table in `system.verbs.apps`.

The main thing about the app verbs is not to try to use them for driving applications other than those for which they are intended. You can tell if an application responds to the app verbs by consulting the `app1supported` entry in its `appInfo` table (in its glue table, in `system.verbs.apps`). For example, `cardEditor.appInfo.app1supported` (cardEditor is MacBird) is `true`.

---

* Because they lack `'aete'` resources, which are discussed later in this chapter. For technical information, see the Applet Toolkit in the Frontier SDK.

Object Model

Most commands and queries to applications involve specifying some object in that application's world which is what you want to affect or ask about. Thus, you need a way to specify the object. For example, suppose the application is a scriptable word processor, and we want it to report what text appears somewhere in one of its windows; but *what* window, and precisely *what* stretch of text do we want to know about?

Furthermore, some commands involve placing or creating things. Thus, you need a way to specify a location. For example, our word processor can be scripted to insert some text in a document; but precisely *where* in the document should this text go?

In UserTalk, such specifications are made using the *object model*. The object model is implemented through two components.

First, the object model is implemented through UserTalk syntax (*object model syntax*). UserTalk itself allows you to construct object specifiers. The syntax involved is analogous to other familiar syntactical features of UserTalk. For example, to refer to the first window of an application, you might say:

```
window[1]
```

This should remind you of an array reference. To refer to the second word in the first window, you might say:

```
window[1].word[2]
```

The concept of containment is expressed by a dot, which should remind you of how the dot is used to express containment in a database object reference.

There are two kinds of object that can be contained in another object. Some objects represent features of which the container has (or might have) many, and therefore need an index to specify which one is meant. Other objects represent features unique within the container, and therefore get no index; for instance, you might say:

```
window[1].name
```

Something like **window** or **word**, which requires an index, is called an *element*; something like **name**, which requires no index, is called a *property*.

Second, the object model is implemented through constants and scripts contained in **system.macintosh.objectmodel**. For example, in the phrase **window[1]**, the constant **window** is defined in **system.macintosh.objectmodel**, so that when you use it, it is translated into a string4 value suitable for sending as part of an Apple event.

There is a search path to `system.macintosh`, but there is no search path to `system.macintosh.objectmodel`. Therefore, a line of UserTalk that uses the object model should generally appear in a **with** whose domain is `objectModel`:

```
with objectModel
    window[1].name
```

Furthermore, `system.macintosh.objectmodel` defines constants for only a few universal object types. An application is free to invent other object types specific to its own functionality, and constants for these are defined in the glue table for that application. For example, the Finder glue table includes a string4 definition for `folder`. The Finder's glue table is at `system.verbs.apps.finder`, and there is a search path to `system.verbs.apps`; so, you might say:

```
with objectModel, Finder
    get( window[1].folder[1].name )
```

And that is the form of a typical UserTalk command that drives another application.

The **with** structure isn't merely a way of giving access to constants; it is also how you say what application you want to talk to. In our example, the verb `get()` is also defined in `system.verbs.apps.finder`; and it is defined so that it sends its Apple event to the Finder, and not to some other application.

Containment and with

As an alternative to dot-notation, containment can be expressed using **with**. For example:

```
with objectModel, Finder, window[2]
    get( folder[1].name ) « means window[2].folder[1].name
```

But an object specifier in the **with** bundle that is not contained in the **with** domain can cause an error:

```
with objectModel, Finder, window[2]
    get( folder[1].name ) « fine, means window[2].folder[1].name
    get( window[1].folder[1].name ) « error!
```

One solution is to begin such specifiers with the container `null`. This constant is defined in `objectmodel`, and refers to the top of the containment hierarchy. So:

```
with objectModel, Finder, window[2]
    get( folder[1].name ) « fine, means window[2].folder[1].name
    get( null.window[1].folder[1].name ) « fine too
```

You can refer to the domain of the **with** from inside the **with** bundle by using the special constant `it`. So:

```
with objectModel, Finder, window[2]
    count( it, folder ) « that is, folders of window[2]
```

In these examples, the with syntax may seem to have little advantage over the dot-notation syntax as a way of expressing containment. But it comes in very handy for simplifying commands that make repeated reference to a particular container.

Indexing

Given a type of object that's an element (such as window or word), there are various ways to say *which* one(s) you mean, corresponding to various kinds of items that can go into the square brackets.

A number

For example, word[2]. Unlike normal UserTalk, negative indexing is permitted; for example, word[-2] is the next-to-last word.

A name, which must be a string

For example, window["Chapter 3"].

A range

Indicated by two indexes separated by to. For example, word[2 to 4]. This specifies more than one object (in our example, word[2], word[3], and word[4]).

An application-specific unique index

If an application defines its own index type, you can use the name of the index followed by a colon and then the index value. For example, FileMaker Pro assigns each field an ID number, called, in the glue table, its id1; thus one can speak of field[id1:2].[*]

A relative position

Defined in objectmodel; they are next and previous. Containment shows what object the relationship is to. For example, selection.word[next] means the word that is next after the current selection.

An enumerative constant

Defined in objectmodel; they are all, any, first, last, and middle.

A boolean expression

Discussed separately, in a moment.

Note that you don't have to use a literal as an index; what's in the square brackets is evaluated before it is used. So you can put into the square brackets a variable (its value is the index), a verb call (its result is the index), and so forth.

---

[*] The reason this index is called id1 in the glue table is that the name id is already taken: it refers to FileMaker's own application ID, the creator code 'FMP3'.

Boolean Indexing

An object specifier with an index which is a boolean expression refers to all objects of the given type of which the boolean is true. Obviously, the boolean should be something which can be true of this type of object.

For example, the boolean might say something about one (or more) of the things that the object can contain. Suppose, for instance, that in a word processing application one can speak of window[x].word[y].character[z]. Then, since character is an element of word, one can test words in terms of their characters. One might say:

```
window[1].word[character[1] == 'f']
```

This means: "Within window 1, every word of which it is true that its first character is 'f'." Notice that it is permitted to use indexing within the boolean index expression.

This is all very well, but it doesn't handle every situation. How would you ask for all words that contain an 'f'? In the boolean expression, we don't want to speak of any element or property of a word; we want to speak of the word itself. Therefore, a special constant it is defined:

```
window[1].word[it contains "f"]
```

Any UserTalk boolean operator may be used:

```
window[1].word[it beginswith "f" and it endswith "s"]
```

The boolean index does not have to be on the last index of the object specifier. The following is perfectly legal:

```
with objectModel, Finder
    get( window[name contains 'e'].folder[1].name )
```

An object specifier generally cannot contain more than one boolean index. This is not because it is syntactically forbidden, but because no application (so far as I know) implements an ability to respond. The following returns an error:

```
with objectModel, Finder
    get( window[name contains 'e'].folder[name contains 'e'].name )
```

Locations

Some glue verbs require a location parameter. Locations specify where something is to be put. They are expressed using one of five verbs defined in objectmodel: before(), after(), beginningOf(), endOf(), and replace(). The parameter is an object specifier.

As an example, we drive Microsoft Word to insert the string `"hello"` in its front-most window at the point where the third word is now:

```
with objectModel, MSWord, window[1]
    make( text, replace(word[3]), "hello" )
```

Property Lists

Some glue verbs take one or more parameters that are *property lists*. These are actually not lists, but records. I usually find it more legible to create these records as local variables in separate lines, but this is purely a matter of taste.

This example is the same as the previous one, except that we specify the style and size of the new word as we create it. We drive Scriptable Text Editor because Microsoft Word doesn't implement property lists:

```
with objectModel, STE, window[1]
    theProps = {style:{bold, underline}, size:14}
    make( text, replace(word[3]), "hello", theProps )
```

Note that one of the items in the property list was itself a list. The example shows how UserTalk lists and records convert directly to Apple event lists and records.

Glue

The usual way to drive an application with UserTalk is by means of the application's *glue*. This term refers to the verbs and constants in the application's glue table, which is a table in **system.verbs.apps**. These verbs and constants give the UserTalk programmer access to an application's full range of scriptability without ever having to deal directly with an Apple event. Glue provides a level of abstraction which makes scripts easy to read and write.

Consider, for example, the following script, which drives Eudora. It runs through all mail messages in the In box, looking for digests from the Frontier-Beginners mailing list (identified by their subject line); any such messages are copied out to a certain folder on disk for later reading, and then moved to Eudora's Trash folder.

Example 32-2. Eudora session

```
local (theFolder = "Desire:digests:frontierBeginners:")
local (theList = {}, x, subj, theMessage)
on zOrP (n) « zero or positive
    return n * (n > 0)
on niceName (s) « make legal filename
    s = string.delete (s, 1, zOrP(sizeOf(s)-31))
    return string.replaceAll (s, ":", ";")
on gatherSubjects (whatMessage) « add one message subject to theList
    theList[0] = get (whatMessage.subject)
    return true
```

Example 32-2. Eudora session (continued)

```
on trashMessage (whatMessage) « move one message to the Trash
    moveMessage (whatMessage, endOf(mailbox["Trash"]))
« --- the action starts here ---
with eudora, eventInfo
    visitMessages (mailbox["In"], @gatherSubjects)
« now theList contains the subjects
for x = sizeof(theList) downTo 1
    subj = theList[x]
    if subj beginsWith "FRONTIER-BEGIN"
        with objectmodel, eudora, eventInfo
            theMessage = mailbox["In"].message[x]
            saveAs (theMessage, theFolder + niceName(subj))
            trashMessage (theMessage)
```

Knowing UserTalk and the object model, even someone who has never scripted Eudora can see just what's going on here. Abstraction is achieved by the use of glue verbs such as `eudora.visitMessages()` and `eudora.saveAs()`, by the constants `mailbox` and `message` in Eudora's glue, and by a series of local handlers at the start of the routine. After the point where the action starts, the structure and purpose of the script is clear. The routine was easy to write, it is easy to read, it executes quickly, and there isn't an Apple event in sight.

Where Glue Comes From

Glue for some applications is included in the database as distributed by UserLand. Some applications, such as QuarkXPress, ship with a packed object containing their glue. A user can create a glue table and distribute it to other users.

However, you do not have to rely upon an outside source for glue. It is possible to have Frontier generate glue for a scriptable application automatically. This process relies upon the presence in the application of an `'aete'` resource, a kind of map describing the Apple events and object types recognized by the application.

To make a glue table in this way, start by choosing **Commercial Developers** from the **Suites** menu, and then choose **Enter Your App's Name** from the **Glue** menu that appears. The name you enter will be the name for the glue table (not the name of the application on disk); decide carefully, because it is not possible to change the name later on. After one more chance to back out, you are asked a series of important questions. Should an `appInfo` table be created? (You should say **OK**; the `appInfo` table is explained below.) Does this application support menu sharing? (If you don't know that it does, say **No**.) Next, you're asked to locate the application on disk using a StandardGetFile dialog. Finally, should glue be generated from the application's `'aete'` resource? (You should say **OK**; that's what you're here for.)

Tweaking Glue

Glue generated by the Commercial Developers suite may be perfectly satisfactory and ready to use; but in some cases it will not be. This is not Frontier's fault; it's because the `'aete'` resource is an imperfect vehicle for communicating the syntactical variations accepted by the application. It may be necessary to correct the glue (called *tweaking*) with the help of written documentation where it exists.

As an example, let's take the image conversion utility Clip2Gif. Glue automatically generated from Clip2Gif includes a verb **save()** which says, in part:

```
on save (x, as, saveIn = nil ... )
    return (appleEvent (clip2gif2.id, 'core', 'save', '----', x, \
        'fltp', string4 (as), 'kfil', filespec (saveIn) ... )
```

This verb converts an image and saves the result as a file on disk. But the same verb, we are told by Clip2Gif's written documentation, can also be used to hand back the converted image as the result of the call—if we supply `'TEXT'` as the value of the parameter **saveIn**. But that's impossible as the script stands, because that parameter is being coerced to a fileSpec. So the glue is faulty, and needs some tweaking.

Precisely how to tweak it is a stylistic and architectural decision. One possibility is just to remove the fileSpec coercion. Personally, I don't like that idea, because it throws upon the calling script the onus of explicitly coercing the parameter to a fileSpec when we do want to save to a file. I feel that the purpose of glue is to provide abstraction, so that the programmer doesn't have to bother with the nitty-gritty of Apple events.

My preference, therefore, is to write separate glue verbs for the different functionalities, so that the glue, not the caller, does the work. One verb might save the converted image as a file to disk:

```
on saveToFile (x, as, saveIn = nil ... )
    return (appleEvent (clip2gif2.id, 'core', 'save', '----', x, \
        'fltp', string4 (as), 'kfil', filespec (saveIn) ... )
```

Another verb hands back the converted image as the result; it's the same Apple event, but I've removed the **as** and **saveIn** parameters and hard-coded their values for this particular functionality:

```
on returnGIF (x, ... )
    return (appleEvent (clip2gif2.id, 'core', 'save', '----', x, \
        'fltp', 'GIFf', 'kfil', 'TEXT' ... )
```

In the special case where what we want to convert is a `'PICT'` in the clipboard, the documentation tells us that the direct object (the parameter **x**) should be `null.clipboard`; again, that's an unintuitive thing for the caller to have to say, so I prefer to write yet another glue verb, without even an **x** parameter:

```
on returnClipAsGIF ( ... )
    return (appleEvent (clip2gif2.id, 'core', 'save', \
        '----', objectmodel.null.clipboard, \
        'fltp', 'GIFf', 'kfil', 'TEXT' ... )
```

Here is a utility that I use to include screen shots in Web pages written with Frontier's Web site management tools. GIF images need to be stored in the database as binaries of type `'GIFf'`. I use Snapz Pro to take a screenshot and leave it as a `'PICT'` in the clipboard; then I run the following utility script to convert the clipboard contents to a GIF in the database.

Example 32-3. clipToGifInDatabase()

```
on clipToGifInDatabase()
    local (p, theNewName = "newGIF")
    with clip2gif
        bringToFront() « need clip2gif in front to use clipboard
        p = returnClipAsGIF() « PICT clipboard is now GIF in p
    frontier.bringToFront()
    if dialog.ask("What would you like to call it?",@theNewName)
        pack(p, @user.html.images.[theNewName])
        setBinaryType(@user.html.images.[theNewName], 'GIFf')
        table.goToAddress(@user.html.images.[theNewName])
```

Now, when this script drives Clip2Gif, it uses a nice legible call with an informative name; it's easy to write the script, and easy to read it and understand what it does:

```
with clip2gif
    bringToFront()
    p = returnClipAsGIF()
```

But suppose we had never created `returnClipAsGIF()`; suppose we had instead tweaked the glue for `clip2gif.save()` by just deleting the fileSpec coercion. Then the script could achieve the same functionality, but it would have to say this:

```
with objectModel, clip2gif
    bringToFront()
    p = save (null.clipboard, 'GIFf', 'TEXT') « huh?
```

The call to **save()** is difficult to write, and no one reading it could possibly guess what it does. The example shows clearly why I prefer the glue, not the caller, to take care of the messy details.

Anatomy of a Glue Table

A glue table typically has the following components. (I find it easiest to sort glue tables by Kind so as to view them easily; if you are following along by examining a glue table, you might wish to do the same.)

An appInfo *subtable*

The information in this table helps Frontier identify the application. Note in particular:

appInfo.name

This initially matches the name of the glue table as a whole, but you are free to change it. Frontier will use this in reporting error messages sent by the application.

appInfo.id

This is the application's creator code.

appInfo.path

This the pathname of the application's location on disk. Frontier will use this to launch the application if you start driving it when it isn't already running. If the application is not there (perhaps you've moved the application, or you've copied the glue to a database on a different computer), Frontier will prompt you to locate the application, and will update path; so you needn't (and normally wouldn't) alter it manually.

The id *entry*

Distinguish here between, e.g., eudora.id and eudora.appInfo.id; we are speaking now about the former. They often have the same value, but they sometimes don't, and they are used for different purposes. The latter is invariant. The former is the current addressee of Apple events sent from this glue table. We'll talk more about this in a moment.

Basic verbs

These are common to most applications, and are generated even before consulting the application's 'aete' resource. They are:

openDocument(), printDocument(), quit()

Implemented by calling the corresponding required verbs (listed earlier in this chapter, under "Apple Event Suites").

isRunning()

Calls sys.appIsRunning() to return a boolean.

bringToFront()

Launches the application if necessary, then makes it frontmost with sys.bringAppToFront(). Returns false if either is impossible.

```
launch()
```
> Gets the application running if it isn't running already, but does *not* bring it to the front. This verb performs some other tasks, too, in connection with the `id` entry; we'll talk about this in a moment.

Other glue verbs
> These are the scripts that you call in order to drive the application.

Constants
> These are string4s and enums. They are sometimes in the glue table itself, and sometimes in a subtable. When they are in a subtable, it makes the glue table easier to read, but you have to remember to include the subtable as a domain in the **with** when you call a glue verb.

> For example, Eudora's glue table has its constants in a subtable, `eudora.eventInfo`. That's why in Example 32-2 we address Eudora like this:

```
with Eudora, eventInfo
    visitMessages(mailbox["In"], @gatherSubjects)
```

> Otherwise, Frontier wouldn't know what **mailbox** means.

The `objectHierarchy` *outline*
> This is created when glue is generated from an application's `'aete'` resource. It can be a helpful map for understanding what types of object an application knows about, and what their elements and properties are.

Documentation
> If a glue table has been tweaked by hand and then publicly distributed, it may contain documentation that helps explain how to use the glue. See, for instance, `eudora.readme` and `eudora.examples`; `finder.baxter`; and `fileMaker.examples`.

Addressing an Application

When you use a glue verb to drive an application, how does Frontier know which application you want to talk to? It uses the `id` entry in the glue table. If you examine a glue verb, you will see that the `appleEvent()` call uses the `id` entry from the same table as its first parameter. For example, the first parameter in the `appleEvent()` call in `eudora.get()` is `eudora.id`.

So, given a particular glue table, you can change its `id` value to change what application glue verbs in that table will talk to. If you want to talk to a local copy of the application, `id` should be the application's string4 creator code (or it can be the pathname of the application file). If you want to talk to a remote copy of the application running elsewhere on an AppleTalk network, `id` should be the process's network string pathname, or the binary returned from **sys.browse-Network()**.

However, you should not change the `id` entry directly. Instead, you should call a verb which manages it for you. Typically, this will be the glue table's `launch()` verb.

When you call a glue table's `launch()` verb, `app.start()` is called, which in turn calls `app.startWithDocument()`. This verb checks `app.idNetworkApp` to see if it is defined. If it is, it is copied into the glue table's `id` entry; if not, the local version of the application is started if it isn't already running (using the information in `appInfo`), and its creator code is copied from `appInfo.id` into the glue table's `id` entry.

So whether the glue table will address a remote or a local copy of the application depends upon `app.idNetworkApp`. To define and destroy this object, you call `app.linkToNetworkApp()` and `app.clearNetworkApp()`.

Thus, the mechanism for establishing a remote or local instance of an application as the addressee of its glue table is to call `app.linkToNetworkApp()` or `app.clearNetworkApp()` followed by its `launch()` verb. Then you can use the other glue verbs.

In the following example, we tell first a remote Microsoft Word, then the local Microsoft Word, to type "Hello from Frontier!".

Example 32-4. Remote addressing with launch()

```
app.linkToNetworkApp("*:Matt's Other Computer:Microsoft Word")
with objectModel, MSWord
    launch() « establish remote Word as addressee
    doScript ("Insert (\"Hello from Frontier!\")")
app.clearNetworkApp() « go local for next session
with objectModel, MSWord
    launch() « establish local Word as addressee
    doScript ("Insert (\"Hello from Frontier!\")")
```

Some glue tables, such as the Finder's, lack a `launch()` verb. In this case, you must call `app.start()` yourself; its parameter is the string name of the glue table entry in `system.verbs.apps`. So in the following example, we tell first a remote Finder, then the local Finder, to open its About This Macintosh window.

Example 32-5. Remote addressing with app.start()

```
app.linkToNetworkApp("*:Matt's Other Computer:Finder")
app.start("Finder") « establish remote Finder as addressee
with objectModel, Finder
    open(aboutThisMacintosh)
app.clearNetworkApp() « go local for next session
app.start("Finder") « establish local Finder as addressee
with objectModel, Finder
    open(aboutThisMacintosh)
```

Faithful adherence to these procedures will ensure that you won't accidentally try to communicate with a remote application when you intend a local one.

Note that communication with remote applications requires that the application be running and available on the network, that program linking on the remote machine be turned on, and that the user have access privileges for that machine. If the user has not already linked to the remote application, the Mac OS Link To dialog will appear. This can be disruptive,* so if you know the correct ID and password, you can prevent the dialog from ever appearing by first calling `loginAs.loginAs()`, which sets the userID and password values for any subsequent remote communications.

Using Glue

Suppose now that you possess correctly working glue for some application, such as the glue included in the Frontier database. Ideally, you should be able to figure out how to drive that application by examining its glue table. Unfortunately, the matter is not always so easy.

The fault lies not in Frontier but in the nature of scripting. The problem is that both the degree and the style of scriptability vary greatly from one application to another. Thus, your experience of driving one application may not help you much in driving a different application. Learning to drive a new application isn't like learning to drive a new car; it's more like learning an entirely new language. You're at the mercy of the developers of the application; everything depends upon how intelligibly they have implemented its scriptability.

There are two main kinds of stumbling-block which application developers can introduce: inadequate exposure of the application's functionality, and unintuitive syntax.

Inadequate exposure of an application's functionality means that you may search in vain for a particular verb or object. For example, in Eudora the user can save a message as a textfile on disk, by choosing the **Save As** item of the **File** menu; but there is no Apple event to make Eudora do the same thing.† Nisus Writer can't tell you the font or style of the currently selected text. And so on.

Unintuitive syntax means that even though the glue table shows you all of the verbs and constants for an application, this isn't sufficient for putting those verbs and constants together into the kinds of expression that the application expects.

---

* Especially because if the Link To dialog appears and the user cancels, script execution will break off completely.

† Luckily, Frontier can make a textfile, so the authors of the Eudora glue have been able to write a `saveAs()` verb anyway.

As with learning a natural language, you may know a lot of vocabulary, but that doesn't mean you know how the language expresses a particular idea idiomatically. How often I have seen people on the Net stumped as to how to get Eudora to move the currently selected message to the Trash mailbox:

```
with objectModel, eudora, eventInfo
    move (message[""], endOf(mailbox["Trash"]))
```

Who would have thought that "the currently selected message" would be `message[""]`, or that you would have to move a message, not to the Trash, but to the end of the Trash? And we have already seen, in connection with Example 32-3, how obscure Clip2Gif's syntax can be if its glue is not tweaked.

So, how can you learn to drive an application? Here are some tips.

- Study the application's glue table. Look at its verbs, and even more important, look at its `objectHierarchy` outline.* Some of the glue included with Frontier has been hand-tweaked in a very helpful way. Some applications have secondary glue tables; for instance, Brent Simmons's `suites.fileMakerLib` is meta-glue that makes scripting FileMaker much easier.

- Supplement this with whatever documentation is supplied with the glue or with the application. I wouldn't have known how to tweak Clip2Gif's glue without the documentation that comes with it, but thanks to that documentation, it was easy.

- Nothing is as helpful as examples. Sometimes the documentation includes examples; sometimes there are additional examples hidden in the database (such as `suites.samples.basicStuff.quarkDocServer`, for Quark).

- Also, some scriptable applications are recordable. With such an application, you can perform actions by hand, and the application will generate the User-Talk code that would have performed those actions. If you don't know whether an application is recordable, just try it. Open a script edit window and press the **Record** button. Switch to the application and do something there; then return to Frontier and press the **Stop** button.

Tips and Pitfalls

In this section, I have distilled years of experience into nuggets of wisdom intended to jump-start your learning efforts and to save you from heading up unnecessary blind alleys.†

---

* If a glue table has no `objectHierarchy` outline, have Frontier generate a second glue table, without overwriting the first; you can then copy the outline from the second table into the first, and throw away the second table.

† There may be a prize for figuring out how many metaphors were mixed in that sentence. Or maybe not.

- If you have the string4 value of a constant, you can make Frontier look up its name for you with `displayString()`. Use **with** to dictate the region of the database in which Frontier should search; the **objectModel** table is searched automatically.

  For example, recording Fetch generates string4s instead of constant names, because the constant names are hidden in an **eventInfo** subtable. So recording the renaming of a remote file generates this:

  ```
  set (['cFHB'] ["bookfolder.sit"].name, "bookfolder2.sit")
  ```

  To learn the name of the constant that has **'cFHB'** as its value, run this in Quick Script:

  ```
  with fetch.eventInfo {displayString('cFHB')}
      « "remoteFile"
  ```

  Now we know that the syntax for renaming a remote file is like this:

  ```
  with objectModel, Fetch, eventInfo
      set (remoteFile["bookfolder.sit"].name, "bookfolder2.sit")
  ```

- Similarly, `displayString()` can often decode mysterious garbage returned from a query, and help you understand the structure of a piece of data. For example, if you say to QuarkXPress:

  ```
  with objectmodel, QXP, defs
      local
          theStyle = get(document[1].currentbox.paragraph[1].style)
      msg (theStyle)
  ```

  you see garbage in the Main Window. But if you change the last line to:

  ```
  msg (displayString(theStyle))
  ```

  you see this in the Main Window:

  ```
  {onStyles:{plain}, offStyles:{bold, italic, underline ... }}
  ```

  and now you understand the structure of a style record.

- The way to obtain or change an object's value is with `get()` and `set()`, not the assignment operator. Forgetting this is probably the beginner's most common error. For instance, you might mistakenly say:

  ```
  with objectmodel, finder
      local (x = disk["Power"].folder[1].name)
  ```

  This does not cause a runtime error, but **x** is not the name of a folder—it's an objSpec that refers to the name of a folder. What you probably meant to say was this:

  ```
  with objectmodel, finder
      local (x = get(disk["Power"].folder[1].name))
  ```

Of course, sometimes an objSpec *is* what you want. See Example 32-2 and the way the variable `theMessage` is used.

- When asking for data, consider using `getAs()` instead of `get()` in order to avoid receiving some confusing private datatype. For example, this is not a good idea:

```
with objectmodel, QXP, defs
    local
        theSize = get(document[1].currentbox.paragraph[1].size)
```

You'll find `theSize` fairly useless. What you probably mean is:

```
theSize = getAs(document[1].currentbox.paragraph[1].size, longType)
```

- Just the other way, you may need to coerce a parameter to the right datatype. For example, this does not work:

```
with objectmodel, finder
    set(folder["book disk:book folder"].view, name))
```

You have coerce **name** to an objSpec:

```
with objectmodel, finder
    set(folder["book disk:book folder"].view, objspec(name)))
```

Some glue verbs take care of this sort of thing for you, others don't.

- Experiment with object specifications to learn what an application wants. For instance, in Eudora, normally you have to specify a message as contained in a particular mailbox. But if you try to learn the subject of the currently selected message like this:

```
with objectmodel, eudora, eventinfo
    local (x = get(mailbox["In"].message[""].subject))
```

you get an error; in this case, the mailbox must be omitted.

- Similarly, experiment with object specifications to learn what an application returns. For instance, suppose the first sentence of the first window of Scriptable Text Editor says "This is a test". You might suppose that saying this:

```
with objectmodel, STE, window[1]
    get (word [1 to 4])
```

would return:

```
"This is a test"
```

but it doesn't; you've referred to four separate objects, and therefore you receive four separate answers, in a list:

```
{"This", "is", "a", "test"}
```

What you need to say is:

```
with objectmodel, STE window[1]
    get (text [word[1] to word[4]],'TEXT'))
```

Another simple but subtly tricky example arises when you don't know how many objects your expression refers to. This might return a string or a list of strings, depending on whether one or several items are selected:

```
with objectmodel, finder
    get (selection.name)
```

- Error messages from applications are often just informational, and not worth aborting execution for. Once you have your commands debugged, embed them in a **try** bundle. This enables the script to catch the error messages from the application and proceed usefully.

For example, referring to a non-existent object often generates an error message. Instead of testing the object's existence first with **exists()**, you can save time with a **try**. Here, we wish to tell QuarkXPress to select the textbox called "theStory"—if there is one:

```
with objectmodel, QXP, defs
    with document[1]
        try
            set (currentbox, textbox["theStory"])
            return true
        « if we reach this point, there is no textbox "theStory"
```

If it works, we're done. If it doesn't, the script proceeds, and we've learned (without the extra overhead of an **exists()** call) that there's no such textbox. When you have a **try**, errors become your friends.

- Experiment with what modes of indexing an application will accept for a particular object or a particular command. Where you can use boolean indexing, it is a far faster and more elegant way to gather information than looping within Frontier, because you're sending just one Apple event.

Here, I obtain from Quark a list of all textboxes whose first paragraph is in a style the name of which starts with "Body Copy" or "Body Sub":

```
local (boxList)
with objectmodel, QXP, defs, document[1]
    try
        boxList = list(get( \
        textbox[paragraph[1].styleSheet.name beginswith "Body Copy" or \
        paragraph[1].styleSheet.name beginswith "Body Sub"]. \
        objectreference))
```

There's a lot of power packed into this single command.

- Verbs that move or create objects require a location parameter, and locations have to be provided with one of the five functions provided in **object-Model**—**before()**, **after()**, **beginningOf()**, **endOf()**, and **replace()**. These may all seem unnatural given what you want to do: too bad, you have to pick one. If there doesn't seem to be a meaningful object to relate it to, the parameter of the function might be `null`, or `it`. For example, here's how to make a new bookmark list window in Fetch:

```
with objectmodel, fetch, eventInfo
    create (bookmarkListWindow, at:beginningOf(null))
```

- Be prepared for the consequences of operating inside a `with`. Remember Lawton's Law (Chapter 7, *The Scope of Variables and Handlers*). The domains of the `with` might contain entries with names that will throw off your scripts. For example, the following innocent-looking script doesn't work, because `delete()` doesn't mean what you think it does:

```
with objectModel, filemaker
    local (cellList = get (record[1]))
        « {"dvorak", "antonin", "quintet", "a major", "81"}
        « now I want to get rid of the "81" from this list
    delete(cellList[5]) « error!
```

The problem is that `filemaker.delete()` exists.

33

AppleScript

AppleScript, like UserTalk, is a programming language which sends Apple events to drive other applications. It is a product of Apple Computer, Inc. Frontier scripts can be written in AppleScript. That's possible because Frontier is an Open Scripting Architecture (OSA)–compliant scripting program: it can run scripts in any OSA dialect present on the machine, and AppleScript is an OSA dialect. (For more about the OSA, see Chapter 34, *Driving Frontier from Outside*.)*

This chapter describes how to write and run AppleScript scripts in Frontier. It does not teach the AppleScript language, because someone who doesn't know AppleScript already won't derive, from learning AppleScript, any benefit that Frontier doesn't already offer through UserTalk.†

This chapter also discusses how to call an OSAX, which is an AppleScript language extension. An OSAX can add functionality to UserTalk without the use of AppleScript, so this part of the chapter may be of use even to readers who do not know AppleScript.

AppleScript Scripts

The language of a script object is revealed in a rectangular box at the bottom of its edit window. That box is actually a popup menu, from which you can choose a different language. On most machines, the choices will be **AppleScript** and **UserTalk**.

---

* Other OSA dialects besides UserTalk and AppleScript are QuicKeys Script (present if QuicKeys is loaded at startup), tclScript, and MacPerl. This book does not discuss these dialects.

† To learn more about AppleScript, see both *http://www.applescript.apple.com/default.html* and *http://devworld.apple.com/dev/techsupport/insidemac/AppleScriptLang/AppleScriptLang-2.html*.

In an AppleScript script, lineation is meaningful according to the usual rules of AppleScript, but indentation is not; indent as desired for legibility. You can check a script's correctness by pressing the **Compile** button; if there is an error, you can find it with the **Go To** button in the error message dialog. There is no Debug mode.

AppleScript scripts run within Frontier just as they would in another context. But note that properties are not static after execution terminates; instead, store static values in the database, as explained next.

UserTalk and the Database

You can call an AppleScript script from a UserTalk script or from another Apple-Script script.

- If you call an AppleScript script from a UserTalk script, or if you press the **Run** button in an AppleScript script edit window, top-level commands are executed, or a `run()` handler is executed if there is one—there cannot be both, but there can be other top-level handlers, since a handler is not an executable command.

- If you call an AppleScript script from another AppleScript script, its eponymous handler is called—if there isn't one, an error results. There can be other top-level commands, but they won't be executed.

So, for example, suppose `workspace.testing` contains an AppleScript script that goes like this:

```
on testing()
    display dialog "You called me from AppleScript."
end
display dialog "You called me from UserTalk, or with the Run button."
```

Now suppose we have another script, perhaps `workspace.testing2`, which says:

```
workspace.testing()
```

Let the language popup in `workspace.testing2` say **UserTalk**. If we press the **Run** button in `workspace.testing2`, we see: "You called me from UserTalk, or with the Run button." But if we now change the language popup in `workspace.testing2` to be **AppleScript** and press its **Run** button again, we see: "You called me from AppleScript."

UserTalk script objects in the database can be called from an AppleScript script. There is no need for a `tell` block; you are already inside Frontier, so you can call UserTalk verbs directly.

AppleScript scripts cannot speak of any other database objects, so two special verbs are provided to allow database entry values to be manipulated.

To set or get a database entry's value from an AppleScript script:

```
db.set (entryPathString, value)
db.get (entryPathString)
```

In general, UserTalk datatypes and AppleScript datatypes match, so you can hand values between AppleScript and UserTalk as parameters in a verb call, or get and set database entries, without difficulty.

AppleScript on the Fly

Up to this point, we have been assuming that the AppleScript script is prepared and living in the database in advance. Sometimes, though, this is too much trouble or is not feasible: it is desired, instead, to generate and execute a few lines of AppleScript on the fly, from within a UserTalk script. This is easy to do with a utility script.

The approach taken here is: given a string, to create an AppleScript script object, put the lines of the string into it, and call it.

Example 33-1. doAsAppleScript()

```
on doAsAppleScript(s)
    local (sc); new(scriptType, @sc)
    script.setLanguage(@sc, "AppleScript")
    target.set (@sc)
    local (x)
    for x = 1 to string.countFields (s, cr)
        op.insert (string.nthField(s, cr, x), down)
    return sc()
« here's a little test
doAsAppleScript ("display dialog \"Howdy.\"")
```

OSAXen

An OSAX is an AppleScript language extension. AppleScript scripts can call OSAXen transparently;[*] when an OSAX lives in the *Scripting Additions* folder, its commands become part of the AppleScript language.

UserTalk scripts, too, can call OSAXen. They do this by way of glue. This extends the UserTalk language by permitting it to take advantage of the many existing OSAXen. (For other ways of extending UserTalk, see Chapter 23, *Extending the Language*.)

To generate UserTalk glue for an OSAX, do this. First, make a temporary copy of the OSAX file, and rename the copy as you want its database entry to be named.

---

[*] "OSAXen" is the correct plural of "OSAX." Don't ask me how I know; I *know*.

The new name should contain no spaces or other nonstandard characters. Now, in Frontier, choose **Load an OSAX** from the **AppleScript** submenu of the **Main** menu, or drop the OSAX file onto Frontier's icon. This will create a glue table for the OSAX in `system.extensions`, and will also load the OSAX itself as a binary into the glue table. Now you can delete the renamed copy of the OSAX.

At this point you have a choice. You can simply proceed to use the OSAX by calling the glue verbs. But there are disadvantages to this approach. Some OSAXen, for one reason or another, won't work properly unless they live as files in the *Scripting Additions* folder: for example, they expect themselves to be there, or they need to be able to modify themselves, or they require additional resources contained in the files.

So you might want to tweak the glue to call the copy of the OSAX in the *Scripting Additions* folder rather than the one in the database. To do so, go through the script objects in the glue table, changing the first parameter in their `appleEvent()` calls to 0. This, as explained Chapter 32, *Driving Other Applications*, sends the Apple event to the system, which passes it to the system-level copy of the OSAX.

If you distribute scripts that call OSAXen, don't forget that they won't work unless those OSAXen are installed on the user's machine.

Import/Export

An AppleScript script object can be exported as a file which can be opened or run by other AppleScript scripting programs, such as Script Editor, Script Debugger, and OSA Menu. To do so, choose **Export** from the **Main** menu, and choose **Script Editor Format** from the first popup in the Export dialog.

An AppleScript file which is already in Script Editor format (because, for instance, it was created by Script Editor) can be imported into the database directly. Just choose **Load AppleScript** from the **AppleScripts** submenu of the **Main** menu, or drop the file onto Frontier's icon. Frontier can import an uncompiled script, a compiled script, or an AppleScript "application"; the object is loaded into the `workspace` table. A folder of such files can be imported all in one move with the **Load Folder** command; a subtable is created in the `workspace` table, and the files are imported as its entries.

If you own Script Debugger, you can use it to edit (and debug) AppleScript script objects in the Frontier database. See Chapter 35, *External Editors*.

34

Driving Frontier from Outside

In an earlier chapter, we saw how Frontier can make other applications do its bidding by sending them an Apple event (Chapter 32, *Driving Other Applications*). In this chapter, we turn the tables: other applications can make Frontier do their bidding by sending it Apple events. And this aspect of Frontier, like so many others, is open to user modification; when Frontier receives an Apple event, it calls a script within the database, and you can modify that script. In fact, you can determine what Apple events Frontier can respond to, and how it will respond to them.

There is also a second mechanism by which other applications can drive Frontier. We have seen that if AppleScript is present in the system, Frontier can run Apple-Script scripts internally (Chapter 33, *AppleScript*). By the same token, if Frontier is running, certain applications can run a UserTalk script internally, almost as if the application contained Frontier. The mechanism is parallel to menu sharing (Chapter 26, *Menus and Suites*), except that with menu sharing, it is the user who is calling Frontier, by choosing from a menu. In this case, the applications themselves are able to initiate UserTalk scripts.

We divide the discussion into three parts. First, we explain how certain applications can run UserTalk scripts internally. Then, we talk about some universal Apple events to which Frontier is prepared to respond, especially the `'dosc'` (doScript) Apple event, which even an application with limited scripting abilities may be able to send. Finally, we explain how Frontier can be configured to respond to custom Apple events.

OSA Scripting Applications

The Open Scripting Architecture (OSA) is a system feature making certain scripting languages (called "OSA dialects") available systemwide, so that applications can ask the system to compile and run scripts written in those languages. UserTalk is an OSA dialect, so it is available as a scripting language from within applications that are OSA-savvy in the right way. Such applications include the Script Editor and HyperCard. Nisus Writer, too, with its `Frontier Do Script` command, compiles and runs UserTalk scripts through the OSA.

A UserTalk script written in these environments must use the semicolon and curly-braces syntax. When it runs, it runs in the host application's context (so that, as with menu sharing, Frontier dialogs can appear in the host application); the application is not driving Frontier by way of Apple events, but is running the UserTalk script internally through the OSA, almost as if it were Frontier.

Nisus Writer

One convenient way to use Nisus Writer's `Frontier Do Script` command is to construct the UserTalk script in a clipboard, and then execute the contents of that clipboard with the `Frontier Do Script` command.

Here's an example. Frequently, I use Nisus Writer to edit the contents of Frontier wptext objects (such as Web pages to be rendered with Frontier's Web site management tools). Then I need a way to send the text back into the Frontier database. In the Nisus macro in Example 34-1, a Nisus document is broken into pieces and each piece is stored as a wptext object in the Frontier database. The script looks for lines of this form:

```
#entrySource user.websites.mySite.gerbils
```

to tell it where the pieces of the document are and where they should go; everything after such a line, up to the next such line or the end of the document, is placed into a wptext at the stated location.

Example 34-1. sendToFrontier, a Nisus macro

```
0
jump "1"
// quote and single-quote
q = numtochar(34)
sq = numtochar(39)
loop:
// get a title, make a new wptext there
find next "^\#entrySource\s\(.*\)\r" "gT-W"
eval 'clipboard = "\1"'
setselect (selectend, selectend)
entry = clipboard
```

Example 34-1. sendToFrontier, a Nisus macro (continued)

```
clipboard = "new(wptexttype,@" + entry + ")"
frontier do script '\CC'
// get the whole content and copy it into clipboard 0
theStart = selectstart
find next "^#entrySource" "gE-W"
setselect (theStart, selectStart)
copy
// in clipboard 1, formulate commands...
// ...to put clipboard into our new entry
1
clipboard = "target.set(@" + entry + "); "
clipboard = clipboard + "wp.setText(clipboard.getvalue("
clipboard = clipboard + sq + "TEXT" + sq + ")); "
clipboard = clipboard + "target.clear()"
0
frontier do script '\C1'
goto loop
```

The complementary "other half" of this Nisus macro—the UserTalk script that sends a wptext object to Nisus for editing—is given as Example 35-2.

HyperCard

HyperCard is a program in which one can design a custom graphical user interface whose components respond to user actions by running scripts and sending one another messages in an object-oriented manner. HyperCard scripts can be written in HyperTalk or in any OSA dialect present on the computer when Hyper-Card starts up, including UserTalk.

HyperCard is a particularly compelling environment in which to run UserTalk scripts, because it effectively wraps Frontier in a custom-designed application-like interface. The user, instead of encountering Frontier's menus, edit windows, and database, sees instead what appears to be a single-purpose application with such familiar features as clickable buttons, editable fields of text, meaningful menus, and so forth—and is unconscious of the fact that Frontier may be running in the background, supplying some of this "application's" functionality.

To make a HyperCard script whose language is UserTalk, make sure Frontier is running before HyperCard starts up. Then open a HyperCard object's script and change the popup menu at the top of the script window to **UserTalk**.

UserTalk scripts inside HyperCard can receive HyperTalk messages. For example, if a HyperCard button script written in UserTalk says this:

```
on mouseup() { « notice the parentheses -- it's UserTalk!
    dialog.notify ("Frontier, at your service.") }
```

clicking the button will bring up the Frontier dialog within HyperCard.

If a message passes a parameter, a UserTalk script can receive it; this is a good way to communicate values between HyperTalk and UserTalk. Suppose, for example, a stack's background script is in UserTalk:

```
on dialogNotify(s) {
    dialog.notify (s)
```

Then you can pass a message up to it from a HyperTalk button script, sending along a parameter:

```
on mouseup
  dialogNotify ("This stack has" && ¬
  the number of cards of this stack && "cards.")
end mouseup
```

UserTalk scripts inside HyperCard can drive HyperCard in the same way that User-Talk drives any application—by calling verbs in its glue table. For example, this HyperCard button script written in UserTalk obtains the value of the HyperCard global variable **userName**. This illustrates another way to communicate values between the two languages:

```
on mouseup() {
    local (userName);
    with objectmodel, hypercard {
        userName = get(variable["UserName"]) };
    dialog.notify(userName) }
```

HyperTalk lacks UserTalk's wide range of datatypes—for the most part, HyperTalk values are strings, or numbers—but this is not usually problematic when passing information between them.

As a rather less trivial example, imagine using HyperCard as a database-like store for email received in Eudora. Figure 34-1 shows a possible background layout.

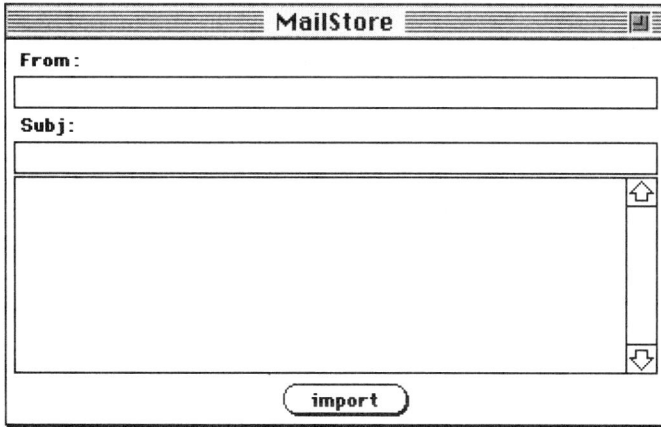

Figure 34-1. HyperCard front-end to Frontier actions

The three editable background fields are named "fromFld", "subjFld", and "textFld". The **Import** button might then have a UserTalk script to import parts of the currently selected Eudora message into the fields of the current card:

```
on mouseUp() {
    local (theFrom, theSubj, theText);
    with objectModel, eudora, eventInfo {
        theFrom = get(message[""].sender);
        theSubj = get(message[""].subject);
        theText = get(message[""].body)};
    with objectModel, hyperCard {
        set (backgroundField["fromFld"], theFrom);
        set (backgroundField["subjFld"], theSubj);
        set (backgroundField["textFld"], theText)}}
```

In real life one would probably have the button script call a script object in the Frontier database, relegating the actual functionality to that script object, which will be much more convenient to edit than a UserTalk script within HyperCard.

Standard Apple Events

Frontier can receive and respond to Apple events; therefore, applications that are not full-fledged OSA scripting applications but can send even a limited range of Apple events can drive Frontier.

How Frontier Responds to Apple Events

When Frontier receives an Apple event, it automatically routes it to a script somewhere within **system.verbs.traps**. The exact location of this script depends upon the name of the Apple event.

Every Apple event is a kind of verb, identified by two string4s, its type and its subtype (Chapter 32). Each entry in **system.verbs.traps** is a table whose name is an Apple event type; each entry in each of those tables is a script whose name is a subtype of that type. In these names, case is significant (because it is significant in Apple event identifiers). Frontier calls the script whose location corresponds to the type and subtype of the Apple event it has received; the parameters of the Apple event are handed as parameters to the script. The script must therefore have an eponymous handler.

So, for example, if Frontier receives an Apple event of type 'core'/'setd', it calls the script at **system.verbs.traps.core.setd**. A 'core'/'setd' Apple event has two parameters: the "direct object," specifying the object whose value is to be set, and the 'data', which supplies the new value for that object. So **system.verbs.traps.core.setd()** takes two parameters.

Frontier responds to five Apple events from the Standard Suites (Chapter 32; see also Chapter 47, *Apple Event Suites*):

- `'core'/'delo'`, corresponding to `core.delete`

- `'core'/'doex'`, corresponding to `core.exists`

- `'core'/'getd'`, corresponding to `core.get`

- `'core'/'setd'`, corresponding to `core.set`

- `'misc'/'dosc'`, corresponding to `misc.doScript`

The first four are about database entries, and to implement them, Frontier defines an object type `cell`. So for example, an AppleScript script running in Script Editor can say this:

```
tell application "UserLand Frontier™"
    get Cell "user.name" -- "Matt Neuburg"
end tell
```

The `'misc'/'dosc'` Apple event is surely the most important, because it can be used to tell Frontier to evaluate any UserTalk expression—which is to say that it can make Frontier do anything whatsoever.

I shall illustrate how to drive Frontier with these Apple events from two commonly used applications, Microsoft Word and FileMaker Pro.

Microsoft Word

Microsoft Word can send Apple events by running an AppleScript script. We have already seen how to use AppleScript to send a `'core'/'getd'` Apple event. We shall now make Microsoft Word do the same thing.

The following WordBasic macro puts up a dialog within Word, asking for the name of a Frontier database entry; it then types into the current Word document the value of that database entry. The strategy is to build the AppleScript script as a string and then send it:

```
Sub MAIN
On Error Goto done
cell$ = InputBox$("What database entry would you like?")
q$ = Chr$(34)
s$ = "tell application " + q$ + "userland frontier™" + q$
s$ = s$ + " to get cell " + q$ + cell$ + q$ + " as string"
result$ = MacScript$(s$)
Insert result$
done:
End Sub
```

If, for example, when the dialog appears, the user types **user.initials** and clicks **OK**, the macro constructs and executes this line of AppleScript:

```
tell application "userland frontier™" to get cell "user.initials"
```

This sends a `'core'/'getd'` Apple event to Frontier. The result **man** comes back from Frontier (on my machine), and is inserted into the Word document.

FileMaker Pro

FileMaker Pro can run an AppleScript script, but it can also send a `'misc'/` `'dosc'` Apple event directly, without passing through AppleScript at all. The second method being much faster and rather more interesting, we choose to illustrate it.

Suppose we have a FileMaker database in which one of the fields, "FilePath," is to contain a file pathname. To set its value, we wish to be presented with a dialog that will let us choose a file; the pathname of the chosen file should then go into the "FilePath" field. We'll get Frontier to implement this functionality.[*]

First, we'll prepare Frontier. A script at **people.[user.initials].setFMFile-Path** goes like this:

```
on setFMFilePath()
    local (thePath)
    if file.getFileDialog("Pick a file:", @thePath, 0)
        with objectModel, filemaker
            set (currentRecord.cell["FilePath"], string(thePath))
            bringToFront()
```

Now we create a FileMaker script that initiates the Frontier script by way of a `'misc'/'dosc'` Apple event, as shown in Figure 34-2. The second checkbox at the bottom is unchecked, and this is very important; the script will "hang" if we don't make sure to do this.

The reason is that FileMaker is not "re-entrant": Frontier can't talk to FileMaker while FileMaker is waiting for a response to something it has already said to Frontier. Remember, we're going to initiate the Frontier script from FileMaker; FileMaker is going to send Frontier an Apple event, and as long as it is waiting for a response, Frontier cannot send an Apple event to FileMaker. But the `set()` command does try to send an Apple event to FileMaker. Unchecking the second checkbox guarantees that when this happens, FileMaker will be idle and drivable, and not still waiting for a response to the Apple event that *it* sent.

---

[*] The example is slightly artificial, since an AppleScript-based solution without Frontier would be possible; but I find Frontier easier to program and snappier in execution than AppleScript. In any case, the *form* of solution presented here is very commonly used.

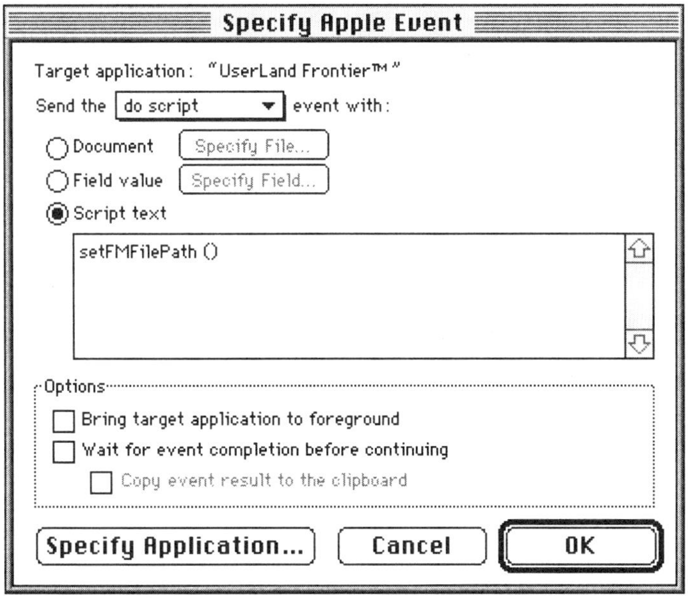

Figure 34-2. Making FileMaker send an Apple event

Now all we have to do is launch the FileMaker script from within FileMaker, in any of the usual ways (for example, by pushing a button in the layout, or choosing from the **Script** menu).

A commonly asked question is how to get FileMaker Pro to send Frontier a `'misc'`/`'dosc'` Apple event in which part of the UserTalk expression is calculated at the time of sending. The solution is to select the **Field value** radio button in Figure 34-2 instead of the **Script text** button, and have the field in question be a calculation field which constructs the UserTalk expression.

One reason why this is important is so that the Frontier script needn't rely upon the `currentRecord` object to specify what record contains the field in question. Suppose we were to launch our FileMaker script, which in turn initiates our Frontier script, and then, in the moment between the `thread.evaluate()` call and the `set()` call, the user, or a received Apple event, were to cause FileMaker to switch to another record. The `set()` call would then set the field in the wrong record. No matter how unlikely we think this is, it is a possibility that cannot be ruled out; and it would become much more likely if our Frontier script performed some time-consuming task between the two calls.

To handle this possibility, we might have a FileMaker calculation field whose formula is:

```
"setFMFilePath(" & Status(CurrentRecordNumber) & ")"
```

The UserTalk expression for our `'misc'`/`'dosc'` Apple event would come from this field. Now we simply modify our UserTalk script so that it accepts one parameter, which is the record number:

```
on setFMFilePath(whatRecord)
    local (thePath)
    if file.getFileDialog("Pick a file:", @thePath, 0)
        with objectModel, filemaker
            set (record[whatRecord].cell["FilePath"], string(thePath))
            bringToFront()
```

Custom Apple Events

Since it is possible to add objects to `system.verbs.traps`, it is possible to configure Frontier to respond to any desired Apple event. The various reasons why this might be done are exemplified by the existing entries in `system.verbs.traps`.

You might be an application developer wanting your application to be able to send Frontier some custom message or class of messages. Thus, MacBird and droplets are applications specifically developed by UserLand to speak to Frontier in a special way; this is implemented through the `IOWA` and `OHIO` tables. Recent versions of BBEdit include a feature added through cooperation between User-Land and Bare Bones, so that BBEdit can act as an external editor for wptext and string objects in the database; this is implemented through the `R*ch` table and the `odbeditor` suite (see Chapter 35, *External Editors*).

An application might establish a custom message that any other application can receive, and you might want Frontier to be able to take advantage of this. Thus, Eudora permits any application to "register" with it, indicating a desire to be notified of such events as the arrival or sending of mail; Frontier can register itself with Eudora, upon which it receives communications from Eudora through the `CSOm` table and the `mail` suite. WebStar (formerly MacHTTP) was the first Web server to implement CGIs on Mac OS, which it did by sending a custom Apple event; Frontier was configured to respond to this through the WWWΩ table and the `webserver` suite, and thus became a CGI application (see Chapter 40, *CGIs*).

You might devise a custom Apple event to be sent from one copy of Frontier to another as a kind of private communications link. That's how the mini-application NetFrontier works; it lets copies of Frontier speak to one another over an Apple-Talk network through a set of custom `'netf'` Apple events (see Chapter 39, *NetFrontier*).

Driving Frontier from a Web Page

A peculiar application of custom Apple events makes it possible to drive Frontier by clicking a link in a Web page in Netscape Navigator. (This method doesn't work with Microsoft Internet Explorer.) It is easiest to describe by a hands-on example.

In Frontier, select and copy `webBrowser.protocols.otherhandlers.usrtlk`; now open `webBrowser.protocols.activehandlers` and paste into it. Save the database and quit Frontier.

In a word processor such as SimpleText, create a Web page that includes the following HTML:

```
<html><head>
<title>Wow</title></head><body>
<p>Click
<a href="usrtlk:frontier.bringToFront();dialog.notify(%22Hi!%22)">me</a>.
</p></body></html>
```

Save the page as *wow.html* and quit the word processor. Start up Navigator if it is not already running; make sure it is the only browser running. Start up Frontier. Switch to Navigator and open *wow.html* with it. Click the link in the sentence "Click me." Frontier will come to the front and display "Hi!" in a dialog.

The way this works is as follows. Navigator can call a helper application to handle particular protocols when a link is clicked. When Frontier starts up with a `usrtlk` script in the `activehandlers` table, if it finds a browser running, it tells the browser that any `usrtlk` links should be passed to Frontier (this is called "registering a protocol"). When you click the link in the Web page, Navigator sees the `usrtlk` protocol in the `href` and sends Frontier an Apple event of type `'WWW?'`/ `'OURL'`, with a parameter consisting of everything after the colon in the `href`. Frontier dispatches the parameter to the `usrtlk()` script, which passes it through `toys.urlDecode()` and then evaluates it. Thus, Frontier ends up executing a script:

```
frontier.bringToFront()
dialog.notify("Hi!")
```

In this example, we were just playing, so the Web page was a local file. Obviously, though, it is possible to serve up, across a network, a Web page containing a `usrtlk` link. If a person who downloads that Web page has Frontier, and is using as browser a copy of Navigator with the `usrtlk` protocol registered to Frontier, *any* desired UserTalk script can be made to run on that person's computer in response to their clicking the link.[*]

---

[*] To represent a space in the script, use `%20`; to represent a double quotation mark, use `%22`. If you generate the Web page with Frontier's Web site management tools, you can create the link by way of a macro call to `embeddedUserTalk()`, and this URL-encoding is taken care of for you.

This scenario may sound far-fetched over the Internet, but it is not at all far-fetched in an intranet situation (in a corporate setting, for instance) where there is coordination as to what's on client computers. I know of system administrators who are using this client-side scripting feature to powerful effect. Those concerned (rightly!) with security should consult `webBrowser.protocols.safetyCheck()`.

35

External Editors

Previous chapters (Chapter 32, *Driving Other Applications* and Chapter 34, *Driving Frontier from Outside*) have described Frontier's ability to drive other applications and to be driven by other applications, through the sending and receiving of Apple events. This chapter is about one particular integrated use of these abilities: editing a database object within another application.

Certain types of object, though editable within Frontier, might sometimes be more conveniently edited in another application. It is possible to copy such an object from Frontier, paste it into the other application, edit it there, copy it, and paste it back into Frontier; but instead of this, the process is automated, so that Frontier and the other application appear seamlessly integrated. From within Frontier, the object is directly opened for editing as a document in the other application; when, within the other application, the document is saved, its data replaces that of the original object in the Frontier database.

This integration requires that the other application be written in such a way as to notify Frontier automatically when saving a document that came from Frontier. At present, the following applications and object types support such integration: MacBird for card binaries; Script Debugger for AppleScript scripts; and BBEdit or PageSpinner for wptexts and strings.* Nevertheless, we shall also suggest in this chapter how to extend the mechanism to other applications.

---

* For Script Debugger, see *http://www.latenightsw.com/*. For BBEdit, see *http://www.barebones.com/*. For PageSpinner, see *http://www.algonet.se/~optima/pagespinner.html*.

The odbEditor Suite

As we have already seen (Chapter 25, *Dialogs*), to edit a MacBird object, you select it and choose **Edit with App** from the **Main** menu; MacBird opens and comes to the front, displaying that object in its edit window. When you save a MacBird card from within MacBird, the card binary in the Frontier database is updated and the Frontier database is saved.

This model is generalized to other object types and other applications by way of the `odbEditor` suite. When you select a database object and choose **Edit with App**, the object's address is handed to `odbEditor.edit()`, which proceeds to seek an editor appropriate to the object's type from the `odbEditor.editors` table. Each entry in this table is itself a table, with four entries:

- `apps`, a list of the creator codes of applications that can edit this type of object

- `canEdit()`, a script which, when handed the object, decides whether it is of a type that this editor can edit

- `edit()`, which can be called to hand the object to an application for editing

- `traps`, the scripts to go into `system.verbs.traps` to receive notifications from the editing application

`odbEditor.edit()` begins by polling the `canEdit()` scripts in order, looking for one which will return `true`, meaning that its editor can edit the object. If a `canEdit()` script returns `true`, `odbEditor.edit()` looks in `user.odbEditors` to see if the user has already registered a choice of editor for this type of object. If so, and if that application is at the specified location on disk, the corresponding `edit()` script is called; otherwise, the list of creator codes in `apps` is used to try to set the pathname to an appropriate editor application, and if that doesn't work, the user is offered a chance to select an application in a file dialog.

So, to establish Script Debugger as your external editor for AppleScript scripts, select an AppleScript script and choose **Edit with App**; usually, Script Debugger will simply open the script directly (if Frontier has trouble finding a copy of Script Debugger it will ask to be pointed in the right direction). Similarly for BBEdit (or PageSpinner) and wptexts. If you were using BBEdit for wptexts and now wish to use PageSpinner instead, delete `user.odbEditor.textEditor` and change `odbEditor.editors.Text.apps` to read:

```
{'JyWs', 'R*ch'}
```

The important thing here is the order of the two items in the list. In fact, I switch between using PageSpinner and BBEdit often enough that I have a utility script to do it for me.

Example 35-1. switchTextEditor()

```
on switchTextEditor ()
    try{delete(@user.odbeditors.texteditor)}
    local (addrList = @odbeditor.editors.text.apps)
    local (x = addrList^[1])
    delete (addrList^[1])
    addrList^[0] = x
    dialog.notify ("The current text editor is " + \
        file.fileFromPath(file.findApplication(addrList^[1])))
```

In the rest of this chapter, I discuss some further specifics of the implementation, and then talk about ways in which the mechanism can be extended.

Script Debugger

The `edit()` script hands the AppleScript script to Script Debugger by way of an Apple event, and receives it back by way of **system.verbs.traps.asDB.EDsv**. The process is simple because the OSA serves as a bridge between Frontier and Script Debugger.

I have found that an error message in Script Debugger when the window is closed can be avoided if this line, in **suites.odbEditor.editors.AppleScript.edit()**:

```
local (requestCloseEvents = true)
```

is changed so that **true** reads **false**.

When an AppleScript script from the database is exported to Script Debugger's context, calls to script objects in the Frontier database appear in "pipes" in Script Debugger's window, like this:

```
set x to |db.get|("user.name")
```

This is to prevent Script Debugger, which is not inside Frontier's context, from seeing such expressions as a syntax error. Nonetheless, the script can still be run and debugged in Script Debugger. If you add new calls to Frontier verbs in Script Debugger, surround the verb names with "pipes" manually. The "pipes" will be copied back into Frontier when the script is saved, but the script will still run.

BBEdit and PageSpinner

The `edit()` script saves the object out to a textfile and tells the application to open the textfile. The textfile is placed in the *Frontier Text Files* folder in the *UserLand* folder inside your system folder's *Preferences* folder. The textfile is not deleted later on; it is up to you to clean up the folder from time to time.

Frontier maintains a correspondence between database objects sent to the external application for editing and their textfiles, by pairing them in lists kept at `user.odbEditors.data.openfiles`. If your computer crashes, or you quit Frontier, while a database object is being edited externally, these lists may get out of synch with reality, and your application may seem no longer to work correctly as an external editor for Frontier. Simply quit the application, clean out `user.odbEditors.data.openfiles` and the *Frontier Text Files* folder, and all will be well.

Once BBEdit has opened a textfile, you may use **Save As** to save it to a different location. Frontier is able to keep track because it is notified through `system.verbs.traps.["R*ch"].FMod`. This does not work with PageSpinner.

BBEdit implements syntax coloring based on the suffix of the filename (and the way you have set up its preferences). Frontier assumes that you want HTML syntax coloring, and establishes this by setting `user.odbEditors.textFileExtension` to `".html"`; this causes Frontier to add `".html"` to the name of the textfile. You can change this value, for example to `""` if you don't want syntax coloring. If you delete this database object, Frontier will restore it and set it to `".html"` again.

A weakness in the mechanism that maintains the correspondence between a database object and its textfile is that the name of the textfile is generated, in `odbEditor.editors.Text.edit()`, by concatenating the name of the object's parent table with the object's own name (with a dot between them) and then truncating the result to 26 characters before appending `".html"`. The truncation is necessary because of the limit on the length of Mac OS filenames, but because it is performed by lopping off the end of the name, two database objects with names that differ only in their final characters can wind up being represented by the same textfile. This was first brought to my attention when a user complained that editing her database object `Action_Learning1996.Actionlearning_abs12` caused a different and previously edited database object, `Action_Learning1996.Actionlearning_abs11`, to be overwritten. An obvious workaround was to shorten considerably the name of the table `Action_Learning1996`.

Scalars

A scalar might be too long to edit comfortably in a table edit window. It is possible to edit a string in BBEdit, but this might be regarded as overkill, and doesn't solve the problem for other types of scalars. We will describe how to create an entry in `suites.odbEditor.editors` that will set a dialog as the "external" editor for scalar types.

The entry is a subtable, `suites.odbEditor.editors.string`. (The name is important only in that it precedes `text` alphabetically; that way, our dialog will be used in preference to BBEdit for strings.) It has four entries:

- `apps`, a one-element list: `{'LAND'}`
- `biggerAsk`, a MacBird card binary
- `canEdit`, a script
- `edit`, a script

Here is `canEdit()`:

```
on canEdit (addr)
    local (typeList = \
        {stringType, doubleType, longType, listType, recordType})
        « extend the list as desired
    local (type = typeOf (addr^))
    return (typeList contains type)
```

You can edit more types by adding them to the `typeList` list; they need to be scalars that can be coerced to a string and back, because the dialog will deal only with strings.

Here is `edit()`:

```
on edit (addrObject)
    scratchpad.tempObj = addrObject^
    card.run(@odbEditor.editors.string.biggerAsk)
    delete (@scratchpad.tempObj)
```

The card `biggerAsk` consists of four items: an **OK** button (`"item1"`); a **Cancel** button (`"item2"`); a large text edit field (`"item3"`); and a non-editable text (`"item4"`). The non-editable text has a recalculation script that runs when the card starts up:

```
nameOf (addrObject^) + ":"
```

The card table has a `startcard()` script:[*]

```
card.setObjectText ("item3", scratchpad.tempObj)
```

The **Cancel** button has an action script that runs when it is pressed:

```
card.close()
```

---

[*] Why are we setting the text edit field's text with a `startCard()` script instead of a recalculation script that runs when the card starts up? Because there seems to be a bug whereby a recalculation script truncates the text to 255 characters, defeating the whole purpose of the dialog. This is a pity, because `addrObject` is visible to a recalculation script, but not to a `startCard()` script; thus, `odbeditor.string.edit()` must copy `addrObject` to a global from which `startCard()` can retrieve it.

The **OK** button has an action script that runs when it is pressed:

```
try {addrObject^ =
coerceTo (card.getObjectText("item3"), typeOf (addrObject^))};
card.close()
```

coerceTo() is given in Example 10-2.

Now when you select a scalar of one of the specified types and choose **Edit with App**, a dialog appears in which you can edit the scalar. Figure 35-1 shows the dialog in action.

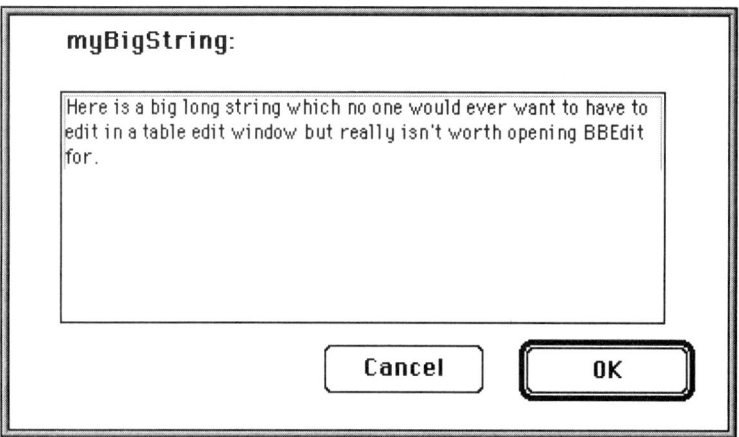

Figure 35-1. The scalar edit dialog

Other External Editors

Can any other applications act as external editors for Frontier database objects? Yes, but not so smoothly and automatically.

The external editing link to Script Editor and BBEdit works because the developers of those applications cooperated with UserLand, agreeing upon a notification mechanism whereby Frontier would be told through an Apple event when a file has been saved or a window has been closed. Other applications lack this notification mechanism. So even if you can drive an application to open a database object for editing, you have no automatic way to hand the edited result back to Frontier.

Still, if the other application has a macro language of its own and can drive Frontier, you can sometimes devise a satisfactory means of communicating data between them.

For example, I sometimes like to edit wptext objects in Nisus Writer. So I have a Frontier script (Example 35-2) which appends the currently selected wptext to the end of the current Nisus Writer document.

Example 35-2. wpToNisus()

```
on wpToNisus ()
    if not sys.appIsRunning('NISI')
        dialog.notify("Nisus isn't running.")
        return
    local (whatToSend = toys.getcursoraddress())
    if typeOf(whatToSend^) ≠ wptexttype
        dialog.notify ("You need to select a wptext first")
        return
    local (s = "#entrySource " + whatToSend + cr + string(whatToSend^))
    clipboard.putvalue(s)
    nisus.bringToFront()
    nisus.doscript("setSelect (endcharnum, endcharnum)")
    nisus.paste()
    frontier.bringToFront()
```

A special line, beginning with the word **#entrySource**, is prefixed to the text, so that Nisus will know later on where this piece of text goes in the Frontier database. When I'm done editing in Nisus, I run the **sendToFrontier** Nisus macro that was discussed as Example 34-1; it puts the pieces of text back into their wptext objects in Frontier.

Even if an application doesn't have a macro language, you may still be able to automate communication between it and Frontier. Consider, for instance, using NoDesktopCleanup and KeyQuencer[*] along with its AESend extension. NoDesktopCleanup can "attach" a KeyQuencer script to any application's **Save** menu item, so that when you choose it, the script is triggered. Thanks to the AESend extension, that script can send an Apple event to Frontier. Since that is almost identically what choosing **Save** does in BBEdit or Script Debugger, you've effectively turned the application into an external editor for Frontier.

---

[*] For NoDesktopCleanup, see *http://persoweb.francenet.fr/~alm/*. For KeyQuencer, see *http:// www.binarysoft.com/kqmac/kqmac.html*.

VI

Applied Frontier

Earlier sections have described what Frontier is; the question now is what to do with it. Frontier is a tool for building tools; what *those* tools should do is up to you. But you do not have to build all your own tools from scratch; many have been included in the Frontier database. In fact, many users come to Frontier explicitly to take advantage of these included tools.

Some of the tools are small and simple, little more than illustrations or suggestions of the kind of thing you might have Frontier do. But others, such as Frontier's famous Web site management facilities, are powerful, flexible frameworks whose functionality rivals that of commercial programs.

In this section, most of the included tools are described. You can use them "straight out of the box," or let them serve as inspiration for further UserTalk programming of your own. Frontier is an open system: the UserTalk scripts that lie behind these tools are there for you to study.

The section starts with some basic tools, the ToDo List and the Scheduler, before proceeding to network-related materials. Frontier's abilities to serve as a TCP/IP client or server application are described, and there is a chapter on NetFrontier, which lets copies of Frontier talk to one another across an AppleTalk or TCP/IP network. Then the use of Frontier as a CGI application is explained, and, at long last, we'll examine the Web site management tool that lets you construct and manage Web sites of any desired size and complexity. Finally, CGIs and Web site management are combined to make dynamic Web sites, sites whose content changes automatically over time or in response to commands given through Web pages.

These tools are all suites. On how a suite is implemented, see Chapter 26, *Menus and Suites.*

Besides the tools included with Frontier, there are many others written by users all over the world and available on the Internet. Space and time don't permit these to be documented here, so your journey of discovery is only beginning. As you become an adept UserTalk programmer, you too may contribute to the ever-growing library of Frontier tools.

36

ToDo List

The ToDo List is a simple, minimal tool, no more than an example, really; but, precisely as an example, it suggests how quickly and easily you could assemble your own Frontier tools to make your life more convenient. It is a self-contained suite with clear functionality and a convenient menu interface.

The `todo` suite maintains two lists of tasks, in the form of outlines (at `todo.lists.todo` and `todo.lists.done`). Show these lists by way of a menu interface, which appears as the **ToDo** menu when you choose **ToDo List** from the **Suites** menu: **Edit To Do List** shows the one, **Edit Done List** shows the other. Add items to the ToDo list manually. When you've completed an item, select it in the ToDo list and choose **Move To Done List**; the item is cut from the ToDo list and filed in the Done list under the date and time when it was completed.

The suite also comes with an agent which, if copied to `system.agents` and renamed `To Do Monitor`, and if given priority in the Main Window, posts there every 60 seconds the next item on the ToDo list.

That's all! But before you dismiss this as trivial, ask yourself: what would it take for this to be a practical, valuable ToDo list application? Categories of things to do? Priorities? Extra notes that can be appended to each thing to do? How would you implement all of that? Such things would make good exercises for the reader...

37

Scheduler

The Scheduler is a suite that triggers the performance of tasks associated with a certain time, or with the passage of a certain time interval since the previous performance. You might want to perform a backup every night in the middle of the night, or put up a dialog every couple of hours telling you to take a break, or put up a dialog once telling you that your dentist appointment is in an hour.[*] True, a recurring task could also be implemented simply as an agent; but the Scheduler brings all such tasks together under one roof, as it were, and avoids the problem of all agents being triggered when Frontier first starts up.

Every task waiting to be performed is represented by a table in `user.scheduler.tasks` (a "task table"), consisting of four entries:

error
> Temporary storage for an error message in case a runtime error arises during execution of the task. Its initial value is the empty string.

taskTime
> The date-time when the task should next be performed. It is pointless to specify a time more precise than to the minute, and even then there is a possibility that the task won't be performed until other tasks have cleared the queue.

minutesBetweenRuns
> The number of minutes after `taskTime` when the task should be performed again. When the Scheduler performs a task, it automatically increases its `taskTime` by adding `minutesBetweenRuns` to it repeatedly until `taskTime`

---

[*] Those who, after studying the Scheduler, find it inadequate to their needs, may wish to consult Preston Holmes's cron suite, at *http://siolibrary.ucsd.edu/Preston/scripting/root/suites/cron.html.*

is after the current time (this is called "rescheduling"). If a task is to be performed only once, `minutesBetweenRuns` should be 0; when the Scheduler performs such a task, it deletes its task table.

script

> A string, which will be evaluated when `taskTime` comes; this is the action of the task. `script` will probably represent a verb call, so performing the task will consist of calling that verb. Tasks are performed with `evaluate()`, not `thread.evaluate()`; this is why a task may have to wait past its `taskTime` until other tasks have finished performing. (Of course, the task itself could involve spawning a thread; that's up to you.)

The heart of the suite is an agent, `system.agents.["Scheduler Monitor"]`, which calls `scheduler.monitor()` to do all the work. `scheduler.monitor()` first checks the current time; then, for every task whose `taskTime` is at or before that moment, it reschedules the task and performs it; finally, it calls `clock.sleepFor()` in such a way that the agent will run again at the start of the next minute.

While a task is running, `scheduler.currentTask` contains the address of its task table. The task, as it runs, can use this information if desired; it might be used to maintain a global between calls, or a task could change its own `taskTime` or `minutesBetweenRuns`—even its own `script`.

Suppose, for example, you have a PTA meeting on the third Tuesday of every month, so you want a dialog to appear each month on that day. That's a complicated concept, but the Scheduler can handle it because the task can reschedule itself. The task table might be set up with an initial `taskTime` of `clock.now()+60` and a `minutesBetweenRuns` value of 60; these values aren't very important because the first performance of the task will change them (without putting up the dialog). This works because `scheduler.monitor()` reschedules a task *before* running it; so the Scheduler reschedules our task, but our task then reschedules itself and has the final say.

If `script` is `"pta()"`, then `people.[user.initials].pta()` might look like:

```
on getThirdTue(mo, yr)
    local(aDate, aDay = 1)
    aDate = date.set(aDay, mo, yr, 1, 0, 0)
    while date.dayOfWeek(aDate) != 3
        aDate = date.set(++aDay, mo, yr, 1, 0, 0)
    aDate = date.set(aDay+14, mo, yr, 1, 0, 0)
    return aDate
local (theTask = scheduler.currentTask)
local (day, month, year, hour, min, sec, aDate, today)
date.get(clock.now(), @day, @month, @year, @hour, @min, @sec)
today = date.set(day, month, year, 1, 0, 0)
aDate = getThirdTue(month, year)
```

```
if today < aDate « reschedule for this month
    theTask^.taskTime = aDate
if today == aDate
    dialog.alert ("You have a PTA meeting today.")
    today = today + 1 « fall thru to next test
if today > aDate « reschedule for next month
    theTask^.taskTime = getThirdTue(month+1, year)
```

Missed Tasks

There is some question of what to do about tasks that may have been missed because Frontier was not running. You get some say in this by setting `user.scheduler.prefs.reschedule`. If you set it to `true`, then when Frontier starts up, if it finds that a task was missed and that its `minutesBetweenRuns` is greater than 60, a dialog appears letting you know that the task is ready to run; if you click **OK**, the task runs at the top of the next minute. Otherwise—that is, if you set `user.scheduler.prefs.reschedule` to `false`, or if a missed task's `minutesBetweenRuns` is less than 60, or if you click **Cancel** in the dialog—the task is simply rescheduled, as it would have been if it had just been performed.

Thus, regardless of how you set `user.scheduler.prefs.reschedule`, a one-time task missed during the night will be performed at the top of the first minute after startup. If this isn't what you want, you may have to write your own version of `scheduler.monitor()`.

Logging

The performance of tasks (including, if there was a runtime error, the error message) is logged in an outline at `user.scheduler.log`—if it exists. The Scheduler comes with a default hourly task; so if you've experienced a mysterious situation where, after a save, you do nothing with Frontier, and yet when you quit Frontier, it asks whether you want to save changes (*what* changes?), it might be that the change is in `user.scheduler.log`. You may wish to check this outline from time to time; it's up to you to see that it doesn't get ridiculously long.

To turn off logging, just delete or rename `user.scheduler.log`.

Making a Task

There are two chief ways to make a new task. Tasks to be performed every hour on the hour and tasks to be performed at 2:00 A.M. each morning are so common that there are menu items to create them in the **Scheduler** menu that appears when you choose **Scheduler** from the **Suites** menu. Otherwise, the easiest way is to call `scheduler.addTask()`. The syntax is:

```
scheduler.addTask (taskTime, script, minutesBetweenRuns)
```

So, to install our task that puts up a dialog on the third Tuesday of each month, one might say:

```
scheduler.addTask(clock.now()+60, "pta()", 60)
```

38

TCP/IP

Frontier speaks and listens across a TCP/IP network, such as the Internet, by way of NetEvents, a scriptable application with a very small RAM footprint. The glue verbs for NetEvents are at `system.apps.verbs.netEvents`. The user can call any of these directly; there are also higher-level verbs that hide the details, and this chapter concentrates mostly on the latter.

NetEvents basically just manages TCP/IP connections, referred to as *streams*. One may picture streams as analogous to telephone calls: NetEvents is like a telephone operator, managing a bank of phones. It can open a stream (like calling a server elsewhere on the network), write to a stream (speak over the phone), read from a stream (hear over the phone), and close a stream (hang up the phone). It can also listen for an incoming stream, which means that clients elsewhere can call *it* on the phone. And it can perform a few utility functions, such as obtaining domain name resolution.

However, NetEvents knows nothing of what to say over the phone, or how to interpret what it hears; that's up to your scripts. The higher-level verbs described in this chapter do all the hard work. Thanks to them, Frontier and NetEvents can easily be made to function as either a client or a server on a TCP/IP network.

Server

A *server* is an application which receives and obeys commands over a network. To continue our telephone conversation analogy: with TCP/IP, a computer's IP address is its phone number. A server is assigned to a particular *port*, which is like an extension at that phone number; one computer might have several servers running, so when a client program phones in, it asks for the particular port whose server performs the desired function. By convention, a few standard server types are assigned standard port numbers; other port numbers are free for custom servers.

A server doesn't want clients to get a busy signal, so it implements enough phones on its port so that at least one will pretty surely be free at any given moment. It is then up to the server to manage talking on all the phones at once if need be. This can be a challenge, but Frontier and NetEvents are multithreaded, so they are up to the job if the scale is reasonably contained.[*]

So, for Frontier to act as a server, it first tells NetEvents to implement a bank of phones and to do the actual "listening" to see if there's a call. When a call comes in, NetEvents patches it through one of the phones and hands the phone to Frontier. Frontier must know how to be a good server on this port: it will be expecting some particular command or commands, and it should obey and reply in whatever the proper manner may be. When the conversation is over, Frontier tells NetEvents to hang up the phone.

inetd

Frontier's basic server functionality is implemented through the `inetd` suite. The name "`inetd`" comes from the UNIX command which sets up a server, or "Internet daemon." A simple menu interface to the suite appears as the **Inetd** menu when you choose **Internet Daemon** from the **Suites** menu.

To use the `inetd` suite, you first describe your server as a table (the "daemon table"); this table will usually be an entry in `user.inetd.config`, but, as we shall see, it doesn't have to be. The daemon table must have three entries:

port
> The TCP/IP port number for your server: if a client phones up asking for this port, your `daemon` server script will be called.

count
> The number of listeners you want to open on this port (i.e., the size of your server's phone bank). This will be the number of listening threads that Frontier will spawn, and represents the maximum number of clients your server will be able to respond to simultaneously.

daemon
> The server script which will be called when a client phones up.

---

[*] Experience will determine what the limits are. Each thread takes up memory in NetEvents's heap space and memory in Frontier's heap space, and slows down other operations slightly.

The `daemon` script should take one parameter; this parameter will be the address of a table. The table will contain an entry `stream`, the stream ID identifying the particular client who has just phoned up; you will need `stream` in order to speak with this client by way of NetEvents. The table also contains an entry `client`, the IP address of the client, encoded; you can decode this, if the information is needed, by handing it to `netEvents.nameToAddress()`. Once your script is called, it should find out what the client is saying, by calling `netEvents.read-Stream()`, and then reply, by calling `netEvents.writeStream()`. Your script does not close the connection; that is handled automatically when your script returns.

To start your server running, call either `inetd.startOne()` or `inetd.start()`. `inetd.startOne()` takes as parameter the address of your daemon table. `inetd.start()` takes no parameters: it creates listeners for all daemon tables in `user.inetd.config`. (**Start Daemons** in the **Inetd** menu does the same thing.) These verbs call a utility, `inetd.supervisor()`, in a new thread, `count` times for each daemon. Each supervisor thread calls `netEvents.listenStream()` to create a listener (a "phone"), and then loops waiting for a client to appear on that listener. If a client does appear, the supervisor thread calls the daemon script. The daemon script does whatever it does, and when the daemon script returns, the supervisor closes the stream, and returns to the listening state. Your daemon script may need to consider that it could be called simultaneously in different threads.

To shut down all listeners, set `user.inetd.shutdown` to `true`. (**Stop Daemons** in the **Inetd** menu does the same thing.) This causes all supervisor threads to stop looping, and this in turn causes each NetEvents listener to time out (because the supervisor loop was what was keeping it alive); this may take some time, but after a while you should see all the "state" squares in the NetEvents status window turn gray.[*]

Web server example

To illustrate, we use the `inetd` suite to turn Frontier into a primitive Web server. (This is basically the example included with the `inetd` suite.) Set up a daemon table; since we're just testing, set `count` to 1, and the `port` must be 80, because this is where browsers will expect the server to be. The `daemon` script is shown as follows.

---

[*] It seems incredible that the only way to shut down one inetd server is to shut all of them down; but that's the way it is, as far as I can tell.

Example 38-1. Minimalist Web server

```
on daemon (addrParams)
    local (request, f, filetext, url, serverfolder, htmltext)
    request = netEvents.readStream (addrParams^.stream, 10000)
    serverfolder = "HD:qpq:" « fix this!
    url = string.nthField (request, ' ', 2)
    url = string.delete (url, 1, 1) « pop off leading /
    f = serverfolder + string.replaceAll (url, "/", ":")
    try
        filetext = toys.readWholeFile (f)
    else
        filetext = user.webserver.fileNotFoundPage
    htmltext = "HTTP/1.0 200 OK\r\n\r\n" + filetext
    netEvents.writeStream (addrParams^.stream, htmltext)
```

The fourth line hardcodes a folder which will serve as the root folder of our "Web site"; you'll need to rewrite this line, designating a folder on your hard disk. Put into that folder an HTML file called *default.html*. Now call `inetd.start-OneDaemon()`, handing it the address of the daemon table. Wait until NetEvents shows `LISTENING` in its status window. Then go to another computer, start up a browser, and ask it to open the URL *http://otherComputer.com/default.html*, where *otherComputer.com* is the name or IP number of the server computer. The HTML file will be displayed in the browser!

When our daemon script is called, it calls `netEvents.readStream()` to obtain the browser's request. `10000` is just an arbitrary number of bytes, larger than the expected size of the request. We assume that the request begins like this:

GET /default.html HTTP/1.0

This is followed by a lot of other stuff which we ignore; remember, this is a *primitive* Web server! We parse the second word to get the file's pathname, read the file from disk, and send it back with `netEvents.writeStream()`.

When you're finished playing, set `user.inetd.shutdown` to `true` on the server computer. While Frontier would probably not substitute for an industrial-strength dedicated Web server application, this is a convincing demonstration of its potential.

odbServer

The `odbServer` suite sets up Frontier as a server when another copy of Frontier is to talk to it across a TCP/IP network; for this purpose it is easier to use, and more flexible, than `inetd`.

The `odbServer` architecture uses a "dispatcher." There is a table, `user.odb-Server.commands`, containing scripts ready to respond to commands arriving over the network. When a command arrives, it is routed by name to the correct script in `user.odbServer.commands`. The receiving script should take one parameter, the address of a table. The entries in this table are strings; the script should know what to do based on their names and values. Whatever the script returns will be sent back over the network to the client.

There are only four commands to know:

`odbServer.serverStart()`
> Given on the server side. Calls `inetd.startOne()` to create listeners in accordance with the specifications in `suites.odbserver.daemon`. To stop the threads and time out the listeners, set `user.inetd.shutdown` to `true`.

`odbServer.commandEncode(`*commandName, addrAttributeTable*`)`
> Used on the client side to prepare a command for sending; *commandName* is the name of a script on the server machine in `user.odbServer.commands` to be called and handed as parameter a copy of the table at *addrAttributeTable*.
>
> May also be used on the server side to return a table as a reply; in this case, *commandName* must be the empty string.

`odbServer.commandDecode(`*encodedReply, addrTable*`)`
> Used on the client side to decode a reply, if it was encoded on the server side. The encoded reply *encodedReply* is decoded into the table at *addrTable*.

`odbServer.commandSend(`*encodedCommand, server*`)`
> Used on the client side to send to the server the *encodedCommand*, which is the result of `commandEncode()`. *server* is the IP address of the server computer; if omitted, `user.odbServer.prefs.serverAddress` will be used.

The architecture sets up just one server, on port 1997; but that server can easily handle numerous commands, because each command is dispatched to a different script.

The server is given an opportunity to preprocess the attribute table before it is handed to the command script; to take advantage of this, modify `user.odb-Server.commandFilter`.

remoteGet example

As an example, we will use `odbServer` to implement a rudimentary version of the NetFrontier `remoteGet()` functionality described in Chapter 39, *NetFrontier*. The idea is that a client copy of Frontier will be able to obtain from a server copy of Frontier the value of any entry in the latter's database.

At the server end, we have a script `user.odbServer.commands.remoteGet()`.

Example 38-2. An odbServer server script

```
on remoteGet (addrTable)
    local (whatToGet = addrTable^.whatToGet)
    local (replyTable); new(tableType, @replyTable)
    if defined (whatToGet^)
        pack(whatToGet^, @replyTable.value)
        replyTable.type = typeOf(whatToGet^)
        replyTable.error = ""
    else
        replyTable.error = "Not defined."
    return odbserver.commandEncode("", @replyTable)
```

The script expects the parameter table to contain one entry, **whatToGet**, a string that can be coerced to the address of a database entry. The script returns a table with three entries: **error**, to communicate error information, if any; **value**, a binary version of the requested database entry's value; and **type**, the requested database entry's datatype. The reason for this elaborate encoding is that **odbserver.commandEncode()** will coerce every entry in our returned table to a string; therefore, we pack the database entry's value to a binary to preserve its contents, and send along its datatype separately because that information will otherwise be lost when the binary is coerced to a string.

The following is a script on the client computer that asks the user for the name of the database entry to get, sends the command to the server computer (whose IP address is assumed to be in **user.odbServer.prefs.serverAddress** already), receives the reply, checks for errors, and deposits the requested entry's value at **scratchpad.reply**.

Example 38-3. An odbServer client script

```
local (t); new (tableType, @t)
if not dialog.ask ("Database entry to get:", @t.whatToGet)
    return
local (packet = odbServer.commandEncode ("remoteGet", @t))
local (reply = odbServer.commandSend(packet)) « talk to the server
odbServer.commandDecode(reply, @t)
if t.error != ""
    dialog.alert(t.error)
else
    local (value = binary(t.value))
    setBinaryType (@value, t.type)
    unpack (@value, @scratchpad.reply)
    edit(@scratchpad.reply)
```

The **value** is coerced from a string back to a binary, assigned the correct internal datatype, and then unpacked, to reconstruct the original value.

Client

In order to act as a client to a remote server, Frontier needs to know the protocol for talking to that server. A *protocol* is a well-defined grammar of command and response. For instance, in Example 38-3, we knew in advance that the `remoteGet` command was expecting a table with one entry called `whatToGet` and that it would return a table with three entries called `error`, `value`, and `type`. A protocol can get quite complicated, because it may involve various commands, where the server must parse the request and do different things in different cases, and then the client must do different things depending on what reply comes back.

The reason we knew the protocol for `remoteGet` is that we wrote the server ourselves. But apart from this sort of situation, how are we going to know any protocols? The standard Internet server types, such as FTP, HTTP, NNTP (Usenet "news"), and POP and SMTP (mail), each have standardized protocols which are widely available.* Some simple servers with unique protocols respond to a command HELP by returning a description of their protocol. Also, if you already have a client application that "knows" a protocol, you may be able to intercept its conversation with a server and deduce the salient features of the protocol.†

tcpCmd

In Frontier, the most important standard protocols are implemented already, in `system.extensions.tcpCmd`. The salient features of this table are as follows:

useUCMD
> A boolean. If `true`, the UCMD at `tcpCmd.code` will be used for talking to the network; if `false`, NetEvents will be used. It is generally agreed that using NetEvents is now better for most purposes.

interfaces
> The bottom-level commands for talking to NetEvents.

ftp, gopher, *etc*.
> Tables containing scripts for each command in a protocol, along with scripts that combine these into larger commonly needed structures.

examples
> A table of scripts and outlines that show some of the commands in action.

---

* For example, the HTTP protocol is available at *http://www.w3.org/Protocols/*.

† On Mac OS, using Open Transport, you can do this with Peter N Lewis's OTSessionWatcher application, available at *ftp://ftp.stairways.com*.

Scripts that perform high-level self-contained sessions with a human interface

For example, `sendMail()` gets your SMTP server to send a mail message; `download()` obtains a file from an FTP server; and so forth.

The `tcpCmd` scripts are far too extensive to document fully here.[*] They provide the same sort of logic and functionality that a full-fledged client application has to provide—a remarkable body of work, especially considering that they must juggle the details of conforming to protocols, interfacing with NetEvents or the UCMD, interacting with the user, and so forth.

FTP example

As an illustration, here is a utility script that relies upon `tcpCmd`. As part of a Web site I manage, I sometimes have to upload an *.hqx* file to the site via FTP. But the FTP server at the far end always unBinHexes these, which isn't what I want; the workaround is to give the file a meaningless suffix such as *.hqxx*, upload it, and then rename the remote copy so that it ends in *.hqx*. The operation is common and tedious, so a scripted version is clearly in order.

To learn to write the script, I first performed the whole operation with OTSession-Watcher running, so that I could capture the TCP/IP messages and figure out the protocol. Then I explored the scripts in `tcpCmd`. Some of these are very high-level, such as `tcpCmd.upload()`, which manages an entire session and was clearly going to be the model for the structure of my script. Others perform large, high-level tasks as part of a session, such as `tcpCmd.ftp.logon()`, `tcpCmd.ftp.putfile()`, and `tcpcmd.ftp.getHostDirectory()`. Some implement sending a single command in the protocol, such as `tcpCmd.ftp.dele()`, which sends the DELE command to delete a file. These are supported by slightly lower-level utilities such as `tcpCmd.interfaces.sendCommand()`, which sends commands not already in the scripted repertory; I would need this for RNFR and RNTO, which rename a remote file.

Example 38-4. uploadAndRename()

```
on uploadAndRename(f)
    local (mySession, dir = "/Servers/WebSTAR/matt/ftp")
    local (fnam = file.filefrompath(f))
    if not (string.lower(fnam) endsWith ".hqxx") {return}
    fnam = string.delete(fnam, sizeOf(fnam), 1) « ".hqx" version
    try
        tcpCmd.ftp.logon(@mySession, "ftp.mySite.com", "neub", "haha")
    else
        scriptError ("Problem logging on: " + tryerror)
    try
        tcpCmd.ftp.cwd(@mySession, dir) « change directory
```

---

[*] See *http://www.erols.com/asg/tcpcmdDocs/*. The scripts are mostly the work of Alan German.

Example 38-4. uploadAndRename() (continued)

```
        tcpCmd.ftp.putfile(@mySession, f) « upload the file
        scratchpad.theList = tcpCmd.ftp.getHostDirectory(@mySession, dir)
        if scratchpad.theList contains (" " + fnam)
            tcpCmd.ftp.dele(@mySession, fnam)
        tcpCmd.interfaces.sendCommand(@mySession, "RNFR " + fnam + "x")
        if mySession[tcpCmd.properties.response] beginswith "350"
            tcpCmd.interfaces.sendCommand(@mySession, "RNTO " + fnam)
        tcpCmd.ftp.logoff(@mySession)
    else
        tcpCmd.ftp.logoff(@mySession)
        scriptError ("Problem after logon: " + tryerror)
uploadAndRename("power:testing.hqxx") « testing stub
```

We first perform a reality check: if the file isn't prepared by ending the name with *.hqx*, the user has made a mistake. Then we log on to the FTP server, get to the right directory with `tcpcmd.ftp.cwd()`, and upload the file. Then we get a directory listing, to see if the directory already contains an earlier version of the file with the name we are about to give ours; if so, we delete it. Now we rename the file, which the protocol decrees is a two-step process (say what file is to be renamed, then say the new name); notice the technique for checking the response from one command before proceeding to the next. That's it, so we log off; the whole operation is also embedded in a `try` so that we can log off if anything goes wrong along the way.

Further Studies

For a splendid example of a nonstandard protocol client implemented from the ground up, obtain the **webster** suite,[*] which gets word definitions from the University of Michigan's dictionary server.

We have seen Frontier act as a rudimentary HTTP server; we have seen it perform an entire FTP session without human intervention and without the use of FTP client software. We have also seen how copies of Frontier can easily communicate over the network as custom client and server. The examples given are elementary, but they all perform quite significant tasks, and should serve to suggest Frontier's potential power as a TCP/IP network citizen.

---

[*] By Preston Holmes; see *http://siolibrary.ucsd.edu/Preston/scripting/root/suites/webster.html*.

39

NetFrontier

NetFrontier is an ingenious suite which makes it easy for Frontier to communicate with other copies of Frontier running remotely. Of course, it would be possible to arrange this yourself: for AppleTalk networks, you could build a Frontier glue table and send a remote copy of Frontier one of the five standard Apple events that it already knows how to receive; for TCP/IP, you could write a server with `odbServer` and communicate with it through a script like the one in Example 38-3. But NetFrontier does it all for you, and more:

- The difference between the AppleTalk and TCP/IP implementations is transparently handled.

- It lets you assign each remote Frontier a convenient nickname (called a "node"), and you can group nicknames into sets so that commands can be sent to multiple copies of Frontier. (For example, Frontier might be used to monitor and configure an array of laboratory computers.)

- It provides both a menu interface and a UserTalk interface, for manual or programmed transactions.

Programming Example

Let's suppose we want a remote Frontier to report to us on the contents and file sizes of everything in its machine's system folder. We will use an approach which allows the scripts that do this to be developed and tested on our own machine.

We begin with a simple adaptation of `samples.basicStuff.buildFolder-Outline()`.

Example 39-1. workspace.outlineFolder()

```
on outlineFolder (path)
    new (outlineType, @scratchpad.otl)
    target.set (@scratchpad.otl)
    op.setLineText (path) « path in main head
    on traverse (path)
        local (f)
        fileloop (f in path)
            op.insert (file.fileFromPath (f) + " " + file.size(f), dir)
            dir = down
            if file.isFolder (f)
                dir = right
                traverse (f) «recurse
                if dir != right «at least one item added from the folder
                    op.go (left, 1)
                dir = down
            rollBeachBall ()
    local (dir = right) «first headline to right of summit
    traverse (path)
    target.clear()
    return true
```

`workspace.outlineFolder()` creates at `scratchpad.otl` an outline of the folder which is its parameter. But initially we don't necessarily know the pathname of the remote system folder, so here's a routine which queries the Finder about that, and hands the result to `workspace.outlineFolder()`.

Example 39-2. workspace.outlineMaster()

```
local (path)
with objectmodel, finder
    path = get (startupdisk.systemfolder, 'TEXT')
workspace.outlineFolder(path)
```

Once we confirm that these routines work locally, our strategy is to transfer them both to the remote Frontier's database, call the remote `workspace.outline-Master()`, retrieve the resulting remote `scratchpad.otl`, and then delete the remote `scratchpad.otl`, `workspace.outlineMaster`, and `workspace.outlineFolder`. This approach, called "put-run-delete," is very common.

NetFrontier's UserTalk interface provides four actions: `remoteGet()`, `remotePut()`, `remoteDelete()`, and `remoteRun()`. One should also clean up after a transaction by calling `remoteQuit()`. The script we shall use illustrates all of these verbs.

Example 39-3. outlineRemoteSysFolder()

```
with netFrontier
    remotePut(@workspace.outlineFolder)
    remotePut(@workspace.outlineMaster)
    remoteRun("workspace.outlineMaster()")
```

Example 39-3. outlineRemoteSysFolder() (continued)

```
    remoteGet("scratchpad.otl", @scratchpad.otl)
    remoteDelete("workspace.outlineFolder")
    remoteDelete("workspace.outlineMaster")
    remoteDelete("scratchpad.otl")
    remoteQuit()
edit(@scratchpad.otl)
```

Literal addresses may be used in referring to the local database, but it is wise to use strings for addresses in the remote database, so that their validity as addresses will not be checked against the local database (see Chapter 8, *Addresses*, and compare our similar concerns in Chapter 30, *Multiple Databases*).

To prepare for execution, it is simplest to use NetFrontier's menu interface. First, choose **NetFrontier** from the **Suites** menu to bring up the menu interface. Now choose **Edit Nodes** from the **Nodes** submenu of the **NetFrontier** menu to designate the remote computer as a node.

All of the commands in `outlineRemoteSysFolder()` can take an optional parameter saying what node is to be communicated with, but if this is omitted, the "current node" is used. So, make sure that the remote computer is the "current node," with **Select Current Node** from the **Nodes** submenu. Now it is sufficient to click the **Run** button in the edit window of `outlineRemoteSysFolder()`. After a while, the outline of the system folder of the remote machine appears on the local machine.

Behind the Scenes

If the remote computer is linked to the local computer by an AppleTalk network, NetFrontier works by creating and sending custom `'netf'` Apple events to the remote Frontier. To see, for example, the heart of `remoteGet()`, look at `netFrontier.ATclient.get()`; it sends a `'netf'/'valu'` event, and the remote Frontier handles it with the script at `system.verbs.traps.netf.valu`.

But if the remote computer is linked to the local computer by a TCP/IP network, such as the Internet, Apple events are not involved. NetFrontier packages the call information and hands it over to `tcpCmd` (see Chapter 38, *TCP/IP*), which in turn drives NetEvents to send the package across the network. At the remote station, another copy of NetEvents is functioning as a listener, and is being polled in a separate thread by a loop set up in `netFrontier.TCPserver.listener()`; when the package arrives, the polling process picks it up and NetFrontier responds. It is only in this context that `remoteQuit()` does anything: it closes the connection between the two copies of NetEvents, returning the listener to listening and the local copy to an idle state; by including the call in Example 39-3, we ensure that the script will work over either AppleTalk or TCP/IP networks.

Further Exploration

NetFrontier is well documented, at `netFrontier.docs`. The following points are worth noting.

- In the previous example, we used the scripting interface to drive NetFrontier. But it is often more convenient to use the menu interface: see the commands in the **Nodes** submenu of the **NetFrontier** menu, which let you transfer database objects to and from the remote machine, run scripts located on the remote machine, write your own QuickScript and run it on the remote machine, and perform a put-run-delete of a script object located on the local machine.

- The remote machine's copy of Frontier must be configured to receive communications. For AppleTalk networking, this means that program linking must be turned on, and the next-to-last item in the NetFrontier menu must say **Disable AT Server** (meaning that it is now enabled). For TCP/IP networking, it means that the user must be given permissions with **Edit Users** from the **TCP Server** submenu of the **NetFrontier** menu, NetEvents must be started up, and **Start Server** must be chosen from the **TCP Server** submenu (to put NetEvents into "listen" mode).

40

CGIs

The term *CGI* (for "common gateway interface") refers to a means whereby Web pages can be programmatically generated in real time in response to a browser request for a document.

When a Web browser requests a document (because the user clicks on a link, or asks for a URL), a Web server somewhere on the Internet receives the request and sends the text of the document back to the browser, which displays it. Typically, the Web server obtains this text by fetching it from a file on disk. But with a CGI, the Web server passes the document request on to another application (the "CGI application") and receives the text from that application. How the CGI application obtains the text is an open question; typically, it "calculates" the text in some way. The CGI application might also perform other actions before returning the text; for instance, it might write to a database. Thus, CGIs can provide a Web browser interface to the computational and data-processing power of a remote computer over the Internet.

An *ACGI* is an *asynchronous* version of a CGI. With a normal CGI, both the Web server and the CGI application are tied up in conversation with one another from the moment the Web server passes the document request to the CGI application to the moment the CGI application gives the Web server the text. The Web server cannot respond to any other requests from the outside world while this is going on. By contrast, with an ACGI, the Web server passes the CGI application a document request and then goes on about its business: it assumes that the CGI application will send a response whenever it is ready.

On Mac OS, the way a Web server and a CGI application speak to one another is with Apple events.

Frontier is a natural choice as an ACGI application. It is programmable; it can calculate text; it can read text from files on disk or from its own database; and it can drive other applications, so they can perform auxiliary functions such as storage, lookup, and calculation. It can respond to custom Apple events; it can send Apple events. It is multithreaded, so it can receive an Apple event while others are still being processed. It implements semaphores, so threads won't trip over one another accessing data. Plus, with NetFrontier, you can easily develop CGI scripts locally and then upload them to a remote copy of Frontier.

Configuring

By convention, the name of a document whose text Frontier is to provide usually ends in *.fcgi*. Let's presume we're going to adopt this convention. The Web server must be told in advance that requests for documents named in this way should be passed on to Frontier. We want the Web server to see Frontier as an ACGI application, so an alias to Frontier should be placed where the server will find it[*] and given the name *frontier.acgi*. A new action needs to be defined, called FRONTIER, which corresponds to the alias `frontier.acgi`. And a new suffix mapping needs to be defined, which associates the action FRONTIER, the suffix `.FCGI`, and the MIME type `text/html`.

Frontier itself also must be configured. When you choose the **WebServer** menu item from the **Suites** menu, the first item in the **WebServer** menu that appears contains an important configuration choice: it says **Use Script-based Trap**, and you can check or uncheck it. This determines whether a script or a UCMD will be placed at `system.verbs.traps.WWWΩ.sdoc`. In the typical situation, where just one Web server program will be calling Frontier, and it is WebStar or a WebStar-compatible Web server, you should check this item.[†] You should also use the next two items in the **WebServer** menu so that Frontier knows what Web server you are using and where your Web site folder is.

Thereafter, whenever the Web server receives a request for a document whose name ends in *.fcgi*, it sends Frontier an Apple event of type `'WWWΩ'/'sdoc'`. Frontier routes the Apple event to `system.verbs.traps.WWWΩ.sdoc`, which calls scripts in the `webserver` suite. There is a "dispatcher" architecture: a script in the `webserver` suite takes the name of the requested document, removes the *.fcgi* suffix, and puts `"suites.webserverScripts"` in front of it to get the name

---

[*] This might be, depending on the server, in the same folder as the server, or in the default folder of the Web site.

[†] Unchecking **Use Script-based Trap**, so that the UCMD is used, restricts the flexibility of your CGI scripts and may cause some occasional bugs; still, in certain unusual situations it may be necessary to do so.

of a script, which is then called.* That script—your CGI script—is handed the information that accompanied the document request; it does whatever it does, and returns a complete piece of HTML text, which is sent back to the Web server and from there across the Internet to be displayed in the requester's browser.†

Making a CGI Script

One way to make a CGI script is to choose **New Script** from the **Main** menu and set the popup menu to **CGI Script**. This generates a script in `suites.webserver-Scripts` with some shell statements and information to jog your memory as you write the script. The script has one parameter, `adrParams`; this is the address of a table holding information from the document request, and the names of its entries are shown in a comment.‡ The text to be returned has been named `htmltext`; the text's construction has begun with a call to `webServer.httpHeader()`, and a local handler, `add()`, has been provided so that you can continue its construction. All you do is call `add()` repeatedly until you've constructed the Web page, and return `htmltext`.

As a first example, we write a CGI script that makes a Web page consisting basically of nothing but the current time. We don't need to refer to any of the parameters in `adrParams`, so our script might look like the following.

Example 40-1. webserverScripts.theTime()

```
on theTime (adrParams)
    local (htmltext = webServer.httpHeader ())
    on add (s)
        htmltext = htmltext + s + cr
    add ("<html><head>")
    add ("<title>What Time Is It</title>")
    add ("</head>")
    add ("<body>")
    add ("<p>The time is:</p>")
    add ("<center>")
    add ("<h1>" + string.timeString() + "</h1>")
    add ("</center></body></html>")
    return htmltext
```

---

* I'm simplifying here; as we shall see in a moment, the object in `suites.webserverScripts` doesn't actually have to be a script.

† The computer with the Web server and Frontier running does not have to be your main Web server. Indeed, your main Web server can be an NT or UNIX computer, provided it can see your Frontier computer's Web server and route CGI requests to it; this lets you take advantage of Frontier for handling CGIs and NT or UNIX for handling Web traffic. One reason for keeping your CGI computer separate from your Web server computer is so that one can crash without taking down the other.

‡ To learn more about the information passed to a CGI application in the Apple event sent by WebStar and WebStar-compatible servers, see *http://www.biap.com/datapig/mrwheat/appleevents.html*.

Testing

To test the above script, ask your browser for *http://yourServer.com/theTime.fcgi*, substituting for *yourServer.com* the IP number or name of the computer where the Web server and Frontier are running. On Mac OS, if your computers don't have assigned IP numbers, you can fake a miniature Internet using two computers networked with AppleTalk: use the TCP/IP control panel to make both computers use TCP/IP networking and give them artificial IP numbers, such as 255.255.255.254 and 255.255.255.255. Open Transport lets you switch conveniently between a fake configuration like this and your normal configuration.

A brilliant feature of the `webserver` suite is its robust handling of errors. Errors do not cause execution to stop or put up an error dialog—it could be a serious problem if they did, since your server computer might be remote and unattended. Instead, errors are trapped and logged in a file, and also are reported back to the browser in an HTML page, which displays the error message that would have appeared in Frontier. This means that calling `scriptError()` in a CGI script is a good way to return a simple error message to the browser.

You can customize the HTML framework of the error page by editing `user.webserver.errorPage`; you should probably leave the `<!--#errorMessage-->` tag alone, because this is what causes the text of the error message to be inserted into the page. (This expression is a "macro"; macros are explained in a moment.) Similarly, you can customize a page at `user.webserver.fileNotFoundPage` which is displayed if the Web server asks Frontier for a "document" that doesn't exist as an entry in `webserverScripts`. You can also increase the error information in the log file by setting `webserver.preferences.verboseLogging` to `true`.

Errors often depend on what was in `adrParams`, which is a local variable and therefore has gone out of existence by the time you learn of the error. A useful debugging script appears at `webserverScripts.samples.tellParams`; it returns to the browser a Web page showing all the entries in the `adrParams` table. To access it, have your browser request *samples.tellParams.fcgi*; or, alternatively, have the first statement of your CGI script simply be this:

```
return webserverScripts.samples.tellParams(adrParams)
```

(Later, we illustrate a trick which lets your CGI script either make this call, for debugging purposes, or perform its normal function.) Also, turning on debugging in the **WebServer** menu causes the `adrParams` table to be saved in the `scratchpad` table each time the Web server calls Frontier.

Macros

Example 40-1 works, but it represents an inelegant and impractical programming model. One will scarcely be motivated to create good HTML through a series of clumsy `add()` calls. A far more pleasant and flexible approach is to store the HTML as a wptext; an eligible place for it is `user.webserver.utilities`, which is open for customized storage.

But how will a call to `string.timeString()` take place from the middle of a wptext? The solution is to use a *macro*. A macro is an HTML comment, of the form `<!--#xxx-->`, where *xxx* refers to a database entry: if *xxx* names a string or wptext, its value will be substituted for the macro; if *xxx* is a verb call, its result will be substituted for the macro.

Here is the syntax for the verb that performs macro substitution:

```
webserver.utilities.processMacros (string, addrTable)
```

where *addrTable* is optional. This verb returns *string*, except that any macros within *string* are replaced by their substitute. Resolution of object references in a macro is by means of a nest of `with`s; Frontier looks first in the table pointed to by *addrTable* if present, then in `webserver.macros`, then in the database as a whole (we will illustrate this in a moment).

So, modifying Example 40-1, we could move the HTML to a wptext at `user.webserver.utilities.timeHTML`, as demonstrated in Example 40-2.

Example 40-2. user.webserver.utilities.timeHTML

```
<html><head>
<title>What Time Is It</title>
</head>
<body>
<p>The time is:</p>
<center>
<h1><!--#string.timeString()--></h1>
</center></body></html>
```

Then, `webserverScripts.theTime()` could look like Example 40-3.

Example 40-3. webserverScripts.theTime() revised

```
on theTime (adrParams)
    local (htmltext = webServer.httpHeader ())
    with webserver.utilities, user.webserver.utilities
        return htmltext + processMacros(string(timeHTML))
```

This method presents the prospect of easily generating powerful, interesting Web pages through a CGI script. The bulk of our HTML text is now a wptext, so it can be developed with any convenient tool and copied into the database, or the wptext object itself can be edited externally with BBEdit or PageSpinner (Chapter 35, *External Editors*).

Form Example

An HTML form requires a CGI application that will process and respond to the form when its Submit button is pressed; we now illustrate how Frontier's CGI framework can deal with a form.

Here is the HTML for a simple form; the idea is that the user registers by giving a name and an email address:

```
<html><head>
<title>Registration Form</title>
</head><body>
<p>Please register.</p>
<form action="register.fcgi" method=post>
<p>Your name:</p>
<p><input type=text size=40 name=name value=""></p>
<p>Your email address:</p>
<p><input type=text size=40 name=email value=""></p>
<p><input type=submit value="Register">
<input type=reset value="Reset Form"></p>
</form>
</body></html>
```

Notice the use of the **post** method; this is important because it causes Frontier to parse the form input items for you.

The following is a CGI script that responds when the user presses the **Register** button in the form. Because the **post** method was used, the form input items arrive as a subtable of the table pointed to by **adrParams**, called **argTable**.

Example 40-4. webserverScripts.register()

```
on register (adrParams)
    local (htmltext = webServer.httpHeader ())
    local (f = file.folderFromPath(frontier.getProgramPath()) + "regs")
    with adrParams^
        if not defined(argTable)
            scriptError ("Data missing.")
        with argTable
            if not defined(name) \
            or not defined(email) \
            or name == "" \
            or email == ""
                scriptError ("Form not fully filled out.")
```

Example 40-4. webserverScripts.register() (continued)

```
        if email == "@@@@" « secret code, call tellParams() for debugging
            return (webserverScripts.samples.tellParams(adrParams))
        if not file.exists (f)
            file.new (f)
            file.setType (f, 'TEXT'); file.setCreator (f, 'ttxt')
        semaphores.lock (f, 3600)
        try
            if not file.writeLine (f, "Name" + tab + name) \
            or not file.writeLine (f, "Email" + tab + email)
                scriptError("Could not write to file.")
        else
            file.close (f) « just in case
            semaphores.unlock (f)
            scriptError (tryError)
        semaphores.unlock (f)
    with webserver.utilities, user.webserver.utilities
        return htmltext + processMacros(string(registerHTML), @argTable)
```

We first check to see that both fields were filled out; if not, we cause an error message to be returned. If all is well, we append the data to a file; because each CGI request generates a new thread, a semaphore is used to make sure that two simultaneous instances of this script don't clash over access to the file. We also include a test for a secret code as the **email** value; this code causes the **tell-Params()** page to be returned, for debugging purposes, instead of performing our script's normal functions.

Finally, we return a response. In this case, it's a Web page acknowledging the registrations; once again, the page is a wptext in **user.webserver.utilities**. The wptext is passed through a call to **processMacros()**, and this time the call includes a second parameter, namely **@argTable**; this enables the macros in the wptext to access the name-value pairs in **argTable**, as follows.

Example 40-5. user.webserver.utilities.registerHTML

```
<html><head>
<title>Registration Complete</title>
</head><body>
<p>You have been registered as
<!--#name-->, email <!--#email-->.</p>
</body></html>
```

An important trick frequently used in form submissions (though not in this example) is to have one or more **hidden** type input items in the form. This allows information which the user doesn't see in the browser to be communicated from the Web page to the CGI script in **argTable** entries. For example, you might have several forms, all of which are submitted to a single CGI script: the script knows which form it is receiving by a **hidden** input whose value identifies the form.*

Other Servables

A CGI script, instead of returning HTML text itself, can ask the Web server to serve up a different document (e.g., another *.fcgi* "document," a document on disk, or a URL elsewhere on the Internet) as its response to the browser request. This is called *redirection*, and is done by returning the result of a call to **webserver.httpHeader()** with **"302 FOUND"** as the first parameter and the URL of the desired page as the second. For example, to return the text of a file on disk after registering the user in Example 40-4, one might put this in place of the last two lines:

```
    return webserver.httpHeader("302 FOUND", "regDone.html")
```

This example uses a relative (partial) URL, but of course you can use a full URL if needed.

The entry in **webserverScripts** requested by the browser does not have to be a script. For one thing, it can be a binary, in which case it is simply returned, with its internal datatype used to determine the MIME type. A table at **user.webserver.mimeTypes** is consulted for correspondences between datatypes and MIME types; you are free to augment this table. Thus, for example, a PDF or a QuickTime movie could be served up from within the database.

Another possibility is that the entry might be a string object, a wptext object, or an outline object. In these cases, the object is coerced to a string, passed through **processMacros()**, and returned, as an HTML document if it contains <html>, as plain text if it does not. The call to **processMacros()** includes **adrParams** as its second parameter, so the entries in the parameter table are available to any macros within the object.

Thus, since Example 40-3 performs no actions of its own—it just passes **user.webserver.utilities.timeHTML** through **processMacros()** and returns the result—we can skip it altogether; instead, we can move the wptext of Example 40-2 to the **webserverScripts** table and name *it* **theTime**, and it will work all by itself!

We can use the same technique, though, even if a CGI does perform actions. In Example 40-4, the **webserverScripts** object was a script, which performed some actions and then called **processMacros()** to return a wptext. But it might

---

* If you use this trick, bear in mind that certain characters cannot appear as the value of an input item. You can get around this by encoding the value with string.urlEncode(); the value is automatically decoded with string.urlDecode() before the CGI script gets hold of it.

have worked the other way round: the `webserverScripts` object might be a wptext, containing a macro which calls a script to perform some actions. This arrangement is illustrated in the next example.

Search Example

When a CGI document request is submitted by a browser to a Web server, two kinds of ancillary argument can be passed along with it, by using this format for the URL:

```
myCGI.fcgi$pathArgs?searchArgs
```

The $ and ? function as delimiters telling the server that what follows, up to the other delimiter or the end of the URL, are path arguments or search arguments, respectively. Both types of argument are optional, but their order is fixed. With the Frontier CGI framework, such arguments arrive as `adrParams^.pathArgs` and `adrParams^.httpSearchArgs`, respectively.

To illustrate one way of taking advantage of this feature, we follow a suggestion given in various tutorial examples.[*] Imagine that we have a FileMaker Pro database of first names, last names, and ages. The user fills out a form to define a search of this database; in a CGI script, we perform the search, and return an HTML page listing all the matching records. We would like this list to consist of live links, so that the user can click on one to see that individual record in full.

Here's the trick. When we generate the HTML for the list of matching records, we include in the `<a>` tags that make them live links a search argument which is the ID of the corresponding record. So, when the user clicks on a link, a CGI script examines `httpSearchArgs`, asks FileMaker for the record with that ID, and returns the information.

Here is the HTML for a search form that the user fills out in the browser:

```
<html><head><title>Search</title></head><body>
<form action="fm.showMatching.fcgi" method=post>
<p>First Name:</p>
<p><input name="firstname" type=text size=40></p>
<p>Last Name:</p>
<p><input name="lastname" type=text size=40></p>
<p>Age:</p>
<p><input name="age" type=text size=40></p>
<p><input type=submit value="Search">
<input type=reset value="Reset Form"></p>
</form></body></html>
```

---

[*] Such as that by Preston Holmes, at *http://siolibrary.ucsd.edu/preston/scripting/FilemakerDemo.html*, or that by Brent Simmons, *ftp://ranchero.com//software/filemaker/writingFilemakerCgis.sit.hqx*.

When the user clicks the Search button in this form, the CGI framework looks for `webserverScripts.fm.showMatching` and finds it is a wptext (as follows).

Example 40-6. webserverScripts.fm.showMatching

```
<html><head>
<title>Matching Records</title>
</head><body>
<p>Here are the records matching your request;
click one to see the full record:</p>
<blockquote>
<!--#user.webserver.utilities.showMatching(@argTable)-->
</blockquote>
</body></html>
```

The real action is performed by `user.webserver.utilities.showMatching()`. Recall that when the object in `webserverScripts` is a wptext, its macros have access to the entries in the parameter table; one of these is `argTable`, which contains the input from the form, so the macro passes along `argTable`'s address in the verb call. `showMatching()` consults FileMaker Pro and generates a series of live links representing the records that match the user's request.

Example 40-7. user.webserver.utilities.showMatching()

```
on showMatching (addrArgs)
    local (s = "<p>No entries matched your request</p>", x)
    semaphores.lock("FMP", 3600)
    try
        with fileMakerLib, addrArgs^
            findAll(1)
            local (ct=find(1,{firstname,lastname,age},{"first","last","age"}))
            if ct
                s = ""
                for x = 1 to ct
                    s = s + "<p><a href=\"fm.showRec.fcgi?" + getID(1,x)
                    s = s + "\">" + getCellNthRecord(1,"first",x) + " "
                    s = s + getCellNthRecord(1,"last",x) + "</a></p>\r"
    else
        semaphores.unlock("FMP")
        scripterror(tryerror)
    semaphores.unlock("FMP")
    return s
« for instance, if the user put "M" in the Last Name field:
    « <p><a href="fm.showRec.fcgi?5">Jim Morison</a></p>
    « <p><a href="fm.showRec.fcgi?8">Mickey Mouse</a></p>
```

The comment at the end shows the sort of HTML that is returned; each link asks for the same *.fcgi* "document," but appends a different search argument, which is the ID of the record. These links are inserted into the wptext of Example 40-6, which in turn is returned to the browser. If the user then clicks on a link, the CGI framework encounters the following wptext.

Example 40-8. webserverScripts.fm.showRec

```
<html><head>
<title>One Record</title>
</head><body>
<p>Here is the record you requested:</p>
<blockquote>
<!--#user.webserver.utilities.showThisRec(httpSearchArgs)-->
</blockquote>
</body></html>
```

In the call to `user.webserver.utilities.showThisRec()`, we pass the search argument, which tells `showThisRec()` which full record to retrieve.

Example 40-9. user.webserver.utilities.showThisRec()

```
on showThisRec (theID)
local (theRec, cellNames = {"First Name","Last Name","Age"})
    semaphores.lock("FMP", 3600)
    try
        with fileMakerLib
            theRec = getRecordByID(1, theID)
    else
        semaphores.unlock("FMP")
        scripterror(tryerror)
    semaphores.unlock("FMP")
    local (x, s)
    for x = 1 to 3
        s = s + "<p><b>" + cellNames[x] + ":</b> "
        s = s + theRec[x] + "</p>\r"
    return s
« for example...
    « <p><b>First Name:</b> Mickey</p>
    « <p><b>Last Name:</b> Mouse</p>
    « <p><b>Age:</b> 60</p>
```

When the `get` method is used and not the `post` method (because a form is not being submitted), there is no `argTable` parameter: the CGI script has to parse `httpSearchArgs` itself. If `httpSearchArgs` is in the standard form of `name=value` pairs, `string.parseHTTPargs()` can parse it (see Chapter 14, *Strings and Chars*); the most convenient method is to call `webserver.parseArgs()`, which passes its first parameter to `string.parseHTTPargs()` for parsing and then builds the result into a table at the address which is its second parameter, thus returning the exact equivalent of an `argTable`.

Security

CGIs can be dangerous, and Frontier CGIs, although they do take some precautions, are no exception. Frontier has great power over your computer, and a CGI can cause it to exercise that power. To give an extreme example, if you were to install a script in `webserverScripts` containing this line:

```
evaluate( string.parseHttpArgs (adrParams^.httpSearchArgs ))
```

then someone could construct a Web page containing a link which, when clicked, would execute on your server computer absolutely any UserTalk script they desired. The possible results are positively unthinkable.

The key to security-mindedness is to assume that someone will figure out how your CGI scripts work. Do not imagine that you can keep the scripts secret; on the contrary, pretend that they are published for the world to see. Don't put into a CGI script anything which could be dangerous if misused.

Even if your CGI scripts contain no security holes, you should assume that if they perform interesting or useful functions of universal application, someone will attempt to "hijack" them. For example, a Frontier user once wrote a Web form containing an input item which told the CGI script what page to show next; the input item was called `newPage`, and the CGI script ended with a redirection to whatever URL `newPage` contained:

```
return (webserver.httpHeader ("302 FOUND", adrParams^.argTable.newPage))
```

As a result, other people who could not do CGIs on their own server machines put up Web pages of their own which implemented the same functionality by way of the Frontier user's server! The server, and Frontier, responded to these "foreign" pages just as they did to the Frontier user's own pages. No harm was done, but Frontier was kept very busy handling requests from a lot of other people's Web sites.* The solution in a case like this is for your script to check `adrParams^.referer` to make sure that the request is coming from one of your own pages.

Keep It Up

The UserLand folks have distilled the wisdom of years into some tips for keeping a Macintosh healthy and quick when it runs unattended with a Web server and Frontier. I've very little experience with this myself, so the following just summarizes advice from a UserLand Web page.†

---

* I owe this story to Brent Simmons.

† *http://www.scripting.com/contentserver/servertips.html.*

Put an alias to Frontier, and nothing else, in the *Startup Items* folder.* Have a Frontier startup script launch any other applications (such as the Web server) or open any documents (such as FileMaker databases), spacing out the timing of the commands with `clock.waitseconds()` or some similar device. Have an agent in Frontier maintain the Web server as the frontmost application.† Don't open any unnecessary applications, and certainly not a Web browser, on that machine; avoid other client applications, such as email or FTP, if you can (NetEvents is fine). Turn off file sharing if possible. Don't edit live on the machine; edit elsewhere. Close all windows: Finder windows, the Web server's status window, everything; you could even hide the Frontier Main Window.

Make sure you're using good system software. Have the latest version of AppleScript. Have the latest version of QuickTime, as it includes important Thread Manager components. But don't use QuickTime MPEG (with the new Thread Library), QuickTime Conferencing, or QuickTime TV. Run Autoboot (restarts the computer in case of a crash) and OkeyDokey (clicks **OK** buttons in modal dialogs).‡ Remove *all* unneeded system software.

Assign Frontier and the Web server plenty of memory. Have plenty of real RAM so you can do so; don't use RAM Doubler or similar. Avoid unnecessary plug-ins for the Web server, but you can use an image map plug-in (for older browsers that don't support client-side image maps). Turn off DNS lookup on the Web server.

Don't forget to back up the root now and then. And the same for the hard disk as a whole.

If you are not using the script-based trap, it may be unwise to save the database while a CGI thread is executing; a crash can result.§ If you find this to be the case, install as an agent the agent script at `webServer.data.agentScript()`. Notice the values in `user.webServer` on which it depends. In recent versions of the CGI framework, installation of the agent is taken care of through the menu interface (the **Auto-Save** submenu of the **Webserver** menu).

---

* Exception: if you're using Open Transport, it can be slow when your Web server talks to a slow modem client. The solution is an unsupported application, OTTCPSlowLinkTuneup2. Get it from *http://www.macintouch.com/otslow.sit.hqx* and drop it in your *Startup Items* folder.

† My personal experiments suggest that Frontier CGIs run a lot faster if Frontier is frontmost, but these were not under real-world conditions, so perhaps the results are misleading.

‡ As of this writing, the latest versions were *http://hyperarchive.lcs.mit.edu/HyperArchive/Archive/gui/okey-dokey-pro-203.hqx* and *http://hyperarchive.lcs.mit.edu/HyperArchive/Archive/cfg/auto-boot-15.hqx*.

§ The problem is that saving while a UCMD is executing can cause a crash. If `user.webserver.classicTrap` is `true`, then a UCMD is executing so long as a CGI script is executing.

Further Explorations

The material we have covered will suffice for the vast majority of CGI needs. Various `readme` wptexts are scattered about the `webserver` suite, giving further information about its use.

Earlier, we had the Web server define an action and a suffix mapping so that the suffix `.FCGI` would correspond to an action `FRONTIER`, which is passed on to `frontier.acgi`, an alias of Frontier. It is possible to define other actions corresponding to other suffixes, and pass these to `frontier.acgi` as well. In this case, Frontier looks for the requested "document" not inside `suites.webserver-Scripts`, but in `user.webserver.actions`. A script there should match the name of the action.

Your Web server may also come with certain built-in actions that don't depend upon the name of the requested document at all. For example, you can have Frontier preprocess every file that the Web server serves up, by configuring its built-in `PREPROCESSOR` action to be sent to Frontier.

The CGI framework includes a macro action which permits files served from disk to be handed to `processMacros()`. This, however, is largely superseded by the `cgiMacros` suite, which defines new macro commands that can be given from *.fcgi* files served from disk.[*]

Frontier includes a powerful tool for generating Web pages; this is the subject of the next chapter. A CGI script can make a Web page by means of the Web page generation tool, taking advantage of its various abilities: to generate HTML from an outline, to apply a header and footer that fit the look of the rest of the Web site, and so forth.

---

[*] The `cgiMacros` suite is not documented in this book, because at the time of this writing it was unfinished.

41

Web Site Management

A Web site is a collection of Web documents (HTML, graphics, and so forth) that you maintain for people to look at over the Internet. You might have your own server, or you might upload documents to a remote server; it makes no difference. Frontier is ideally suited, not just for making an individual Web page, but for creating and maintaining an entire Web site which might consist of many, many pages.

To see why, consider this: HTML is just text—text that's tedious for humans to deal with directly, but easy for a computer to help generate. Furthermore, the different pages of a Web site often need certain things in common to make for a pleasant, efficient browsing experience. Pages may want a similar look. Navigation might require "Next" and "Previous" links to guide the reader through the various pages. There might be a page which outlines the entire site, with links to the other pages, functioning as a hypertext table of contents. Frontier can generate text files computationally, so it can easily take care of these sorts of details.

In this chapter, we describe Frontier's Web site management facilities. These are implemented mostly by the `html` suite, which uses menu items and the database itself as its user interface. At the end of the chapter, we describe an alternative user interface which allows you to take advantage of Frontier's Web site management facilities through BBEdit and menu sharing.

Frontier's Web site management facilities are powerful, flexible, and generous. Whatever your needs, not only is there probably a way to do it with Frontier, but there is probably more than one way. Suggestions and examples given in this chapter are often not the only means of achieving the desired effect.

The Frontier Web site management facilities also provide lots of room for you to add to them and adapt them by writing scripts of your own. In learning about Frontier Web site management, it is easiest to start by building a simple site and confining oneself to the many features offered by Frontier directly "out of the box."* Once familiar with these, one can begin writing scripts that adapt the system to achieve special effects. UserLand also makes available the source code of many of its own Web sites; you can learn from and borrow this code. And there are Frontier mailing lists and Web sites where users share code that they have written to achieve particular effects.

In the unlikely event that the `html` suite cannot meet one's special needs, it is possible to make a copy of it and modify the code at a deeper level; Frontier is an open system, after all, and the `html` suite is just a suite like any other.

Architecture

The architecture presupposed by the `html` suite is that the raw material for the site's pages, including any graphic images, is stored in the database. Frontier will then computationally produce from this the actual HTML files and graphic files that make up the site. This is called *rendering* the site.

Under this architecture, the terms "Web site" and "Web page" become ambiguous. A Web site can be a table in the database, or the folder on disk into which the site is rendered. A Web page can be a database object in a Web site table, or the file that is created from it when the object is rendered. I will speak of the objects in the database as *Web site tables* and *Web page objects*, which represent the folders and files that are created when the site is rendered.

The part of the `html` suite that actually processes Web page objects to turn them into Web page files is termed the *rendering engine*. The rendering engine performs various transformations upon a Web page object in order to produce a Web page file, but it does not actually change the Web page object in the database; only the user will do that. A Web page object is like a set of instructions from the user on how to make a Web page file; clearly, Frontier must not alter the instructions. In this chapter I may sometimes say that the rendering engine does this or that to a Web page object, but I don't mean that it literally changes the object itself: I mean that it transforms a temporary copy of the object, in the course of producing the Web page file.

Keeping the site material in the database takes advantage of the fact that database objects are Frontier's natural domain, and lets the rendering engine derive

---

* An online hands-on tutorial (modesty prevents my mentioning the name of the author) is available at *http://www.scripting.com/matt/webtutorial.*

meaning from such native features as the hierarchy of subtables. Even more important, it is as easy for Frontier to generate a very large number of files as to generate one, whereas the user's work is kept to a minimum because in the database are instructions for making the Web site; a small change in those instructions is enough to change the whole site. Suppose, for example, that for an existing site, it is decided that a certain logo should appear at the top of every page. Instead of editing each page to insert an `<img>` tag, the user introduces one `<img>` tag in one database object, and has Frontier re-render all the pages.

Web Site Tables

A Web site table should be located inside `user.websites`, and should be created by choosing the **New Site** menu item from the **Web** menu, which makes a new table and inserts certain important components into it.* A Web site table may contain the following sorts of entries:

- A subtable called `#ftpSite`.† The entire site should contain exactly one of these, in the table representing the site's root level; in fact, its presence *actually* means, "this is the site's root level." It should have at least these three entries, which must be configured by the user:

  isLocal

  > Should initially be **true**. If **false**, Frontier will render pages and upload them with an FTP client to a remote site, all in one motion; this is discussed in "Previewing and Releasing," later in this chapter.

  folder

  > The pathname of a folder on disk, into which the site is to be rendered. Neither the folder nor any of its containing folders need exist (though of course the volume must); the rendering process will create them if necessary.

  url

  > The URL of the remote directory from which the files and folders in **folder** will be served up over the Net. This value is of diminished importance and for many purposes can be left unset without penalty. If you set it, do not include `"http://"`, and do add a final slash.

---

* When you choose the **New Site** menu item from the **Web** menu, the dialog that appears expects you to type a full path to the Web site table that is to be created. Many users forget this, and type, for instance, **myNewSite** when they mean **user.websites.myNewSite**. The former does not work; it will cause a mysterious-looking error.

† It is permitted, in theory, for the `#ftpSite` entry to be an address, pointing to a table; but in fact some Web site management features break if you do this.

- An entry called `#template`. This is typically a wptext or outline.* Web page objects are embedded into their templates before being written out to disk as files; this embedding is how opening and closing HTML, and header and footer elements common to several pages, are attached. Templates are explained under "Templates," later in this chapter.

- An entry called `#filters`. This is typically a table† containing two scripts named `pageFilter` and `finalFilter`. The use of these scripts is explained under "Filters," later in this chapter.

- Any other entries whose names begin with `#`. These, along with those already listed, are *directives*. Directives set values which will be used during rendering. Many directive names have special meaning: the rendering engine looks specifically to see what value you have assigned them, if any, so that it knows how you want it to behave. You are also free to make up directive names of your own choosing; these are like variables to which you can refer within page objects. Directives are explained under "Directives," later in this chapter.

- A subtable called `glossary` (or, optionally, `#glossary`). This is a table of name-value pairs for substitution at rendering time; its use is explained under "Glossaries," later in this chapter.

- A subtable called `tools` (or, optionally, `#tools`). This may contain anything at all, such as utility scripts associated with the site table, and is a convenient hiding place for objects which might otherwise be mistaken for Web page objects.

- A subtable called `images`. This is for binaries of graphic images referred to by pages in the site; see "Macros," later in this chapter, on the `imageRef()` script.

- Any other wptexts, outlines, strings, scripts, fileSpecs, or addresses, whose names don't begin with `#`. These are the Web page objects representing the HTML files that constitute the site. The name of the entry determines the name of the file; the suffix `".html"` (or any other desired suffix) will be added during rendering, and should not be part of the entry name. The treatment of these entries is explained under "Web Page Object Datatypes," later in this chapter.

- Any other subtables whose names don't begin with `#`. These represent sub-folders of the Web site folder; the name of the subtable determines the name of the folder. These subtables may contain any of the elements listed here, except an `#ftpsite` entry. Web page objects are generally grouped by table

---

* It may, instead, be the string name of a wptext or outline in `user.html.templates`.

† It may, instead, be the address of such a table.

so as to apply directives to them in common, as explained in the next section, "Directives."

The table `user.websites` is itself a Web site table, and may be examined to see some characteristic Web site table entries.

Directives

A *directive* is a user-defined name-value pair.* Frontier knows that you are defining a directive because the directive name is preceded by #. So, for example, if a Web site table contains an entry named `#link` whose value is the string `"0000FF"`, then Frontier responds to the # by associating a name `link` with a value `"0000FF"`. The association takes place in a special table, `html.data.page`, which is used as a scratchpad during rendering; so in this case Frontier will create, during rendering, an entry `html.data.page.link` whose value is `"0000FF"`.

Defining Directives

A directive may be defined, as we have just seen, by making it a Web site table entry, in which case the name of the entry tells the name of the directive (preceded by #) and the value of the entry tells the value of the directive.

Alternatively, a directive may be defined within a Web page object that is a wptext or an outline. To do this, a paragraph (or, in an outline, a summit-level line) must begin with # followed by the directive name and then a space; everything in the paragraph (or summit-level line) after this space is the value of the directive. The rendering engine will remove these directive definitions after obeying them (so that they don't show up in the final HTML of the Web page object).

Here is how the rendering engine obeys directives in a wptext or outline: it constructs a UserTalk assignment statement, placing the directive's value on the right of the equals-sign; the statement is then evaluated. For example, if a paragraph in a wptext Web page object says:

```
#link "0000FF"
```

then the rendering engine will construct a string:

```
"html.data.page.link = \"0000FF\""
```

---

* Actually, some directives are defined, in the course of rendering, by the rendering engine; the most important of these are discussed later.

and evaluate it. So what comes after the space must be something that *can* go on the right side of a valid UserTalk assignment statement. Moreover, it can be anything that can go there. For example, one might say:

    #link blue

or:

    #link workspace.calculateMyColor( )

Thus, there are two important points of difference between the directive value as assigned through an entry in a site table and the directive value as assigned within a wptext or outline object.

First: in a table, the datatype of an entry is fixed beforehand; you can see it in the Kind column. But in a wptext or outline the datatype of the value will depend on the rules of UserTalk as the assignment statement is evaluated. That's why, when we say:

    #link "0000FF"

the quotation marks are needed around `"0000FF"`, to express a string literal when the rendering engine constructs the assignment statement.

Second: in a table, the value of an entry is fixed beforehand; you can see it in the Value column. But in a wptext or outline, the evaluation takes place at rendering time. For example, when we say:

    #link workspace.calculateMyColor()

the rendering engine will assign to `html.data.page.link` whatever is the result of calling `workspace.calculateMyColor()` at the particular moment of rendering.

Directives with large string values

Assigning a directive a lengthy string value can be tricky. That's because there is a limit of 255 characters on the length of a string literal. But a string can be any length, so one solution is to build the string out of shorter literals, like this:

    #meta "a very long string" + "another very long string"

A possibly more convenient solution is to have the value be a wptext (which can be as long as you like) coerced to a string:

    #meta string(user.websites.mySite.tools.myMetaWptext)

This problem arises for very few directives, but when it does, it's nice to know what to do.

The Directive Hierarchy

The chief reason for defining a directive in one location rather than another is that there is a directive hierarchy, as follows:

- A directive defined in `user.html.prefs` applies to every Web page object you ever render. This table is a kind of global directive repository. Directives in this table should *not* have # before their name!

- A directive defined in a Web site table applies to every Web page object in that table or in a subtable of that table.

- A directive defined in a Web page object applies to that page alone.

Where the same directive is defined in more than one place, the one "nearest" the page object being rendered is the one that applies. In effect, for any given directive, we look for its value first in the page object itself, then in the table containing the page object, then in the table containing that table, and so on up to the root; and then in `user.html.prefs`.[*]

The directive hierarchy enables a sort of object-oriented system of inheritance and override. If a directive is defined in a table, all the page objects and subtables in that table inherit the definition, though each of them can override it. This is what I meant earlier when I said that "Web page objects are generally grouped by table so as to apply directives to them in common." The organization of a Web site's folders on disk is irrelevant to the experience of the human reader browsing the site; it is the links within the pages that organize the site for the reader. So the organization of Web page objects in tables (and consequently folders) is available for this use.

Special tables that are really directives are variously handled. For example, the `tools` table is treated like an ordinary directive: the `tools` table "nearest" the page object being rendered is the one automatically associated with that page object. But entries in `glossary` tables are sought in every `glossary` table all the way up the table hierarchy. Exceptional handling, like that of the `glossary` tables, is noted later.

A Directory of Directives

Here is a reference list of the chief directives consulted during the rendering process. Some of these will be more comprehensible when you have read the rest

---

[*] I have simplified by leaving out templates; see "Templates," later in this chapter. There are situations where `user.html.prefs` is not consulted, and there are a few cases where only `user.html.prefs` is consulted. There probably shouldn't be; but there are. To enumerate these would overly complicate the exposition.

of this chapter. Nevertheless, you should probably read through the list now, as a study of the directives is a good way to learn what sort of thing the rendering engine does for you.

Unless otherwise noted, all directives default to `true` if not defined somewhere in the hierarchy. Some directives are automatically given defaults in `user.html.prefs`; those defaults are not listed here (consult the table, or `html.init()`).

Header directives

The template typically begins with a macro that calls the `pageHeader()` script (in `html.data.standardMacros`). This script constructs the opening HTML for the page, including the `<head>...</head>` region. The following directives correspond to material that goes in this region.

#title
> The string value to appear between the `<title>` tags. The default is the name of the page object.

Most browsers use this value in the titlebar of the window.

It would be silly to assign a #title directive anywhere but in a page object, because every page object should have a unique #title—other features of the Web site management system use the #title to identify a page. Nevertheless, I like to assign `user.websites.#title` the value `"Hey, NO TITLE!!!!!!"`; that way, if I forget to assign a page object a title, I see this when I view the rendered object in the browser, and am reminded.

#includeMetaCharset
> A boolean. If `true`, a `<meta>` tag is inserted saying what character set we are using.* The name of the character set comes from another directive, #charset, which is set in `user.html.prefs` as `"iso-8859-1"`.

#meta
> The entire value is inserted in the `<head>` area of the page. Optional.

Despite its name, you can use #meta to put anything you like into the `<head>` area, not merely a `<meta>` tag; nothing is prefixed or appended to the value of the literal when it is inserted (except a return character afterwards), so forming a tag is up to you. You only get one #meta directive per page (if you define two, the second will override the first), so if you have several tags to insert, they all

---

* The presence of such a `<meta>` tag causes some browsers to load the page twice. This is often undesirable.

need to go into this one directive value. See "Directives with large string values," earlier in this chapter.

#javascript

> The value is surrounded with <script> tags and comment delimiters, and is inserted into the <head> area. Optional.

See "Directives with large string values," earlier in this chapter. The surrounding HTML is taken care of for you, so the value need be only the actual JavaScript script. However, you should precede any left curly braces, and perhaps any left double quotation marks, by a backslash, to prevent the rendering engine from seeing these as macros and glossary entries, respectively (as explained later).

Body tag directives

The template typically begins with a macro that calls the **pageHeader**() script (in **html.data.standardMacros**). This script constructs the opening HTML for the page, including the <body> tag.[*] It looks for the following directives, all optional, corresponding to parameters of the <body> tag.

#link

> The hex color value for links in the page

#alink

> The hex color value for a link in the page as the user clicks it

#vlink

> The hex color value for links in the page that have been visited

#bgcolor

> The hex color value for the page background

#text

> The hex color value for text in the page

To type a hex color value as the value of a string object in a table, it may help to include the surrounding double-quotes. Or, you can use any of the predefined constant names in **system.verbs.colors**.

#background

> The URL for an image file to be used for tiling the page's background. The use of this is further described under "Some Utility Scripts," later in this chapter.

---

[*] Presumably this methodology will change as HTML style sheets become universal.

Footer directives

The template typically ends with a macro that calls the `pageFooter()` script (in `html.data.standardMacros`). This script constructs the closing HTML for page, including the `</body>` and `</html>` tags. It looks for the following directives; both are optional, and if the first is `true` the second is not looked for.

#noHintsInHeader

> A boolean. If `false`, `suites.fatpages.buildPageAtts()` is called to generate a machine-readable "hint" comment at the bottom of the page, after the closing `</html>` tag.[*] The fields of the hint can be retrieved from the final HTML into a table with `fatpages.getPageAtts()`.

#adrPageData

> Address of a database object to be encoded into the hint. The presence of such a database object within an HTML comment makes the Web page a "fatPage." Since HTML is a cross-platform network-servable format, a fatPage Web page is a way to distribute small database objects to other users. Since the exported object is inside a comment, it is invisible to the browser, so your Web page needs to state explicitly that it is a fatPage if readers are to know about it.[†]

HTML generation directives

On the whole, the rendering engine is extremely conservative about generating HTML. Frontier is not trying to save you from knowing HTML; it is assumed that you have created most of the HTML yourself, by typing or with BBEdit or Nisus Writer or some similar external editor, or indirectly with a WYSIWYG tool. However, a few extremely common types of HTML are taken care of for you if you wish.

Note that if a Web page object is an outline, its outline structure can also be used to generate HTML. See "Outline Renderers," later in this chapter.

#autoParagraphs

> A boolean. If `true`, any instance of two successive return characters will have a `<p>` tag inserted just before it.

Trying to combine `autoParagraphs` mode with manual insertion of your own `<p>` and `</p>` tags in a single page can result in some pretty skanky HTML, and is not recommended.

---

[*] The directive is so named because this "hint" used to be put in the header!

[†] Other possible directives for putting things in the "hint" are `#menubar` and `#serverNetAddress`, but I believe that these are no longer in general use.

#clayCompatibility

> A boolean.* If `true`, any paragraph beginning with `***` will have the `***` removed and will be surrounded with `<b>` tags. Also, any paragraph consisting entirely of `---` is converted to `<hr>`.

activeURLs

> A boolean. If `true`, any "word" containing `@` or `//` will be assumed to be a URL, and will be surrounded with `<a>` tags to make it a link.

So, for example, if `activeURLs` is `true` and a wptext Web page object contains the following:

```
<p>Write me at matt@tidbits.com.</p>
<p>Visit http://www.scripting.com.</p>
```

then the rendered page will contain this:

```
<p>Write me at <a href="mailto:matt@tidbits.com"> matt@tidbits.com</a>.</p>
<p>Visit <a href="http://www.scripting.com"> http://www.scripting.com</a>.</p>
```

If `activeURLs` is `true` you can still prevent any particular occurrence of `@` or `//` from generating a link by preceding it with a backslash. See "Escape Character," later in this chapter.

isoFilter

> A boolean. If `true`, the page is passed through `string.iso8859encode()` to translate high-ASCII characters; for details, see Chapter 14, *Strings and Chars.*

Rendering engine directives

The rendering engine performs various standard tasks (all described in the course of this chapter). The user can opt out of any of these with directives, but will not typically wish to do so; the usual set of choices is that `#directivesOnlyAtBeginning`, `#tagSubstitution`, `#processMacros`, `#expandGlossaryItems`, and `#useGlossPatcher` will all be `true`, and `threadedRendering` will be `false` unless a CGI or other threaded mechanism might render a page.

#directivesOnlyAtBeginning

> A boolean. If `true`, the rendering engine will stop looking for directives within any wptext Web page object as soon as it encounters a paragraph that doesn't begin with `#`. Clumping all directives at the top of the wptext and setting this directive to `true` can save time during rendering.

---

* The directive is so named because of an earlier incarnation of the Web site management facilities, called ClayBasket.

#tagSubstitution

> A boolean. If true, the rendering engine will perform its default substitutions for the template pseudo-tags <title> and <bodytext>. See "Templates," later in this chapter.

#defaultTemplate

> The string name of an entry in user.html.templates to be used as the template if a page object lacks an associated template. Defaults to "normal". See "Templates," later in this chapter.

#processMacros

> A boolean. If true, the rendering engine will evaluate expressions in curly braces as UserTalk expressions and substitute the result for the expression (unless the left curly brace is preceded by a backslash). See "Macros," later in this chapter.

#expandGlossaryItems

> A boolean. If true, the rendering engine will treat expressions in double quotation marks as invocations of glossary items and will attempt to replace them with the value of those glossary items (unless the first double quotation mark is preceded by a backslash). See "Glossaries," later in this chapter.

#useGlossPatcher

> A boolean. If true, the rendering engine will resolve glossPatch expressions to relative links. See "Relative Links," later in this chapter.

threadedRendering

> A boolean; can be set only in user.html.prefs, and must not be undefined. If true, the rendering engine will raise a semaphore before first using html.data.page, and lower it when finished. Should be set to true if there is a possibility that Web page objects might be rendered at the behest of more than one thread simultaneously (e.g., by a CGI script).

#renderOutlineWith

> Meaningful only when the Web page object being rendered is an outline. Designates a script object which will be called to interpret the page object's outline structure to generate HTML. See "Outline Renderers," later in this chapter.

File directives

These settings mostly determine details about rendered files as files.

#textFileCreator

> The string4 creator code for rendered HTML files. Frequently, a browser's creator code is used.

#imgFileCreator

> The string4 creator code for rendered image files. Frequently, a browser's creator code is used.

#defaultFileName

> The string name (minus any suffix) that the server supplies when a URL specifies no filename. Typically, `"default"` (the default) or `"index"`.

#fileExtension

> The suffix to be added to the names of Web page objects to form the name of the file to which they will be rendered. Typically, `".html"` (the default) or `".htm"`.

#dropNonAlphas

> A boolean. Determines how a filename is formed from the name of a Web page object: if `true`, `toys.dropNonAlphas()` is called to remove nonstandard characters.

#lowerCaseFileNames

> A boolean. Determines how a filename is formed from the name of a Web page object: if `true`, `string.lower()` is called to make the name lowercase.

#maxFileNameLength

> A number. Determines how a filename is formed from the name of a Web page object: a long entry name will be truncated, so that when `fileExtension` is appended the full filename will not exceed `maxFileNameLength`. The default is 31.

#fullTimeNetConnection

> A boolean. If `true`, then when rendering and uploading to a remote site in one operation, the rendering engine will talk to the remote site more frequently. (For example, when a page object is rendered and uploaded, the browser will display the remote version rather than the local version.)

Outline Renderers

When a Web page object is an outline, the rendering engine can use its outline structure to generate HTML. This is can be a great convenience, as it is much easier to arrange an outline than it is to write correct HTML by hand.

How the engine converts outline structure to HTML depends upon the value of the directive #renderOutlineWith (see "Directives," earlier in this chapter). If the directive has no value for a particular outline page object, the object is rendered with the default outline renderer, which results in nested unordered HTML lists. Otherwise, the outline renderer script named by #renderOutline-With is called to generate the HTML. (Therefore, if you define a #renderOutlineWith in a table, it is impossible to specify the default outline renderer for a page object anywhere in that table.)

The value of #renderOutlineWith should be a string, which the rendering engine will try to use as follows:

1. The string is coerced to an address and dereferenced to get an object reference.

2. If that doesn't work (there is no such object), the string is taken to be the name of an entry in the tools table associated with this page object.

3. If that doesn't work (there is no such entry), the string is taken to be the name of an entry in user.html.renderers.

The user is free to create and store outline renderers in any of these locations, and some are included already.

Frontier ships with a number of sample outline renderer scripts, in user.html. renderers. You can learn what they do by studying the scripts, and by rendering an outline Web page object with one of them (see "Previewing and Releasing," later in this chapter). Here is a quick rundown:

newCulture

Levels are undifferentiated. Blank lines have a <p> inserted, and the outline is coerced to a string. This permits the organizational convenience of an outline with minimal effect on the resulting HTML.

twoLevelOutline

Level 0 lines become <h4> paragraphs; lines at all other levels are undifferentiated and become <p> paragraphs.

prettyOutline

Level 0 lines become <h3> paragraphs capitalized with a horizontal rule above. Level 1 lines become <h4> paragraphs, numbered. Other levels are undifferentiated, becoming block-quoted paragraphs.

siteOutliner

Levels are reflected as nested unordered HTML lists, as with the default outliner. The difference is this: if a line contains a comma, what precedes it becomes the basis for a relative URL to turn what follows it into a live link.

daveNetOutline

The whole outline becomes a two-column table: level 0 lines go into the first column, lines at all other levels go undifferentiated into the second column as individual paragraphs. Everything gets <h4> formatting.

`tableOutliner`

> Each level 0 line plus any level 1 lines subordinate to it become cells in one row of a table. Deeper level lines are ignored. The first row of the table is bolded. Optional directives permit some further control:
>
> `#border`
>
> > The value for the `<table>` tag's `border` parameter
>
> `#width`
>
> > The value for the `<table>` tag's `width` parameter
>
> `#cellpadding`
>
> > The value for the `<table>` tag's `cellpadding` parameter
>
> `#centeredCols`
>
> > A list of the numbers of those columns that are to be centered

`cadillac`

> Every line becomes a paragraph whose left margin is indented an amount proportional to the line's level (accomplished through HTML tables); the resulting page looks like the outline. Level 0 lines are bolded.*

The included renderers are good examples for study, to learn to write your own; it's easy.† The outline renderer script is called with one parameter, the address of an outline object. By the time the renderer is called, the outline's directives and summit-level comments have been removed, and the outline is the target, so you can navigate using op verbs immediately. The outline is a copy, so it may be freely modified.

A string must be returned, and there are three typical strategies for deriving it:

- Gather the lines of the outline into a string, editing them as you go, and return the string, like `tableOutliner`.

- Edit the outline in place, and coerce it to a string and return the string, like `newCulture`.

- Edit the outline in place; then send it to `html.ucmds.getOutlineHtml()` to insert some HTML, and return the result, like `siteOutliner`.

`html.ucmds.getOutlineHtml()` takes five arguments—a pointer to the outline, and these:

> a. text to insert before the first line, and before every line indented with respect to the previous line

---

* The `cadillac` renderer does not appear in the Frontier 4.2.3 clean root.

† A number of third-party renderers are also available. My favorite is HALO, available from *http://www.techsoln.com/frontier/HALO/*; it looks for HTML opening tags and puts in all the corresponding closing tags, which makes it very easy to write good HTML while taking advantage of the navigational and organizational features of an outline.

b. text to insert after the last line, and after every line indented with respect to the following line

c. text to precede every line

d. text to follow every line

(c) and (d) are tightest to the line; (a) and (b) go outside of them, like this: ac*xx*db. Nothing is added that you don't explicitly add (such as return characters).

Templates

A *template* is a wptext or outline object which acts as a framework into which a Web page object is embedded as part of the rendering process. A typical template provides at least the opening HTML (the <html> tag, the <head> area, and the <body> tag) and the closing HTML (the </body> and </html> tags), with the Web page object embedded between them; thus, Web page objects typically consist only of the inside of the <body> area.

Multiple Web page objects can share the same template (usually by means of the directive hierarchy). Thus, a frequent use of a template is to provide the beginning and/or end of the inside of the <body> area. All Web pages rendered with that template will have certain material in common, the material provided by the template; the Web page object itself consists of what is different for each Web page. The site as a whole thus ends up with a consistent look.

Furthermore, there are ways to make ·even the part of a Web page that is provided by the template vary from page to page. An example appears in Figure 1-3; the template has put three links at the start of every page; the links are stylistically identical on all the pages, but they are different links on each page. This is typical of the power of templates. (In "Some Utility Scripts," later in this chapter, we shall see how this particular effect was accomplished.)

Every Web page object must be associated with a template. This association is performed by defining a #template directive; the object-template association thus takes advantage of the directive hierarchy. Failure to associate a page object with a template will result in the template named by #defaultTemplate being used; by default, this is "normal" (meaning user.html.templates.normal).

There are two typical modes of #template definition:

• The #template definition is a table entry. In this case, its value is usually the actual wptext or outline object; or it can be the string name of a wptext or outline in user.html.templates.

- The #template definition is inside a wptext or outline Web page object. In this case, its value is usually the string name of a wptext or outline in user.html.templates. However, it is also possible, by a kind of trick, to associate a page object with a template *anywhere* in the database, using this syntax:

      #template user.websites.tools.mySpecialTemplate
      #indirectTemplate false

  The details, if you're going to use this trick, are important: the #template value must be a database object reference, and the #indirectTemplate definition must follow immediately.

By default, a Web page object is embedded by substituting it for the pseudo-tag <bodytext> which appears in its associated template.[*] A template may also contain a pseudo-tag <title>; the value of the #title directive will be substituted for it. Both pseudo-tags may be used only in templates. The two pseudo-tag substitutions take place only if the directive #tagSubstitution is true, which it normally should be. It is possible to obtain other embedding arrangements through the use of custom directives (see "Macros," later in this chapter), but this is not commonly necessary.[†]

The minimal template typically consists of:

- A call to html.data.standardMacros.pageheader() to generate the opening HTML

- The <bodytext> pseudo-tag, to represent the embedded Web page object

- A call to html.data.standardMacros.pagefooter() to generate the closing HTML

The verb calls take place through macros, as explained under "Macros," later in this chapter.

Templates may contain directives. These are handled after directives in the embedded Web page object, so if the same directive is defined both in a Web page object and in its associated template, the template definition will override— unless the directive has already been obeyed earlier in the rendering process. See "Important Routines," later in this chapter, to get a picture of the order of events during rendering.

---

[*] I have coined the term "pseudo-tag" to refer to entities that look like HTML tags but are actually substitution targets for the rendering engine. Instead of <bodytext>, it is permissible to use the pseudo-tag <meat>, but this has gone out of vogue.

[†] For example, the template, instead of consisting of two parts (the beginning and end of the page) with the Web page object sandwiched between them, might consist of three parts, with one half of the Web page object sandwiched between the first two, and the other half of the Web page object sandwiched between the second two.

A template can be either a wptext or an outline. The chief reasons for making a template an outline instead of a wptext are organizational convenience and legibility. It makes no significant difference to the resulting HTML; the outline is coerced directly to a string before the Web page object is embedded (it is not passed through an outline renderer script). Also, making a template an outline lets you take advantage of the `#define` and `#defineScript` directives (see "Macros," later in this chapter).

If you wish to avoid the template mechanism altogether (i.e., a Web page object is to contain or generate all of its HTML, deriving nothing from the template), have the template consist of the single pseudo-tag `<bodytext>` and nothing else; see the template at `user.html.templates.bbedit` for an example.

Web Page Object Datatypes

Here are the datatypes that a Web page object may consist of, and how they are handled by the rendering engine:

An address

> The object pointed to becomes the Web page object. That object must be one of the types listed here. (It can even be another address!)

A table

> The table is rendered into an HTML table, imitating a table edit window, with Name, Value, and Kind columns; directives within the table are not handled. The way to use this feature is to have an object in your Web site table be an address pointing to a table, or to include the table in another Web page object with `renderObject()` (see the next section, "Macros"); that's because a table inside a Web site table will be understood by the rendering engine as representing a folder, not a Web page.

A script

> The script is called, with no parameters; the result is coerced to a string. Directives in this string are not handled, but the script can define directives before returning its result, by inserting them into `html.data.page` directly.

A string, a wptext, or a filespec

> If a fileSpec, the file must be of type `'TEXT'` (or an alias to such a file); the file is read in as a string. If a wptext, it is coerced to a string. Directives in the string are handled.

An outline

> Directives in the outline are handled, and summit-level comments are deleted. The outline is then handed to the outline renderer script specified by the `#renderOutlineWith` directive; if no such directive is in effect, it goes to the default outline renderer. See "Outline Renderers," earlier in this chapter.

Macros

A *macro* is a UserTalk expression in a Web page object; during rendering, the expression is evaluated, and the result substituted for the original expression. Typically, the expression is either a database object reference or a verb call.

A macro provides a way to calculate a stretch of text inside the Web page. Often, the point will be that the text is calculated under the particular conditions at rendering time, such as what page this is, what time it is, what value a certain directive has, and so forth. Also, a macro can be a verb call where the verb's action, not its result, is what's important; if the verb returns the empty string, the original macro will simply be deleted during rendering, with nothing substituted for it.

A macro consists of a UserTalk expression surrounded by curly braces (`{}`). The curly braces are the sign to the rendering engine that this is a macro. A macro may appear in a template or in a Web page object; evaluation takes place after the page object has been embedded in the template. Whether macros are evaluated for a particular template-object pair depends upon the value of the directive `#processMacros`.

Even if `#processMacros` is `true`, an individual instance of curly braces can be exempted from evaluation by preceding the left curly brace with a backslash; this can be important to prevent errors when literal curly braces are intended. Material within double quotes or angle brackets is assumed to be a literal string or an HTML tag, respectively, and is protected from macro evaluation; this protection can be turned off by preceding the left double quote or left angle bracket with a backslash. See "Escape Character," later in this chapter.

The UserTalk expression is evaluated inside a set of nested `with`s so that any object references it contains are sought as follows:

1. Inside the `tools` table associated with the Web page object.

2. Inside `user.html.macros`. This provides a global library of user-created scripts and values that macros can draw upon.

3. Inside `html.data.standardMacros`. This table belongs to UserLand and should not be altered; its scripts are important, though, and are very commonly called in macros. They are discussed under "Standard Macros," later in this chapter.

4. Inside `html.data.page`. The name-value pairs for defined directives are entries here, so it is possible for a macro to pick up directive values simply by referring to them by name.

5. In the database as a whole.

If the expression cannot be evaluated as a legal UserTalk expression in this way, an error message is substituted for it in the rendered HTML.

Custom Directives

Recall that macros are not evaluated until after the Web page object has been embedded in its template; by that time, all directives have been handled. A macro can refer to a directive by name; the directive is an entry in `html.data.page`, and macros are evaluated in a nest of `withs` which includes `html.data.page`, so the value of the directive will be substituted for the name.

For example, the rendering engine creates a `#url` directive, whose value is the URL for the current page; to mention this URL within the page, you could say in the page object:

```
{url}
```

No law says that directives are confined only to those with names that are meaningful to the rendering engine; you are free to define a directive with *any* name. This means that you can insert a name-value pair into `html.data.page` with a directive definition, and retrieve the value later with a macro. This mechanism, the *custom directive* mechanism, is like having temporary variables during the rendering of a Web page object.

A common use for this mechanism is to hand information from a Web page object to its template. Suppose, for example, that every page is to start with a title, different from its `#title`, in large colored letters. So, we have some formatting which is to be true of every page, and which is to appear at the beginning of the page; it makes sense, therefore, to have the title be part of the template. Such an arrangement avoids repeating the same HTML in every page; and if we decide to change the formatting of the title, we change the template once and re-render the site. On the other hand, we have a problem: the text of the title is different for each page.

A custom directive is the answer. We might have this in the template, just before the `<bodytext>` pseudo-tag:

```
<center>
<h1><font size="7" color="#cc9900">{myTitle}</font></h1>
</center>
```

Thus the template sees to the presence and the formatting of the title. Then, in each page object to be rendered with that template, we have a directive definition giving the title's text:

```
#myTitle "History of the Universe"
```

(or whatever the page's particular title might be).

It is also possible to use a directive value inside an HTML tag, but this calls for some trickery because material inside HTML tags is protected from macro evaluation. Suppose we wanted each Web page object to hand up to the template not only the text of the title but its color as well:

```
#myTitle "History of the Universe"
#myTitleColor "#cc9900"
```

The template could contain a macro which constructs the HTML for the <font> tag as a UserTalk expression:

```
<center>
<h1>{"<font size=\"7\" color = \"" + myTitleColor + "\">"}
{myTitle}
</font></h1>
</center>
```

Special Outline Directives

Web page objects that are outlines, and templates that are outlines, can employ two extra directive-defining directives. The reason that only outlines can do this is that these directives take advantage of the outline structure as part of their syntax.

The directives are called #defineScript and #define. Each takes as its value a string that is to be the name of the directive; the bundle subordinate to the definition line is copied into a script or outline object, respectively, to become the value of the directive. The idea is that you can then call the script, or retrieve the outline, by way of a macro or a directive definition.

For example, this is how one of my outline Web page objects establishes its <meta> tag:

```
#define "myMeta"
    <META name="description" content="Free stuff for learning Frontier">
    <META name="keywords" content="Frontier, UserLand, Neuburg, Scripting">
#meta string(html.data.page.myMeta)
```

Except that in my actual page, each content value is much longer; the use of #define thus lets me feed the #meta directive a string which exceeds the 255-character limit on string literals and outline lines.

Standard Macros

Recall that macros are evaluated in a with so that entries in html.data.standardMacros can be referred to by name alone. The following are some of the scripts included in html.data.standardMacros. These scripts, when called from macros, provide some of the most valuable features of the Web site management facilities. The other scripts in the table might prove handy as well, and may be examined at leisure; but the following are the most important.

`renderObject (`*`addrObject`*`)`

> Processes the object at *addrObject*, as described earlier, under "Web Page Object Datatypes." (The normal rendering process calls this script.) The processed object is returned as a string.

`renderObject()` is the standard way of including one Web page object inside another. This feature substitutes for server-side includes. It also allows a Web page object of one type to be embedded in another, such as a rendered outline inside a wptext, or an outline rendered with one outline renderer script inside an outline rendered with a different outline renderer script.

An object to be included in a Web page object is often kept in a `tools` table, so that it will reside in the site table (for easy access) but will not be rendered as an independent HTML file when the whole site table is rendered.

`imageRef (`*`imageObject, alt, hspace, align, usemap`*`)`

> Creates an image file from the binary `'GIFf'` or `'JPEG'` object at *imageObject*, and returns an `<img>` tag referring to it with a relative URL. *imageObject* may be the address of a binary anywhere in the database, or it may be a name string, in which case it is sought in the `images` table in the same table as the Web page object being rendered, and then in each `images` table in each parent table, until the site's root-level table is reached. All but the first parameter are optional; they correspond to parameters in the `<img>` tag.

`imageRef()` is the standard way of dealing with GIFs and JPEGs to which your pages refer. You never have to write an `<img>` tag, because `imageRef()` writes it for you; and you never have to worry about placing an image file among the other files in your site, because `imageRef()` places it for you.[*]

The only hassle is that images must be stored as binaries inside the database. But it isn't much of a hassle, because the **Load Image File** item of the **Web** menu lets you choose a file and load it into the database. Such files are initially loaded into `user.html.images`, which isn't normally what you want; you will probably shift it to an `images` table in your site table. An alternative method of importing is to call `html.loadImageFile()` yourself, providing a file pathname and the address of a destination table.

---

[*] If *imageObject* is an address, the image file ends up in the same folder with the rendered web page file; if *imageObject* is a name string, the image file ends up in an *images* folder in the site, which is created if necessary. Either way, the `<img>` tag points correctly to the image file.

spacePixels (*numPixels, orientation*)

> Creates an image file from the binary 'GIFf' object at html.data.im-
> ages.space, and returns an <img> tag referring to it with a width (or
> height) parameter of *numPixels* and a height (or width) parameter of 1.
> *orientation* is optional; the default is "horizontal", which causes the first
> option, or you can supply "vertical", which causes the second.

The idea of spacePixels() is to make an invisible spacer image. This is particularly useful in HTML tables used for formatting, to help guarantee that a cell will
have a desired width.

outlineSite (*addrTable, width, indent*)

> Returns the HTML for an outline-like table of contents, listing and linking to
> all the Web pages that will be made from all the Web page objects in the site
> table at *addrTable* (to infinite depth). All parameters are optional. If *addr-
> Table* is omitted, the whole current site* is outlined. *width* is the overall width
> of the table, in pixels; *indent* is the multiple of pixel indentation used for
> each hierarchical level of subtable.

The value of the #title directive inside each individual Web page is used as the
listing for that page, from which the link to that page emanates. Each Web page
object may further define a directive #subtext whose string value will be used in
the table of contents as a supplementary description of that page. Further control
over formatting is provided through three optional directives:

#siteOutlineHeadFont

> The complete <font> tag to precede each page listing

#siteOutlineSubtextFont

> The complete <font> tag to precede each #subtext description

#siteDefaultName

> The name of the main (default) page for the site, to prevent it from being
> included in the table of contents

Unfortunately, the titles of Web page objects in each table are sorted alphabetically. It is not hard to write a utility script that makes a table of contents where
you dictate the order; a utility script presented below ("Some Utility Scripts")
suggests one way this might be done.

linkPrev (*linkText*), linkNext (*linkText*)

> Returns the HTML for a relative link to the previous or next page in the site,
> emanating from the text *linkText*. (It is possible for *linkText* to be a call to
> imageRef(), to make the link emanate from a graphic.) To tell Frontier the

---

* As determined by #adrSiteRootTable; see later on how the rendering engine creates this directive.

order of pages in the site, there must be an outline called `#nextPrevs` in the site's root table. The `#useGlossPatcher` directive must be `true`.

To create the `#nextPrevs` outline, choose **Build NextPrev List** from the **Web** menu; specify the site root table in the dialog. (Capitalization counts!) A flat summit-level outline will be generated, listing all page objects at any depth in the table. Rearrange this to the desired order; for any "default" entries, delete them if they don't represent actual page objects, and correct their capitalization (if necessary) if they do.

For reasons explained later, under "Relative Links," it may be necessary to render the whole site twice in succession to make the links work correctly. In many cases this can be avoided with a utility script presented later, under "Some Utility Scripts."

`embeddedUserTalk (`*`scriptText, linkText`*`)`
> Passes *scriptText* through `toys.URLencode()` and returns the HTML for a `usrtlk` link emanating from *linkText*.

This makes a clickable link in a Web page which, if displayed in Netscape Navigator, can run a UserTalk script on the same machine as Netscape. See Chapter 34, *Driving Frontier from Outside*.

Glossaries

A *glossary* is a substitution table consisting of name-value pairs. A *glossary item* is an entry in such a table. During rendering, stretches of text surrounded by double quotes (`""`) and matching the name of a glossary item are replaced by the value of that glossary item. This provides a way for common or boilerplate text to be invoked by a convenient name in a Web page object.

Glossary items may be invoked from within page objects or templates; glossary substitution takes place after the object has been embedded in its template, as part of the same process as macro evaluation. Whether glossary substitution is performed for any particular page is determined by the value of the `#expand-GlossaryItems` directive.

Even if `#expandGlossaryItems` is `true`, any particular stretch of double-quoted text can be exempted from glossary substitution by preceding the first double quotation mark with a backslash; this can save considerable time during rendering. Material within angle brackets is assumed to be an HTML tag, and is protected from glossary substitution; this protection can be turned off by preceding the left angle bracket with a backslash. See "Escape Character," later in this chapter.

Glossary items are sought in a glossary as follows:

1. In the table pointed to by the `#glossary` directive. This directive points by default to the table named `glossary` "closest" to the Web page object being rendered; however, it is permissible to define the `#glossary` directive as the address of any table.

2. In any table named either `glossary` or `#glossary`, looking first in the same table as the page object being rendered, then in its parent table, and so on to root level.

3. In `user.html.glossary`, which thus functions as a kind of global glossary table.

4. If the item is not found by these means, an entirely different tack is taken: the table containing the object being rendered is examined to see if it contains a Web page object with this name; if so, a link to it is formed. This device is largely outmoded by the glossPatch mechanism (discussed later, under "Relative Links").

5. Finally, if none of that works, the original stretch of double-quoted text remains untouched.

A common use of glossary items is to represent links (see `user.html.glossary` for many examples). Therefore, the browser's shared menu contains an item, **Add to Glossary**, which makes a glossary item in `user.html.glossary` from the URL of the page currently being browsed; the glossary item is an absolute link to that page, and it emanates from (and has the same name as) the Web page's title. A utility script provided later, under "Some Utility Scripts," transforms a glossary item whose value is a link so that it emanates from *any* desired text.

Escape Character

We have several times mentioned earlier ("Directives," "Macros," "Glossaries") that this or that function of the rendering engine can be prevented from operating on a particular stretch of text by the use of the backslash character. The backslash character is the escape character for the rendering engine, cancelling the normal treatment of `@`, `//`, `"`, `{`, and `<`.

Since the backslash is just a signal to the rendering engine, the engine removes backslashes as one of its last acts. Thus, a single backslash in a Web page object will never appear in the resulting Web page. To have a literal backslash character appear in a Web page, use two backslashes in a row in the Web page object.

However, material in angle brackets, curly braces, and double quotation marks is protected from escape-character resolution; so, for example, a single backslash within a double-quoted stretch of text *will* appear in the final HTML. But you can remove this protection by preceding the left angle bracket, curly brace, or double quote with a backslash!

It does no harm to escape a character that has no special meaning. The backslash will simply be removed without further effect.

Filters

The `#filters` directive is a table or the address of a table containing two scripts named `pageFilter` and `finalFilter`. Just before a Web page object is embedded into its template, it is placed as a string at `html.data.page.body-Text`, and `pageFilter()` is called, with no parameters; the idea is that the script may now perform whatever action is desired, possibly changing `html.data.page.bodyText` in some way. Similarly, just before the rendered Web page is written out to disk, it is placed as a string at `html.data.page.renderedText`, and `finalFilter()` is called, with no parameters; again, the idea is that the script may perform whatever action is desired, possibly changing `html.data.page.renderedText` in some way.

The default `pageFilter()` in a new site table does two things. First, it enlarges the first character of the page, to make a kind of drop-cap effect. In my own `pageFilter()`, I comment this out, but it is suggestive of the sort of thing the `pageFilter()` is good for. Second, it forms a glossPatch entry in the `glossary` table pointed to by the `#glossary` directive; this is important for the glossPatch mechanism of relative references within the site, as discussed later, under "Relative Links." However, as written, the whole action is in an `if` bundle which prevents it from operating when the object being rendered is named `"default"` (or `"index"` or whatever your default page name is); this restriction should be removed. The default `finalFilter()` in a new site table calls the `glossPatcher()` script; this should be commented out, as the same call will be made automatically elsewhere in the rendering process. By making such changes in the filter scripts within `html.data.newSiteFilters`, you can enforce them for every new site created with **New Site**.

Uses for customized filter scripts are up to your imagination. We suggest one here and another later, under "Some Utility Scripts."

Second Processing

Certain limitations of the rendering process can be overcome by adding to the `finalFilter()` something like this:

```
with html.data.page
    renderedtext = html.processMacros(renderedtext)
```

The routine `html.processMacros()` is where macro evaluation and glossary substitution are performed; it is called once during the rendering process anyway, but we are causing a second round of processing to be performed.

To see why this is useful, it is necessary to understand how `html.process-Macros()` behaves: it passes through the text of its parameter once, looking for characters that signal special processing. If it encounters a left angle bracket, it skips to the next right angle bracket; if it encounters a double quote, it gathers up everything between it and the next double quote and tries to perform glossary substitution upon it; if it encounters a left curly brace, it gathers up everything between it and the next right curly brace and performs macro evaluation upon it. In each case, it then proceeds through the text.

Imagine, now, that your pages frequently use a little GIF that is a colored image of the word "Cool". Suppose that this image is located in an `images` table, where it is called (you guessed it) `cool`. It becomes tedious to have to invoke this image in Web page objects by saying each time:

    { imageRef ("cool") }

It would be nicer to be able to say simply:

    "cool"

where this invokes a glossary item whose value is the `{imageRef("cool")}` macro call. But you can't do this! At rendering time, the rendering engine, having performed glossary substitution on `"cool"`, moves on—and never sees the macro which it has just inserted. Thus a second pass through `html.processMacros()` is needed in order to see the macro and perform macro evaluation upon it.

Similarly, when a macro call to `renderObject()` is used to include one Web page object inside another, macros and glossary item invocations in the included material are not handled. A second processing pass is needed to see and handle them.

So a second round of processing sounds like a pretty good idea. But it can have repercussions that need to be guarded against. For instance, the first round of processing removes all single backslashes (see "Escape Character," earlier in this chapter). So, on the second round, characters that were protected by backslashes are no longer protected, and the actions that those backslashes were intended to prevent, and which *were* prevented during the first round of processing, are performed during the second round of processing. Similarly, if `#autoParagraphs` is `true` during both rounds of processing, an extra set of `<p>` tags will be inserted.

To prevent such repercussions, you may need to plan ahead. If you know that there will be two rounds of processing, put two backslashes where you normally would put one. On the first round of processing, the two backslashes become one backslash; on the second round, the single backslash acts as the escape character and is then removed. (A literal backslash that is to appear in the resulting Web page will have to be coded as four successive backslashes in the Web page object!)

Another way to ease the difficulty is to have the `finalFilter()`, just before it performs the second round of processing, turn off those directives that have already been obeyed and which should not be obeyed a second time. For example:

```
with html.data.page
    activeURLs = false
    autoParagraphs = false
    renderedtext = html.processmacros(renderedtext)
```

You also need to consider how macros will pick up the values of directives. If a directive is defined in an object included in another object with `renderObject()`, the definition will be handled during the first round of processing, as part of the `renderObject()` call. If this is a custom directive which we are picking up with a macro, the macro must be evaluated in the second round of processing rather than the first. This can be attained by postponing evaluation for one round with a backslash:

\{myDirective}

The result is that the first round of processing merely removes the backslash, and the macro {`myDirective`} is evaluated in the second round of processing.

Relative Links

A mechanism called *glossPatch* allows you to generate relative links to other pages in your site in such a way that the links will not break if you move the page objects to a different region of the site table.

There are three parts to the glossPatch mechanism:

1. As each page of the site is rendered, the `pageFilter()` script creates a gloss-Patch expression for it, as an entry in the `glossary` table pointed to by the `#glossary` directive.

2. In the Web page, you have used glossary item invocations wherever you want a link to a page in the site. Glossary substitution now replaces such invocations with the corresponding glossPatch expression that was generated by the `pageFilter()`.

3. Provided the directive `#useGlossPatcher` is `true`, the rendering engine calls `html.data.standardMacros.glossaryPatcher()` to resolve gloss-Patch expressions in the page object into HTML relative links.

A glossPatch expression is computer-generated, computer-readable code; it isn't intended to be useful to human beings. However, it will help our discussion of how the glossPatch mechanism works to know the format of a glossPatch expression, which is like this:

[[#glossPatch My Fourth Page|aSubTable/fourthPage|]]

This glossPatch expression would be generated by a Web page object called `fourthPage` whose `#title` directive value is `"My Fourth Page"`. Besides the outer delimiters (two sets of square brackets), it has three parts: (a) everything up to and including the first space; (b) everything after the first space and before the first "pipe"; (c) everything between the two "pipes."

Part (a) just identifies this as a glossPatch entry.

Part (b) is the text from which the link will emanate. When the `pageFilter()` generates a glossary item which is a glossPatch expression, part (b) of the expression and the name of the glossary item itself are the same: they come from the page object's `#title`.

Part (c) is a path from the root level of the site to the page in question, with slashes showing changes of level; so, in this example, the root level of the site contains a table called `aSubTable`, and in that is our page object, called `fourthPage`.

The result is that in a Web page object, you can say, for example:

"My Fourth Page"

and this will become a link to the Web page whose title is "My Fourth Page". (The link will also emanate from the text "My Fourth Page"; this may not be what you want, so a utility script provided later, under "Some Utility Scripts," lets you change it.) What's more, you can move the object `fourthPage` to a different subtable of your site table, and your links to it will still work, because when the page is next rendered, the `pageFilter()` will change the glossPatch expression for the `"My Fourth Page"` glossary item.

By implication, however, for the glossPatch mechanism to work, `#title` values of all pages in a site must be carefully chosen! They must be unique within the entire site, and they must not be changed. If you change the `#title` of `fourthPage`, there will no longer be a glossary item `"My Fourth Page"` that correctly tracks the location of `fourthPage`; your invocations of "My Fourth Page" in Web page objects will probably break. The `#title` serves as a unique identifier; you're locked into a page object's `#title` as the price of being able to relocate the page.

It may be necessary to render the entire site twice to get relative links to work: once so that step 1 can get all the glossPatch glossary items right, and again so that steps 2 and 3 can resolve invocations of them. In many cases, this can be avoided with a utility presented later in this chapter, under "Some Utility Scripts."

Previewing and Releasing

Let's presume that you have made a Web site table with some Web page objects in it, and wish to render one or more of the objects. We now discuss how rendering is performed.

Previewing

Previewing is a mode of rendering where just one page is rendered for the purpose of viewing the resulting file in the browser. The file goes into a "temporary" folder, *Websites*, in the same folder as the Frontier application; the setting in `#ftpSite.folder` is ignored. The "temporary" folder and its contents are not deleted afterwards; it is up to you to do so from time to time.

Before previewing a page, make sure a browser is running, either Netscape or Microsoft. This is because the `html` suite is going to tell your browser to open the page, by way of the `webBrowser` suite. The `webBrowser` suite provides a uniform programming interface, so that the `html` suite can send commands to the browser without knowing which one it is; but the `webBrowser` suite does need to know which one it is. The agent `system.agents.webBrowserAgent` watches to see what browser is running, and sets `suites.webBrowser.userland.currentid` accordingly; this is how the `webBrowser` suite knows what browser to talk to.

To preview a page, select the page object, either in its edit window or in its parent table's edit window, and choose **View in Browser** from the **Web** menu. Frontier renders the page and orders the browser to display it. (With some browsers, if you previewed the page previously and the earlier version is still displayed in the browser window, you'll have to ask the browser to refresh the window in order to see the new version.)

Releasing

Releasing is a mode of rendering where one page object or an entire site table is rendered into the correct place in the folder designated by `#ftpSite.folder`. This might be preparatory to uploading to the remote site from which the pages are served, or in order to test features involving relative links within the site.

Before releasing, make sure a browser is running. To release a single page, select the page object in its edit window or its parent table's edit window, and choose **Release Rendered Page** from the **Web** menu. To release a table of pages, select a page object within the table's edit window (*not* a subtable!) or select the table itself within its parent's edit window, and choose **Release Table** from the **Web** menu.

After releasing, the browser may not be frontmost, or (worse) may not be displaying a rendered page at all. This is a very inconvenient state of affairs, but selecting a page object and choosing **View Page In Finder** from the **Web** menu can help: it takes you to the icon of the rendered HTML file in the Finder, and you can drag that icon onto the browser to view it.

If changes are made and a page or site is released again, existing files will be overwritten, but nothing will be deleted. It is up to you to clean out (or delete) the site folder manually as necessary.

Releasing Only Changed Pages

The `closeWindow` hook `user.hooks.closeWindow.addToChangedPages()` (see Chapter 24, *Windows*) sees to it that the address of every changed Web page object is recorded in the outline `user.html.lists.changedPages`. This makes it easy to release only those pages that have recently been changed; simply edit the outline and then choose **Release** from the **Changed Pages** submenu of the **Web** menu. This calls `html.buildFromOutline()`. Each object listed in `user.html.lists.changedPages` is released, and that line of the outline is deleted.

The same mechanism could be used to release some standard set of pages automatically. Just call `html.buildFromOutline()` yourself, providing as parameter the address of an outline listing the addresses of the pages. You may want the outline to be a copy, so that the entries in the original outline won't be deleted.

Releasing to a Remote Site

To release and upload to a remote site all in one motion, change `#ftpSite.isLocal` to `false` and add to `#ftpSite` the following entries:

`domain`
> The URL of the remote machine as seen by your FTP client

`account`
> Your FTP username

`password`
> Your FTP password

`directory`
> The path (using slashes as separators) to the directory on the remote machine where you will put the material

For example, Table 41-1 shows what my `#ftpSite` table looks like when I am about to release to a certain remote site (the names have been disguised to protect the innocent).

Table 41-1. An #ftpSite Table

Name	Value
account	mattn
directory	/usr/www/apache/htdocs/
domain	206.31.218.123
folder	HD:local build:
isLocal	false
password	haha
url	www.snarf.com/

An FTP client must already be running. (This is basically for the same reason that a browser must be running before rendering; the `ftpClient` suite is going to be used to send commands to your FTP client program, so the program must be running so that the `ftpClient` suite knows what program it should talk to.) When **Release Rendered Page** or **Release Table** is chosen, the material is first rendered into the local folder, then uploaded with the FTP client. However, image files created with `imageRef()` are *not* uploaded; it is up to you to upload these manually afterwards.

Loading an Existing Site

An existing Web site may be imported into the Frontier database by selecting a destination table and then choosing **Load Existing Site** from the **Web** menu. Both textfiles and image files are imported, maintaining the site structure (folders become tables).

Sites created with FrontPage, PageMill, or HomePage get some intelligent processing: the correct title is separated out, everything outside the `<body>` area is removed, `<p>` tags are cleaned out, and so on.

Object names are assigned by deriving them from file names in accordance with the `dropNonAlphas`, `lowerCaseFileNames`, `maxFileNameLength` and `fileExtension` directives. This process can assist you if you have created a Web site with a different tool and now wish to start managing it with Frontier.

Important Routines

For advanced work, such as debugging, writing utility scripts to be called from macros, and so forth, it helps to possess a mental map of the execution path traversed by the rendering engine. Additionally, some scripts in the `suites.html` table, though intended mostly for use by **Web** menu items and by other scripts, are of sufficient general utility that they might be called from a user script.

Accordingly, here is a brief sketch of some of the chief inhabitants of `suites.html`; this should help guide your further investigations.

`html.buildObject (`*`addrWebPageObject`*`)`

> The rendering engine. Given a Web page object, returns a string consisting of the final HTML for the fully rendered Web page.

> Calls `html.buildPageTable()` to gather table-based directives; then `html.data.standardMacros.renderObject()` to gather object-based directives and to convert the page object to a string (including outline rendering). Calls the `pageFilter()`. Gathers directives from the template by calling `html.runDirectives()` or `html.runOutlineDirectives()`, and embeds the object into the template. Calls `html.processMacros()` to evaluate macros, substitute glossary items, and so forth (see on `string.processHtml-Macros()`) and to translate high-ASCII characters (with `string.iso8859encode()`). Calls the `finalFilter()`. Calls `html.data.standard-Macros.glossaryPatcher()` to resolve glossPatch expressions.

You might call `buildObject()` from a CGI script in order to obtain HTML from a Web page object by way of the rendering engine; thus, `buildObject()` is the link between Frontier's CGI features and its Web site management features.[*]

`html.buildPageTable (`*`addrWebPageObject`*`, `*`addrPageTable`*`)`

> Called by `html.buildObject()`. Gathers table-based directives for the object at *addrWebPageObject*, working its way up the table hierarchy, and storing the results in the table at *addrPageTable* (usually `html.data.page`). Directives that are tables are stored as addresses.

> This routine generates some useful directives that would not normally be set by the user, such as:

`adrObject`

> Address of the object being rendered.

`adrSiteRootTable`

> The table containing the first found occurrence of `#ftpSite`, indicating that this is the root level of the Web site table.

---

[*] For an example, see under "Classified Ads," in Chapter 42, *Dynamic Web Sites.*

subdirectoryPath
> Partial pathname for the folder that will contain the rendered file, relative
> to the site's root folder; thus, it can also be used for the directory of the
> file on the server machine, relative to #ftpsite.directory.

fname
> Filename for the rendered file.

f   Full pathname for the rendered file.

url URL for the rendered file, based on #ftpsite.url, subdirectoryPath
   and fname.

html.runDirectives (*string*)
> Called by html.data.standardMacros.renderObject() to handle string,
> wptext, and textfile objects; also by html.buildObject() to handle wptext
> templates. Gathers object-based directives; each time it finds one, deletes it
> and hands it to html.runDirective() for evaluation.

html.runOutlineDirectives (*addrOutline*)
> Called by html.data.standardMacros.renderObject() to handle outline
> objects; also by html.buildObject() to handle outline templates. Gathers
> object-based directives; each time it finds one, deletes it and hands it to
> html.runDirective() for evaluation, except that #define and #define-
> Script are specially handled.

html.processMacros (*string*)
> Called by html.buildObject(). Calls string.processHtmlMacros() and
> string.iso8869encode() in accordance with the relevant directives.

string.processHtmlMacros (*string, autoParagraphs, activeURLs,
clayCompatibility, addrProc*)
> Called by html.processMacros(). Evaluates macros. Calls html.refGlos-
> sary() to perform glossary substitution. Obeys *autoParagraphs, clay-
> Compatibility*, and *activeURLs* (as directives). Simplifies escaped characters.
> *addrProc* is usually the address of the UCMD html.ucmds.embeddedcode.

The serious work of string.processHtmlMacros() is hidden in the kernel and
in the UCMD, so I have had to guess what the routine actually does.

html.ftpText (*string*)
> Called by **View In Browser** and **Release Rendered Page** menu items. Writes
> *string* to disk as a textfile at the pathname designated by html.data.
> page.f. If #ftpSite.isLocal is false, also uploads the file via FTP to the
> server machine, using #ftpSite.directory, subdirectoryPath, and
> fname to work out where to put it.

`html.getOneDirective (`*directiveName, inString*`)`

    Searches the string *inString* for the first occurrence of *directiveName*; evaluates and returns the rest of the paragraph in which *directiveName* occurs, or returns the empty string if it doesn't occur.

`getOneDirective()` is a brute-force method of reaching inside a Web page object to see how it defines a particular directive; it wastes no time actually handling directives, nor does it check that the rules of directive definition are obeyed—for instance, it doesn't care whether the directive name starts with # or whether it begins a paragraph or summit line. A handy utility.

`html.getFileName (`*nameString*`)`

    Converts *nameString*, typically the name of a Web page object, into a filename, in accordance with the various filename directives such as `#fileExtension`, `#dropNonAlphas`, etc.

`html.getPref (`*prefName*`)`

    Returns the value of the directive *prefName*, looking first in `html.data.page`, then in `user.html.prefs`. In most cases, returns `true` if the directive is undefined.

`html.getSiteTable (`*addrPageTable*`)`

    The parameter is optional, the default being `html.data.page`. Returns the address of the `#ftpSite` table. Needed because in theory `#ftpSite` might be either a table or the address of a table.

`html.traversalSkip (`*addrObject*`)`

    Returns `true` if the name of the database object at *addrObject* shows that it is not a renderable Web page object (i.e., it's a `tools` table, a directive, etc.), `false` otherwise, including if it is a table that might contain renderable Web page objects.

`html.buildFromOutline (`*addrOutline*`)`

    The outline should consist of summit-level items which are references to Web page objects. Each object is rendered to disk and its line in the outline is deleted.

`html.loadImageFile (`*filePath, addrDestinationTable*`)`

    Despite its name, loads *any* type of file at *filePath* as a binary (or, if a textfile, as a wptext) into the database table at *addrDestinationTable*, using the file's filetype as the binary's inner datatype and the file's name to construct the object's name.

Web sites can contain more than just HTML files and image files, and `html.load-ImageFile()` could be used to get them into the database. It would be a simple matter to write a script that does for *.hqx* files (for instance) stored in the database what `imageRef()` does for image files.

Some Utility Scripts

Certain utility scripts that I have developed have come in very handy for me, and I see no reason why the neophyte Frontier Webmaster should be without their benefit. So here they are.

Munging Links

The `glossSub()` utility script, intended to be called from a macro, takes care of a number of limitations having to do with glossary entries and links:

- It is common to store links as glossary items. Unfortunately, this "hardcodes" the text from which the link emanates; for example, if we have a glossary item called `Frontier` whose value is:

  ```
  <a href="http://www.scripting.com/frontier/">Frontier</a>
  ```

  then invoking it in a Web page object with the phrase `"Frontier"` will result in a link to the Frontier Web page emanating from the word "Frontier". But what if we wanted a link to the Frontier Web page, emanating from the words "a cool site"?

- Exactly the same problem arises for glossPatch glossary items generated by the `pageFilter()`. It's fine that the page's `#title` value is used as the name of the glossary item, but we're also stuck with it as the text from which the link will emanate.

- It would be quite easy to do HTML frames if only glossPatch glossary items could be made to include targets.

Example 41-1 contains the script.

Example 41-1. glossSub()

```
on glossSub (whatGlossEntry, whatText = "&&&&", whatTarget = nil)
    local (ss = html.refGlossary(whatGlossEntry))
    on doExceptForTarget(s)
        with system.verbs.builtins.string
            if patternMatch("[[#glosspatch ", lower(s)) == 1
                html.data.page.renderedText = \
                    nthField(s,' ',1) + " " + whatText + "|" \
                        + nthField(s,'|',2) + "]]"
                html.data.standardmacros.glossaryPatcher()
                return html.data.page.renderedText
            return nthField(s,'>',1) + ">" + whatText + "<" + nthField(s,'<',3)
    ss = doExceptForTarget(ss)
    if whatText == "&&&&"
        return string.nthField (ss, '"', 2)
    if whatTarget == nil
        return ss
    return string.replace(ss, ">", ", target=\"" + whatTarget + "\">")
```

glossSub() produces the HTML for a link to the same page as the link or gloss-Patch expression in the glossary item *whatGlossEntry*, but the link emanates from the text *whatText* (which can be a macro call to imageRef(), thus causing the link to emanate from an image). Optionally, a target parameter, whose value is *whatTarget*, will be included in the tag.

If both *whatTarget* and *whatText* are omitted, glossSub() returns just the URL contained in *whatGlossEntry*. This is useful because one might want the URL in constructing a different tag. For example, suppose three of the pages in our site are titled "First Page", "Second Page", and "Third Page"; we can then make an imageMap:

```
<map name="myImage">
\<area shape=rect coords = "0,0,39,33" href={glossSub("First Page")}>
\<area shape=rect coords = "39,0,106,33" href={glossSub("Second Page")}>
\<area shape=rect coords = "106,0,165,33" href={glossSub("Third Page")}>
</map>
```

(Notice the use of the backslash, without which a macro inside a tag would not be evaluated.) The significant thing is that the relative links will not break even if our three pages are relocated within the site, because we're getting our URLs by way of the glossPatch mechanism.

Render Once Instead of Twice

Relative links that depend upon automatically generated glossPatch glossary items typically require the entire site to be rendered twice: once to make the glossary items, and again to use them now that they are correct. The following utility prepares the glossPatch items very quickly; because it uses html.getOneDirective() to peek inside the object and get the #title directive, it works only for string, wptext, and outline Web page objects, but that's what constitutes the vast majority of sites anyway. The script can most conveniently be called from a menu item, just before rendering the site.

Example 41-2. preflightSite()

```
local (whatSite = "user.websites")
if not dialog.ask ("Site table to preflight?", @whatSite) {return}
on traverse(addrT)
    local (x)
    for x = 1 to sizeOf(addrT^)
        msg(nameOf(addrT^[x]))
        if not html.traversalSkip (@addrT^[x])
            local (theType = typeOf(addrT^[x]))
            if theType != tableType
                if theType == wptextType or theType == stringType \
                or theType == outlineType
                    « adapted right from the pageFilter
                    local (stringstrip = string.lower (whatSite) + ".")
```

Example 41-2. preflightSite() (continued)

```
                    local (path = string.lower (@addrT^[x]) - stringstrip)
                    path = string.replaceall (path, ".", "/")
                    path = string.replaceall (path, "[\"", "")
                    path = string.replaceall (path, "\"]", "")
                    local (theTitle = html.getOneDirective("#title", \
                        string(addrT^[x])))
                    whatSite^.glossary.[theTitle] = "[[#glossPatch " \
                        + theTitle + "|" + path + "|]]"
                else « can't preflight this type, let user know
                    dialog.alert ("Could not pre-flight " + \
                        nameOf(addrT^[x]) + \
                        ", sorry. You may have to render " + \
                        "the whole site twice after all.")
            else « it's a table, recurse
                traverse(@addrT^[x])
    traverse(whatSite)
```

Background Images

The **#background** directive lets you specify the URL of an image, but nothing is done to see that the image file is generated or in the right place. This utility remedies that. Define the **#background** directive with a relative URL as if the image file will be in the same folder as the page being rendered—for instance:

```
#background "myBg.gif"
```

When the page is rendered, the **finalFilter()** should contain a call to the following script.

Example 41-3. writeBackgroundImageFile()

```
if defined(html.data.page.background)
    local
        nomad = html.data.page.adrobject
        addrImage
        imagespec = toys.popstringsuffix(html.data.page.background)
    loop « copied from imageRef()
        if nomad == html.data.page.adrSiteRootTable
            scriptError ("Can't locate an image object named \"" \
                + imagespec + "\".")
        nomad = parentOf (nomad^)
        if defined (nomad^.images)
            addrImage = @nomad^.images.[imagespec]
            if defined (addrImage^)
                break
    « write to disk, throw away HTML result
    html.data.standardmacros.imageRef(addrImage)
```

The script imitates **imageRef()**, working its way up through **image** tables looking for the background image. When it finds it, it calls **imageRef()** to write it to disk. Because we supply an address, **imageRef()** writes the file into the same folder as our page, which is just where the `#background` relative URL said it would be.

It does no harm to include in the `finalFilter()` a call to `writeBackgroundImageFile()` even if there is no background image, because in that case the routine does nothing. I have made this call part of my standard `finalFilter()`.

Navigation Links

A common navigation aid is for each page to include a series of links to all pages in the site, or to some common subset of those pages. It is also nicer if none of these links is to the page we are actually in. Here, we illustrate one possible method of generating such links.

In order to know what other pages to link to and in what order to put the links, the script depends upon an outline which the user must already have constructed and arranged. The outline can be the NextPrev list (see "Macros," earlier in this chapter), or any summit-level list of page objects. Because `html.getOneDirective()` is used, the script works only when the page objects are all wptexts or outlines, but this will be the case most of the time. The script depends upon `glossSub()`, which was given previously.

The links do not emanate from the `#title` of each page, but from the value of another directive defined in each page, `#subtitle`. Of course you are free to modify this, and other details of the routine, to fit your own tastes.

Example 41-4. linksToOtherPages()

```
on linksToOtherPages(addr0 = @html.data.page.adrSiteRootTable^.["#nextPrevs"])
    local (s = "")
    on add(t)
        s = s + t
    on doOnePage(pageName)
        local
            theTitle = html.getOneDirective("title", string(pageName^))
            theSub - html.getOneDirective("subtitle", string(pageName^))
        if theTitle == html.data.page.title
            add(theSub)
        else
            add(user.html.macros.glossSub(theTitle, theSub))
        add (" | ")
    target.set(addr0)
    op.firstsummit()
    add ("\r<p>")
    doOnePage(op.getLineText())
    while op.go(down,1)
        doOnePage(op.getLineText())
    target.clear()
    return string.delete(s, sizeof(s) - 2, 3) + "</p>\r"
```

This algorithm could easily be adapted to modify `outlineSite()` so that it orders its table of contents according to the NextPrev list.

BBEdit Front-End

BBEdit can be used as an external editor for wptext Web page objects, as can PageSpinner. This is very convenient, as both these applications are extremely well suited for editing HTML. (See Chapter 35, *External Editors.*)

But there is an entirely different way to use BBEdit in connection with Frontier's Web site management tools: BBEdit becomes a kind of front-end to those tools, or to a simplified subset of them. Frontier and the database lurk in the background, but commands to Frontier are given through a shared menu which appears in BBEdit (the **Site** menu) and are handled through the `bbSite` suite.

Architecture

The architecture is slightly different from what I have been describing up to now. The unrendered Web site is not in the database: it is a folder of textfiles, possibly containing other folders. Templates in the unrendered Web site are files called *#template.html.* The directive hierarchy applies to these: to determine which *#template.html* goes with a file being rendered, Frontier starts in the folder that contains that file, and looks up the hierarchy until it finds the first *#template.html.* At the same time, *#template.html* files help to show Frontier where the unrendered Web site is, as a whole: to determine the root folder of the site, Frontier looks up the hierarchy from the *#template.html* associated with the file being rendered until it finds a folder whose parent folder does *not* contain a *#template.html.*

At the other end of the process, there is a folder into which the site will be rendered—the "output folder," designated by choosing **Choose Output Folder** from the **Site** menu.

This architecture acts as a "wrapper" to the normal Web site management features. A BBEdit file is rendered by bringing its window frontmost in BBEdit and choosing **Render Page** from the **Site** menu. Frontier thereupon reads it and its associated *#template.html* as wptexts into a Web site table, `user.websites.bbsite`, which has been created for you; there, they become a Web page object and template in a Web site table, and now the page object is just rendered in the normal way into the output folder, whose pathname has already been assigned as the value of `#ftpSite.folder` in `user.websites.bbsite`.

When you choose **Render Whole Site**, this same process is repeated for every file in the site, individually, starting at the folder containing the *#template.html* associated with the frontmost window and going down the hierarchy to infinite depth. Files whose names begin with # are not rendered. The contents of user.websites.bbsite are untouched between renderings, and subtables are created to represent folders; so that when the whole site has been rendered, user.websites.bbsite is like a Web site table for it. But the *#template.html* files are all read into user.websites.bbsite itself, not reflecting their place in the folder hierarchy of the site folder.

Further Details

The embedding of the object into its template is not normally performed through substitution for a <bodytext> pseudo-tag; in fact, #tagSubstitution is false. Instead, this line appears somewhere in the template:

```
{html.processMacros (bodytext)}
```

I believe that this difference from the "normal" Web page embedding mechanism is to prevent BBEdit from being confused by any <title> and <bodytext> pseudo-tags in a *#template.html*, when it checks HTML. However, nothing stops you from setting #tagSubstitution to true and using the normal method.

The glossary table is user.html.glossary. However, the glossPatch mechanism is operative, by default, and it places its automatically generated glossPatch glossary items into user.websites.glossary (because user.websites.bbsite has no glossary table).

Images are handled with a macro call to imageTag(), which is in user.html. macros. The syntax is:

```
imageTag (fileName, alt, hspace, align, usemap, height, width)
```

Image files in the unrendered Web site are stored in folders called *images*. image-Tag() looks for such a folder starting in the same folder as the *#template.html* associated with the file being rendered, and then up the hierarchy. It copies the file *fileName* from this folder into an *images* folder at the root level of the output folder, and returns a relative reference to it. All other parameters are optional.

Background images are handled through the finalFilter(). If the value of the background parameter in the <body> tag starts with "images/", the image file is copied out of the unrendered Web site's *images* folder to the correct place in the output folder.

To make it easier for users to leverage their existing AppleScript work, Apple-Script script files in Script Editor format can be kept in a *scripts* folder. This is sought by the same method as the *images* folder. Its contents are loaded into user.html.macros before a page is rendered, thus making them available to be called from macros. The pathname of the file currently being rendered is available at `user.bbSite.prefs.fileBeingRendered`.

Limitations

This system employs the normal Web site management facilities without compelling the user to learn them; instead, the user accustomed to BBEdit and a Web site made up of files on disk is able to stay with this arrangement, while accessing as many of the normal facilities as desired. But as soon as it is desired to go beyond the simplest use of those facilities, it becomes clear that the system is a hybrid. `user.websites.bbedit` must be edited in order to insert table-based directives, modify the filters, add to its `tools` table, and so forth. Features that rely upon the whole site being present as Web page objects in a Web site table, such as `outlineSite()` or the NextPrev list, require rather more work to access than they normally would; in effect, the site must be rendered twice, once to assemble a Web site table and again to use it. `user.websites.bbedit` does not contain multiple templates in different subtables, even if the unrendered site folder contains multiple *#template.html* files in different subfolders. Moreover, despite the clever use of *#template.html* files as markers, in an important sense there is only one Web site, because there is only one output folder unless it is changed manually, and `user.websites.bbedit` must be cleaned out manually when necessary.

In other words, even though the really powerful parts of the Web site management facilities can, with some ingenuity, be accessed, at that point one is effectively using them in the normal manner—from a Web site table—and the file-based aspect of the BBEdit rendering method is merely in the way. A serious Webmaster managing several sites or wanting to take advantage of Frontier's most powerful site management features will therefore probably wish to switch to the normal table-based system.

42

Dynamic Web Sites

Since Frontier generates Web sites from database objects, and since UserTalk scripts can automate the generation and modification of database objects, it makes sense to use Frontier to manage Web sites whose content is to be generated or modified automatically. Such Web sites may be described as *dynamic*: the site is "live" in some way, changing over time, so that readers see different content on different occasions when they browse the site over the Web.

Frontier comes with some suites that illustrate and implement dynamic Web sites; they are the subject of this chapter.

News Page

If you take a look at *http://www.scripting.com* in your browser, you see a News Page in operation: Scripting News. An excerpt is shown in Figure 42-1. Over the course of each day, as Dave Winer browses the Web, he sees pages that he thinks his readers might be interested in, and compiles links to these. He also writes DaveNet essays, and makes links to them. He also transforms email comments from readers of earlier DaveNet essays into Web pages, and generates links to these.

The `newsPage` suite provides a neatly packaged, simplified version of the scripts used to generate Scripting News. It can be used as is, modified to suit your needs, or just studied as a model of basic dynamic Web site generation technique. The basic interface is entirely through menu items in the **NewsPage** menu that appears when you choose **News Page** from the **Suites** menu. The menu items create and maintain a Web site table at `user.newsPage`, and then release it as desired. Of course, it is also possible for the user to customize features of the site table (for

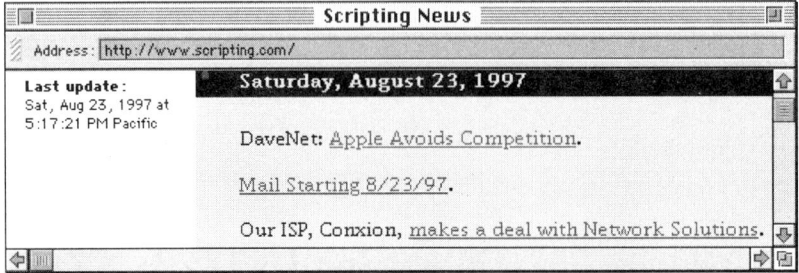

Figure 42-1. A news page

example, you might want to change the template); you can get there quickly with the **Edit Web Site** menu item.

The best way to get to know the `newsPage` suite is to experiment by adding something to each of the pages and then releasing the whole site; you will see immediately how it is structured.

To get started with the `newsPage` suite, choose **Setup** from the **NewsPage** menu. Specify the folder on disk to which the site table will be rendered, and the URL that the site will have when it is served up over the Net; these values will be copied into `user.newsPage.#ftpSite` automatically.

The site table has four page objects, all outlines:

`default`
    (Or whatever the default page name is for your server, according to `user.html.prefs.defaultFileName`.) Links to remote pages you've discovered while surfing the Web with your browser.

`mail`
    Links to received email messages that have been transformed into Web pages.

`downloads`
    Links to files which are downloadable from your Web site.

`links`
    A "free" page, for you to customize as desired, perhaps to link to other parts of a larger site in which the News Page site is embedded.

The pages all have a "diary" structure: a date, followed by all links created on that date. They will be rendered with the renderer at `user.newsPage.tools.defaultRenderer`, which turns level 1 lines into `<h4>` and everything else to an ordinary `<p>` paragraph.

Here is how to add to the default page. While browsing the Web, when a Web page to which you want to add a link is frontmost in your browser, switch to Frontier and choose **News Page** from the **Add To** submenu of the **NewsPage** menu. After you've done this a couple of times, `user.newsPage.default` will look something like this:

```
#title user.name + "'s News Page"
Sun, 24 Aug, 1997
    The <a href="http://info.ox.ac.uk/~classics/">Classics at Oxford</a> page.
    The <a href="http://promo.net/pg/">PROJECT GUTENBERG INDEX</a> page.
```

You are expected to edit this so that each line becomes a more meaningful comment about the link.

On to email. When an email message is frontmost in Eudora, you add it and a link to it to your site by switching to Frontier and choosing **Mail Page** from the **Add To** submenu. The email message becomes a Web page in the table `user.newsPage.messages`; a link to it appears in `user.newsPage.mail`. That link is a reference to a glossary item in `user.newsPage.glossary` whose value will be something like this:

```
[[#glossPatch Msg0001|messages/Msg0001|]]
```

This will cause the link in the rendered version of `user.newsPage.mail` to emanate from the phrase "Msg0001", which is probably not what you want; to fix this, edit the glossary entry's value to something more edifying, such as this:

```
[[#glossPatch An Interesting Comment|messages/Msg0001|]]
```

Be sure that `#autoParagraphs` is `true` or the messages themselves will render poorly.

Finally, downloads. In the Finder, put a file or files into a folder. Now, in Frontier, choose **Downloads Page** from the **Add To** menu, and when the dialog appears, find and choose that folder. The folder is stuffed and BinHexed into the *downloads* folder of your Web site folder (the *.hqx* file is *not* kept in the database),[*] and a link to it is added to `user.newsPage.downloads`; you should then edit that line into something meaningful.

Whenever you want to render parts of the site table, choose the appropriate menu item from the **Build** submenu of the **NewsPage** menu.

Classified Ads

The News Page makes a nice introduction to the concept of dynamic Web sites. The management of the site table and its contents is very largely taken care of automatically through a small number of menu items, which is remarkable and

---

[*] The compression is performed with StuffIt Deluxe, or the StuffIt Engine if you have it. Due to a slight bug, it may be necessary to delete by hand the stuffed, unBinHexed copy in the *downloads* folder.

convenient. But in a sense we are still building the site by hand: making links one at a time, editing them ourselves, and rendering and uploading the site in the normal way.

But now let's crank the "dynamic" concept up a notch. Why should we *ever* put our hands directly on the Web site? The site could just regenerate itself periodically without human intervention. And why should we bother to upload the site? We could just set Frontier going on the same machine that holds the Web server, and let it generate the site in place. But if we do that, how will we communicate with the site? With CGIs, of course.

The `classAds` suite demonstrates these concepts. It is to be used on a Web server machine, where Frontier is acting as a CGI application. It constructs and manages a Web site, which we will call the ClassAds site; this can be a subfolder of a larger Web site. The ClassAds site accepts and displays classified advertisements submitted by readers, by way of a browser, over the Web. You can use the suite as is, customize it, or adapt it as a model for any site where users are to be permitted to publish feedback.

The `classAds` suite combines three major Frontier facilities:

Web site management
> The Web site is stored as a site table in the database, and rendered from there to disk.

CGIs
> Frontier is commanded remotely by way of a browser to make the modifications that will cause the site's content to change.

The Scheduler
> This causes the site to be re-rendered periodically.

Creating the ClassAds Site

The ClassAds site contains four types of pages:

The default page
> This functions as a front page; users come here, read about your classified ads system, and see links to the other pages.

"About This Server"
> A "free" page, for description of and links to the rest of your site as a whole.

The advertisements
> Also called "category pages." There are as many pages as desired, each corresponding to a category which you have predetermined. Readers can come here to read the advertisements in a particular category.

The "New Ad" form

This is the page where the user fills out and submits a new advertisement. Submitting an ad causes a CGI script to return a page showing how the published ad will look, and giving the user an opportunity to edit the ad again, or publish it; publishing the ad causes another CGI script to run, such that when the site is next rendered automatically, the advertisement will be in the correct Web page.

The manual management of the site is performed on the server machine, through menu items in the **ClassAds** menu, which appears when you choose **ClassAds** from the **Suites** menu. Begin by choosing **Setup**; in the dialog, enter the pathname of the folder from which the ClassAds site will be served, and the URL of that folder. It is crucial that the URL be correct, since it will be used in constructing the **POST** action for forms, so that they are submitted to our CGI scripts. If you find you have made a mistake, choose **Start Over** to destroy everything that **Setup** creates, and begin again.

When you click **OK** in the **Setup** dialog, a table `user.classAds` is created, containing the following entries:

`categories`

An outline of summit-level entries stating the categories for the advertisements. You should edit this immediately to specify the categories you want; choose **Open Categories** to reach it easily.

`ads`, `dirtycats`, `prefs`, `serialnum`

Used internally by the `classAds` suite; you should not modify any of these directly. Once a few ads have been created, you might wish to look inside `ads` to understand how information about each one is maintained. The only thing here that might need configuring is `user.classAds.prefs.hoursTillArchive`, which determines at what age an ad will be removed from its page; set this by choosing **Aging**.

`website`

The Web site table.

`templates`

Four templates which determine the look of different Web pages the reader will see. These are not templates in the formal sense of a `#template`; they are text material into which other text material will be embedded in the course of constructing the HTML. You can easily open most of these for editing by way of the items in the **Templates** submenu. They are listed here, along with the site's true `#template`:

For Each Ad—`user.classAds.templates.singleAd`

> Every ad, within its page, is formatted as a table; this is the table. You are free to customize its look. There are seven pseudo-tags for which text will be substituted: `<adnum>`, `<category>`, `<posttime>`, `<headline>`, `<text>`, `<mailaddress>`, and `<url>`. The first two may perhaps be omitted, but the others are information submitted by the user as part of the ad and need to appear somewhere.

For Preview Page—`user.classAds.templates.previewPage`

> This is the complete HTML for the page that the user will see after submitting an ad with the New Ad form; it gives the user a chance to edit the ad some more, or to submit it finally. The `<form>` tag is generated by a macro; the content of the ad is communicated to the CGI script by way of hidden input items whose values are filled in by pseudo-tag substitution. The chief pseudo-tag you'll be concerned with is `<adtext>`, which is the ad as formatted with `singleAd`, just as it will look within its category page.

For Cat Page—`user.classAds.templates.catPage`

> This is the content for each page of advertisements. There are two pseudo-tags: `<category>`, and `<adtext>` which is where all the advertisements for that category will be put, as formatted with `singleAd`.

For Web Page—`user.classAds.website.#template`

> A true template, used for every Web page object: the default page, the "About This Server" page, the advertisement (category) pages, and the New Ad form.

`user.classAds.templates.newPageReturn`

> (There is no menu item for reaching this page.) This is the *complete* HTML for the page that the user will see after clicking the Submit button in the preview page. There is one pseudo-tag, `<url>`; this will be replaced by the value of `#ftpsite.url`.

You may wish to customize any or all of the templates. The easiest thing at first, though, is probably to leave all the templates alone.

Now choose **Build Website**, and the site will be rendered to disk. Test it as you would any CGI; get your Web server running, go to another computer and start up a browser, and ask for the URL that you gave in the **Setup** dialog. Navigate among the pages. Create a few ads; they normally won't appear on the site until it is automatically rebuilt on the hour, but you can give Frontier a nudge on the server machine by choosing **Build Website** again.

At some point, on the server machine, choose **Set Archives Folder** from the **Archives** submenu. When an ad has been displayed for the number of days listed in the **Aging** dialog, the ad table comprising its details is exported to this folder and then deleted from the database; this causes the ad to be henceforth absent from its Web page, yet retrievable from disk if necessary.

Details

You can easily use the `classAds` suite without reading any further; these details are only for those who want to understand better how the suite works, or are thinking of using it as a model for a different kind of site.

There are only two CGI scripts in `webserverScripts.classAds`: when the user submits the form in `newAd.html`, it goes to `previewAd()` which makes and returns the preview page that is returned; and when the user submits the form in *that* page, it goes to `addAd()` which makes and returns a page based on `user.classAds.templates.newPageReturn`. Everything else is served from files on disk, so the reader receives pages as speedily as the Web server can provide them.

When the user submits an ad, no attempt is made to render the category page at that time. The details of the ad are stored in an "ad table," a subtable of the table corresponding to the name of the category inside `user.classAds.ads`. Only when the hour comes around does Frontier actually render any pages; and, to ensure maximum speed, the only thing it renders is those category pages for which the ad content has changed since the last rendering (it keeps track of these in `user.classAds.dirtyCats`).

The real work is done by scripts in `suites.classAds`. Even the CGI scripts call scripts here; this makes for a nicely modularized organization. The shell script for the hourly rendering is `classAds.hourlyTask()`; it archives and deletes any expired ad tables, and then, for each page whose ad content has changed, calls `classAds.buildCatPage()` to render it. This is where we see how clever the system of ad tables and templates is. For each page, `buildCatPage()` has only to cycle through the ad tables, constructing the HTML for each ad with `classAds.getAdHTML()` and the `singleAd` template; embed that accumulated HTML into the `catPage` template; slot the result into the Web site table as a wptext Web page object; and render that object in the ordinary way.

Further Explorations

UserLand distributes various additional suites that can add different sorts of dynamism to your Web site.[*]

---

[*] See *http://www.scripting.com/directory/*.

A particularly powerful extension of the dynamic Web site concept is Content Server, which can be downloaded from UserLand's site.[*] This allows multiple Web page authors to upload material to a server machine via email, FTP, and/or Web page forms; Frontier automatically retrieves this material and transforms it into actual live Web pages.

---

[*] See *http://www.scripting.com/contentserver/*.

VII

Reference

This section is meant to be consulted, not read. Explanations and discussions have come in the preceding sections; here, details are provided for convenient access.

First, there is a chapter on how to write XCMDs and UCMDs. This supplements Chapter 23, *Extending the Language*.

Then there is a chapter on UserTalk operators, and another on UserTalk punctuation. These supplement Part II, *UserTalk Basics*.

Then, all basic UserTalk verbs discussed throughout the course of the book are listed, in alphabetical order, with details describing what they do. For further explanations and techniques of usage, the earlier parts of the book should be consulted.

Finally, some rarely used verbs related to Apple events, mentioned in Chapter 32, *Driving Other Applications*, are listed alphabetically in Chapter 47, *Apple Event Suites*.

43

XCMDs and UCMDs

This chapter provides some technical detail supplementing Chapter 23, *Extending the Language*. It is intended for those thinking of writing a UCMD or XCMD.

Both XCMDs and UCMDs can be written in any development environment you're comfortable with; Metrowerks's CodeWarrior is a likely candidate. For definitive information about XCMDs and how to write them, see the HyperCard documentation. For UCMDs, it is *de rigueur* to download the Frontier Software Developer's Kit (SDK) from *ftp://ftp.scripting.com/userland/*. It contains documentation and examples, and some model projects that you can copy and use as shells; also, it provides the "IAC Toolkit", which greatly simplifies the programming interface for dealing with Apple events. There is also a splendid online tutorial on writing UCMDs, by Brent Simmons, at *http://www.ranchero.com/frontier/ucmds*.

You may be in doubt as to which to write, an XCMD or a UCMD. UCMDs possess several advantages over XCMDs, of which the most salient are:

- UCMDs can receive as parameters and return as results any native Apple event datatype (which means almost any UserTalk datatype); XCMDs receive and return strings only.

- One UCMD can easily act as the repository for a number of different functions.

- The details of getting parameters and returning a result in a UCMD are taken care of for you with functions supplied in the SDK; with an XCMD, these tasks are up to the programmer (and can be rather tedious and tricky).*

---

* However, people have written utility shells that make writing XCMDs much easier; these are available on the Internet. Particularly recommended is Mark Hanrek's shell, available from *ftp://iw.cts.com/public/ InfoWorkshop/Shareware/*. Also, because of the long history of XCMDs, there are commercial utilities that make writing them much easier than is shown here; with these utilities, you can write your XCMD in BASIC or pseudo-HyperTalk and have all the gory details taken care of for you.

XCMDs

Communication with an XCMD depends upon a pointer, called an XCmdPtr, to a
data structure called an XCmdBlock. The XCmdPtr is the argument handed to the
XCMD when it is called.

Example 43-1. XCmdPtr and XCmdBlock

```
XCmdPtr = ^XCmdBlock;
XCmdBlock = RECORD
    paramCount:     INTEGER;
    params:         ARRAY[1..16] OF Handle;
    returnValue:    Handle;
    passFlag:       BOOLEAN;
    entryPoint:     ProcPtr; { to call back to HyperCard }
    request:        INTEGER;
    result:         INTEGER;
    inArgs:         ARRAY[1..8] OF LongInt;
    outArgs:        ARRAY[1..4] OF LongInt;
END;
```

The XCMD can get the actual parameter values by way of **params**; they are all
0-terminated strings. The XCMD ultimately places into **result** a handle to a
0-terminated string, if desired, and returns.

Here is an outline of the C code of a typical XCMD. It is from the Dartmouth
XCMDs, a HyperCard stack by Kevin Calhoun and Roger Brown which includes
many XCMDs and their source code.* It sorts the lines (fields separated by return)
of its first parameter according to the item (fields separated by comma) specified in
its second parameter; but that's not important here. In fact, I have omitted most of
the actual code, reproducing just enough to illustrate typical tasks that an XCMD
performs. ValidStrToNum() shows how you have to worry about converting
string parameters to the desired type. ResultIs() and HandleDoSort() show
ways to construct a result and attach it to the XCmdPtr. Main() shows typical
global and memory management concerns. Comments in italic are mine.

Example 43-2. A typical XCMD

```
/* SortFieldByItem1.0.c */
/* © 1991 Trustees of Dartmouth College */
/* written in THINK C™ © 1991 Symantec Corporation */
/* by Roger Brown 3/2/88  Courseware Development Group 11/19/88 */

#include <Memory.h>
#include <Packages.h>
```

---

* Thanks to Roger Brown, Moonrise Software, *internews@valley.net, http://www.dartmouth.edu/~moon-
rise,* for permission to quote from his code.

Example 43-2. A typical XCMD (continued)

```c
#include "HyperXCmd.h"
#include <stdlib.h>
#include <string.h>
#include "SetUpA4.h"

/* various definitions and globals omitted */

/* change a string to all upper case */
ucase(s)
char *s;
{ /* omitted */ }

char ValidStrToNum(s,n)
char *s;
long *n;
/* check string for valid ASCII and convert if ok. Return validity.
    Input string is changed to Pascal format */
{
    int c,len;

    /* length must be < 32768 */
    len = strlen(s);
    if ((len<1)||(len>32768)) return FALSE;
    /* all characters must be 0..9,-,+ */
    for (c=0;c<len;c++) {
        if ((s[c]<48)||(s[c]>57))
            if ((s[c]!=45)&&(s[c]!=43)) return FALSE;
    }
    CtoPstr(s);
    StringToNum(s,n);
    return TRUE;
}

GetHCItem(inStr,i,outStr)
char *inStr,*outStr;
int i;
{ /* omitted */ }

/* build a return result structure from a string */
ResultIs(paramPtr,theResult)
XCmdPtr     paramPtr;
char        *theResult;
{
    long    len;
    Handle  resultHandle;
    len = 1+strlen(theResult);
    resultHandle = NewHandle(len);
    BlockMove(theResult,*resultHandle,len);
    paramPtr->returnValue = resultHandle;
}

GetLineStarts(source,ucSource)
```

Example 43-2. A typical XCMD (continued)

```
char *source,*ucSource;
{ /* omitted */ }

static int compare(size_t a,size_t b)
{ /* omitted */ }

static void swap(size_t a,size_t b)
{ /* omitted */ }

Handle DoSort()
{ /* mostly omitted, but note this stuff near the end, where
     to build the sorted result each line in the new order
     is grabbed as a C string and concatenated to the result string */

        tempField = theUCASEField;
        next = **tempField = 0;
        for (i=0;i<numLines;i++) {
                sLen = 1+strlen(*theField+*(lineStarts+order[i]));
                for (j=0;j<sLen;j++,next++) {
                        c = *(*theField+*(lineStarts+order[i])+j);
                        if (c==0) c = '\15'; /* change 0s back to CRs */
                        *(*(tempField)+next) = c;
                }
        }
        next--;
        *(*(tempField)+next) = 0; /* put a 0 terminator on the result */
/* the rest is omitted */
}

pascal void main(paramPtr)
XCmdPtr     paramPtr;
{
        Str255 paramStr;

        /* Prepare to use globals */
        RememberA0();
        SetUpA4();

        /* get the input container copy */
        MoveHHi(paramPtr->params[0]);
        HLock(paramPtr->params[0]);
        theField = (Handle)paramPtr->params[0];

        HLock(paramPtr->params[1]);
        strcpy((char*)paramStr,*(paramPtr->params[1]));

        if (ValidStrToNum((char*)paramStr,&sortItem)==TRUE) {
            /* do the sort */
            paramPtr->returnValue = DoSort();
        }
        else {
            strcpy((char*)paramStr,"Error in item number");
```

Example 43-2. A typical XCMD (continued)

```
        ResultIs(paramPtr,(char*)paramStr);
}

/* clean up */
HUnlock (paramPtr->params[0]);
HUnlock (paramPtr->params[1]);
RestoreA4();
return;
}
```

XCMDs feature a callback mechanism whereby the XCMD can ask the calling application to perform certain utility actions or provide information, using a fixed repertoire of functions. This involves setting up the **XCmdBlock** in a particular way and passing it back to the calling application, but the programmer is usually shielded by the API from having to attend to these details. Frontier provides HyperCard-like services in response to some callbacks, but there are many callbacks that are so HyperCard-specific that they just aren't supported (they won't crash Frontier, but they won't be executed either). Table 43-1 shows the supported callbacks.

Table 43-1. XCMD Callbacks Supported by Frontier

Callback Name	Frontier Behavior
EvalExpr	The "expr" must be a UserTalk expression; it is evaluated, and the result is returned.
SendCardMessage, SendHCMessage	The "message" must be a UserTalk expression; it is evaluated, and no result is returned (presumably the message would be a verb call).
SendHCEvent	Should be sent if your XCMD calls WaitNextEvent, so that Frontier can handle unhandled events.
GetGlobal, SetGlobal	The "global" can be any variable in the scope of the calling script. If SetGlobal names a nonexistent variable, an entry is created in the scratchpad table.
ZeroBytes, ScanToReturn, ScanToZero, StringEqual, StringLength, StringMatch, ZeroTermHandle, BoolToStr, ExtToStr, LongToStr, NumToHex, NumToStr, PasToZero, PointToStr, RectToStr, ReturnToPas, StrToBool, StrToExt, StrToLong, StrToNum, StrToPoint, StrToRect, ZeroToPas	Treated normally.

For how to import an XCMD into the database and write glue for it, refer to Chapter 23.

UCMDs

Communication with a UCMD is by way of Apple events. This adds a great deal of overhead to the UCMD, but the Frontier SDK's API shields the programmer from having to attend to the details. The SDK provides a shell script whose `main()` calls `UCMDmain()`, which the programmer must write. This routine obtains the parameters, and returns a result, entirely by the use of IAC verbs from the IAC Toolkit which is included in the SDK. Declarations for these verbs may be found in the file *iac.h*.

Each Apple event datatype (to which UserTalk's datatypes very closely correspond) has a repertory of verbs for manipulating it. `IACgetxxxparam()` obtains a parameter of type *xxx* passed by the caller; the OSType passed to it names the parameter (more about this in a moment). `IACreturnxxx()` causes a value of type *xxx* to be the returned result. `IACgetxxxitem()` and `IACpushxxxitem()` are for obtaining or setting, respectively, the specified item of a list (lists are of type `AEDescList`). For example, in the case of a string:

```
Boolean IACgetstringparam (OSType, StringPtr);
Boolean IACreturnstring (StringPtr);
Boolean IACgetstringitem (AEDescList *, long, StringPtr);
Boolean IACpushstringitem (AEDescList *, StringPtr, long);
```

To understand the workings of a typical UCMD we must start with how the UCMD will be called. A UserTalk "glue" verb will construct an Apple event (see Chapter 32, *Driving Other Applications*) and send it to the UCMD. For example, consider the glue (simplified here) for Brent Simmons's `pbs.deleteListItem()`:

```
on deleteListItem (x, index)
    return (appleEvent (@pbs.code, 'pbsu', 'dlis', \
        '----', list (x), 'indx', number(index)))
```

The first parameter for `appleEvent()` is the address of the UCMD living as a binary in the database. The second parameter is the identification code unique to the UCMD as a whole; it isn't actually important what this is, since the UCMD won't check it (we're just satisfying the rules for Apple events). The third parameter identifies the command you want the UCMD to obey; thanks to this, a UCMD can contain more than one functionality. Then comes the usual series of name-value pairs, of which the first is conventionally the "direct object," whose name is `'----'`.

Now let's look at the source for `pbs.code`, omitting the code for all of its functionality except what is needed for `deleteListItem()` (you can examine the full source at `pbs.source`). Comments in italic are mine.

Example 43-3. A typical UCMD

```
/*
    Brent's Commands 1.0.1
    by Brent Simmons
    <bsimmons@wrldpwr.com>
    © 1996 World Wide Power & Light
    <http://www.wrldpwr.com/>
    Portions copyright © 1996 UserLand Software
    <http://www.scripting.com/>
*/

/*Includes*/
#include <ucmd.h>
#include <appletdefs.h>          /* for some string and Handle routines*/
#include <appletstrings.h>
#include <appletmemory.h>
#include <iac.h>

/*Defines*/
/* mostly omitted, but this one is relevant... */
#define deletelistitemtoken 'dlis'

/*Prototypes*/
/* omitted */

/*Functions*/
/* mostly omitted */

void deletelistitemverb (void) {

    long i = 1, ct = 1;
    long index;
    Handle hlistitem;
    AEDescList list, returnedlist;

    IACnewlist (&returnedlist);
    if (!IACgetlistparam ((OSType) keyDirectObject, &list))
        return;
    if (!IACgetlongparam ((OSType) 'indx', &index))
        return;
    while (true) {
        if (!IACgettextitem (&list, i, &hlistitem))
            break;
        if (i != index) {
            IACpushtextitem (&returnedlist, hlistitem, ct);
            ct++;
            } /*if*/
        else
            disposehandle (hlistitem);
        i++;
        } /*while*/
    AEDisposeDesc (&list);
    IACreturnlist (&returnedlist);
```

Example 43-3. A typical UCMD (continued)

```
        } /*deletelistitemverb*/

    void UCMDmain (void) {
        switch (IACgetverbtoken ()) {
            case deletelistitemtoken:
                deletelistitemverb ();
                break;
            case getlinkstoken:
                getlinksverb ();
                break;
            case listinstringtoken:
                listinstringverb ();
                break;
            case listtostringtoken:
                listtostringverb ();
                break;
            case stringtolisttoken:
                stringtolistverb ();
                break;
            case striphtmltoken:
                striphtmlverb ();
                break;
            default:
                IACnothandlederror ();
                break;
        } /*switch*/
    } /*UCMDmain*/
```

The main routine uses `IACgetverbtoken()` to learn what the third parameter to `appleEvent()` was, and then just switches on that, performing the desired command (and calling `IACnothandlederror()` if it doesn't recognize the command). `deletelistitemverb()` uses `IACgetlistparam()` and `IACget-longparam()` to obtain the parameters by name, `IACgettextitem()` to run through the input list and `IACpushtextitem()` to create the output list, and `IACreturnlist()` to package the output list as the value to be returned.

Like XCMDs, UCMDs can call back to Frontier while they are executing, to ask it to evaluate any valid UserTalk expression, by means of a supplied `runscript()` function. Use of this function is illustrated by a couple of the examples included in the SDK.

Like XCMDs, a UCMD compiled as a resource file can simply be dropped onto Frontier's icon; it will then be imported into the database. It is up to the programmer to write glue.

44

Operators

UserTalk operators are, for the most part, C-like.[*] On datatypes and implicit coercion, see Chapter 10, *Datatypes*. On list, record, and string operations, see also Chapter 11, *Arrays*. On arithmetic operations, see also Chapter 15, *Math*. String comparisons are case-sensitive; see also Chapter 45, *Punctuation*.

=   Performs assignment. The value and datatype of the right side are assigned to the object named on the left side, which may be created if it does not already exist (in accordance with the rules discussed Chapter 6, *Referring to Database Entries* and Chapter 7, *The Scope of Variables and Handlers*).

       Frontier will balk at using assignment to replace an non-scalar type with a scalar, but apart from this, Frontier will happily replace one datatype with another. Use of the assignment operator does not cause implicit type coercion.

+   This operator is overloaded, representing (1) numerical addition, (2) string concatenation, and (3) list/record concatenation. Implicit coercion takes place if the operands are of different types. If neither operand is of any of these types, there may be implicit coercion to a string or a number; if to a number, the result may be coerced back (see Chapter 10).

−   This operator is overloaded, representing (1) numerical subtraction, (2) substring removal, and (3) sub-list/record removal. Implicit coercion takes place if the operands are of different types. If neither operand is of any of these types, there may be implicit coercion to a string or a number; if to a number, the result may be coerced back (see Chapter 10).

---

[*] Amusingly so: for instance, the logical operators && and || are so designated in C because & and | are bit operations; yet UserTalk has the former without the latter.

*   Multiplies two numbers. Implicit coercion takes place if the operands are of different types. If neither operand is a number, there may be implicit coercion to a number, and the result may be coerced back (see Chapter 10).

/   Divides the left operand by the right operand. If neither operand is a number, there may be implicit coercion to a number, and the result may be coerced back (see Chapter 10).

%   Modulus or remainder operator; returns the remainder when the first operand is divided by the second. If both operands are not integers (or integer equivalents), a runtime error is generated. Synonym: `mod()`.

++   Increment. As in C, the operator is either prefixed or suffixed to the name of a numeric object, and the object's value has 1 added to it either before or after evaluating the surrounding expression, respectively. It is not an error to apply this to a non-number, but the results are not meaningful (no implicit coercion takes place) and it can be a bad idea.

--   Decrement. As in C, the operator is either prefixed or suffixed to the name of a numeric object, and the object's value has 1 subtracted from it either before or after evaluating the surrounding expression, respectively. It is not an error to apply this to a non-number, but the results are not meaningful (no implicit coercion takes place) and it can be a bad idea.

@   Address of. When used before the name of an object, returns a pointer to that object. See Chapter 8, *Addresses*.

^   Dereference. Using the name of an address object followed by ^ equates to using the name of the object pointed at by the address object. Using the name of a string object followed by ^ equates to using the string as a name. See Chapter 8.

Operators Used in Boolean Expressions

==   Logical equality. Synonym: `equals`. Used in boolean expressions to test for equality. It is a common beginner's mistake to use = instead:

            if a = 7 « wrong

Fortunately this generates a syntax error and won't compile.* Implicit coercion takes place if the operands are of different types.

!=   Logical inequality. Synonyms: ≠ (typed with Option-= on the Mac), `not-equals`. Note that <>, used by many languages to express inequality, is not

---

* As opposed to C, where it compiles just fine and then you spend hours trying to figure out why the program isn't behaving properly.

valid in UserTalk. Used in boolean expressions to test for inequality. Implicit coercion takes place if the operands are of different types.

< Logical less-than. Synonym: `lessthan`. Used in boolean expressions to test whether the first operand is less than the second. Implicit coercion takes place if the operands are of different types.

<= Logical less-than-or-equals. Synonym: ≤ (typed with Option-period on the Mac). Used in boolean expressions to test whether the first operand is less than or equal to the second. Implicit coercion takes place if the operands are of different types.

> Logical greater-than. Synonym: `greaterthan`. Used in boolean expressions to test whether the first operand is greater than the second. Implicit coercion takes place if the operands are of different types.

>= Logical greater-than-or-equals. Synonym: ≥ (typed with Option-comma on the Mac). Used in boolean expressions to test whether the first operand is greater than or equal to the second. Implicit coercion takes place if the operands are of different types.

! Logical not. Synonym: `not`. Used in boolean expressions to reverse the truth value of the boolean that follows it. Implicit coercion takes place if the operand is not a boolean.

&& Logical and. Synonym: `and`. Operates on two booleans, yielding `true` only if they are both `true`. If the first operand is `false`, the second will never be evaluated ("short-circuiting"). Implicit coercion takes place if an operand is not a boolean.

|| Logical or. Synonym: `or`. Operates on two booleans, yielding `true` if either is `true`. If the first operand is `true`, the second will never be evaluated ("short-circuiting"). Implicit coercion takes place if an operand is not a boolean.

`beginsWith`

Logical begins-with. If the left operand is a list/record, the second operand is coerced to the same and `true` is returned if the elements of the second operand appear, together, in order at the start of the first. Otherwise, both operands are coerced to strings and `true` is returned if the first string begins with the second as a substring.

`contains`

Logical contains. If the left operand is a list/record, the second operand is coerced to the same and `true` is returned if the elements of the second list appear together, in order, in the first. Otherwise, both operands are coerced to strings and `true` is returned if the first string contains the second as a substring.

endsWith

> Logical ends-with. If the left operand is a list/record, the second operand is coerced to the same and **true** is returned if the elements of the second list appear together, in order, at the end of the first. Otherwise, both operands are coerced to strings and **true** is returned if the first string ends with the second as a substring.

45

Punctuation

Spaces between names and operators and punctuation are largely optional in UserTalk. Clearly one cannot run keywords and names together, since this makes a different name; but otherwise spacing is pretty much a matter of style. It makes no difference which of these one says:

```
on myHandler(x=1,y=2)
    « a bit cramped, I think
on myHandler ( x = 1, y = 2 )
    « I like this better
on  myHandler  (  x  =  1 , y  =  2  )
    « legal, but no one writes this way
```

Blank lines in a script are also perfectly acceptable, and are often used to accentuate the script's structure; they do not have any meaning. Sometimes you are required to have at least a blank line, because a loop's bundle cannot be completely empty. Suppose, for instance, you want to pause until the mouse is clicked. It is illegal to say:

```
msg ("Click the mouse, please...")
while not mouse.button()
    « tread water (this is a comment line)
msg ("You clicked it!")
```

This causes a compile error. The **while** loop must contain at least one actual command line, even if it is blank; it cannot consist of just a comment line.

Capitalization is not significant in object names. But capitalization *is* significant in string comparisons. This can be confusing when one is using string comparison in a context where capitalization is normally not significant. For instance, suppose we wish to know whether the currently selected database entry is **workspace.myEntry**. If we say:

```
if table.getcursor() == "workspace.myentry"
```

the test will fail; we have caused implicit coercion of `table.getcursor()`'s result to a string, and are performing string comparison. On the other hand, if we say

```
if table.getcursor() == @workspace.myentry
```

the test will succeed. To construct a case-insensitive string match, call `string.lower()` on both sides of the comparison:

```
if string.lower("MyString") == string.lower("mysTring")
    « succeeds
```

,   Separates parameters in a verb call and in a handler definition (**on** line). Separates names in the parentheses form of a **local** declaration. Separates domain names in a **with** statement. Separates items in lists and in records.

.   In a numeric literal, the decimal point. Separates elements in a reference to a table entry. Separates container from contained in an object specifier in object-model syntax.

:   In a record literal, separates names from values. In a verb call, separates explicit parameter names from values. Separates name from value in an index of an object specifier in object-model syntax.

;   Separates commands when a script is in a text context (as opposed to a script object); a single line of a script in a script object counts as a text context. (See Chapter 4, *What a UserTalk Script Is Like.*) Separates items in the parentheses form of `loop()`.

«   Comment starter. Causes itself and everything following on the same line to be considered a comment.

\   Line continuation marker; at the end of a line, causes the next line to be considered part of the same command. In string and char literals, the "escape" marker, making it possible to type special characters.

()   Dictates order of evaluation; used after the name of a script object to make a call to it. Parameters go inside. Required in a handler definition (**on** line). Parameter variables go inside. Required after `fileloop`. Used in the parentheses form of `loop()`.

{}   Delimits a bundle when a script is in a text context (as opposed to a script object); a single line of a script in a script object counts as a text context. (See Chapter 4.) Delimits a list/record literal.

""   Delimits a string literal.

""   Delimits a string literal. On the interplay between straight and curly double-quotes, see the discussion of string literals in Chapter 10, *Datatypes.*

[]   In an object reference, used to force pre-evaluation of the enclosed expression; the result of the evaluation becomes a single element of the name. Denotes an index in array notation. Denotes an index in an object specifier in object-model syntax.

46

Verbs

For the coercion verbs, see Table 10-1. For UserTalk keywords, see Part II, *User-Talk Basics*. For the core, misc, and required verbs, see Chapter 47, *Apple Event Suites*.

Verbs are listed here in alphabetical order, single-element verbs followed by double-element verbs.

abs

```
abs (number)
```

Returns the absolute value of *number*.

appleEvent

```
appleEvent (appID, type, subtype, name1, value1, ...)
appleEvent (appID, type, subtype, table, ...)
appleEvent (appID, type, subtype, record, ...)
```

Sends to the process specified by *appID* an Apple event of type *type* and subtype *subtype*. The name-value pairs that are the parameters of the Apple event may be expressed in any of three ways: a name parameter followed by a value parameter, a table, or a record. These methods of expression may be combined.

The values in the name-value pairs are sent with whatever datatype they actually have; if binary, the internal datatype is used. The *type*, *subtype*, and the names in the name-value pairs must be string4s.

appID may be either the process's string4 creator code or its string name, or, if the process is remote, either the binary that is returned from `sys.browseNetwork()` or a case-sensitive network address string of the form `"`*zone:machine ID:-processName*`"` (where *zone* can be `"*"` to signify the local zone).

If *appID* is 0 the Apple event will be directed to the system.

callXCMD

    callXCMD (addrBinary, [param1, param2...])

Calls an XCMD. *addrBinary* must point to a binary of internal datatype `'XCMD'` or `'XFCN'`. The other parameters are strings expected as parameters by the XCMD.

close

    close (addrObject)

If the object at *addrObject* is a non-scalar whose edit window is open, closes the window, also returning to the default target situation if the object was the target. Otherwise, returns `false` and does nothing.

complexEvent

    complexEvent (addr, appID, type, subtype, name1, value1, ...)
    complexEvent (addr, appID, type, subtype, table, ...)
    complexEvent (addr, appID, type, subtype, record, ...)

Identical to calling `appleEvent()`, but the result returned from the other application goes into a table of name-value pairs at *addr*.

defined

    defined (objectReference)

Special form. Returns a boolean saying whether the object *objectReference* exists. An object exists if it has a value, even if that value was not explicitly assigned (i.e., it might be `nil`). Never raises an error.

delete

```
delete (addrObject)
```

If *addrObject* is an array specifier, removes the specified item from the array. Otherwise, destroys the object at *addrObject*.

displayString

```
displayString (value)
```

If *value* is a binary or a datatype that can be represented by a literal in UserTalk, returns a string consisting of the characters you would have to type in a UserTalk expression to represent *value* as a literal. Otherwise, returns the result of coercing *value* to a string.

If *value* is a string4 and is the value of an object in the current scope, returns the name of that object.

edit

```
edit (addrObject)
```

If the object at *addrObject* is a non-scalar, explicitly declares it to be the target, opening its edit window visibly and making it frontmost. If the object is a scalar, explicitly declares its parent table to be the target, opening its edit window visibly and making it frontmost, and selecting the entry representing the object.

evaluate

```
evaluate (string)
```

Evaluates *string* as a UserTalk expression and returns the result.

finderEvent

```
finderEvent (appID, type, subtype, name1, value1, ...)
finderEvent (appID, type, subtype, table, ...)
finderEvent (appID, type, subtype, record, ...)
```

Identical to calling `appleEvent()`, but returns immediately without waiting for a response.

gestalt

```
gestalt (selector)
```

Returns the result of calling the Gestalt Manager with string4 parameter *selector.* Knowing what information can be obtained from the Gestalt Manager, and how to interpret the results, is up to you.

getBinaryType

```
getBinaryType (binary)
```

Returns the string4 which is the internal datatype of *binary.*

memAvail

```
memAvail ()
```

Returns the amount of free RAM memory in Frontier's heap, in bytes.

mod

```
mod (dividend, divisor)
```

Returns the integer remainder when the integer *dividend* is divided by integer *divisor.* It is a runtime error if *dividend* or *divisor* is not an integer.

msg

```
msg (string)
```

Displays *string* (which is coerced to a string) in the Main Window message area. If *string* is "", returns control of the Main Window message area to the agent currently selected in the Main Window popup.

nameOf

```
nameOf (objectReference)
```

Special form. If *objectReference* is a defined object with a name, such as an item of a record or a table entry or a variable, returns that name as a string; otherwise, returns the empty string. The name of a table entry is its name within its table, not a path. Never raises an error.

new

```
new (datatype, addrObject)
```

Creates an object at *addrObject* whose value is the result of coercing `nil` to *datatype*. If the object at *addrObject* already exists, its old value is lost.

pack

```
pack (value, addrDest)
```

Coerces *value* to a binary, setting its internal datatype to the datatype of *value*, and stores the result in the object at *addrDest*.

parentOf

```
parentOf (objectReference)
```

Special form. Returns the address of the table of which *objectReference* is an entry; if there is no such table,* returns the empty string. Never raises an error.

random

```
random (integer1, integer2)
```

Returns a pseudo-random integer between *integer1* and *integer2* inclusive.

rollBeachball

```
rollBeachball ()
```

Changes Frontier's cursor to a rolling beachball. Needs to be called at least twice to have an effect. If successive calls occur within 10 ticks of each other, then the rolling beachball cursor persists only until the calls cease. If successive calls occur

---

* This might be because the designated table doesn't exist, or because the object is the root table, or because the object is a local variable. A bug in Frontier 4.2.3 causes `parentOf(x^)` to return `"root"` when x is `"root"`.

more than 10 ticks apart, then the rolling beachball persists until execution ceases. The frequency of calls has no effect upon the speed at which the beach-ball rolls.

scriptError

```
scriptError (string)
```

Throws an error whose error message is *string*.

setBinaryType

```
getBinaryType (addrBinary, string4)
```

Sets the internal datatype of the binary at *addrBinary* to *string4*.

setEventInteraction

```
setEventInteraction (boolean)
```

Causes any subsequent call to `appleEvent()`, `complexEvent()`, `tableEvent()` or `transactionEvent()` within this thread to have associated with it an event interaction value of *boolean*. The default value is `true`.

setEventTimeout

```
setEventTimeout (seconds)
```

Causes any subsequent call to `appleEvent()`, `complexEvent()`, `tableEvent()` or `transactionEvent()` within this thread to have associated with it a timeout value of *seconds* seconds. The default value (`infinity`) is restored when the thread expires or `setEventTimeout(-1)` is called.

setEventTransactionID

```
setEventTransactionID (transID)
```

Causes any subsequent call to `appleEvent()`, `complexEvent()`, or `table-Event()` within this thread to be transmuted before sending to a `transaction-`

Event() call with a *transID* value of *transID*. This situation remains in effect until the thread expires or setEventTransactionID(0) is called.

sizeOf

```
sizeOf (objRef)
```

Special form. If *objRef* refers to an array, returns the number of items in the array. If it refers to an outline, script, or menubar, returns the number of lines. If it refers to a wptext, returns the number of characters. Otherwise, returns the storage space (in bytes) occupied by *objRef.*

syscrash

```
syscrash ([string])
```

Falls into the system-level debugger, optionally displaying *string.*

systemEvent

```
systemEvent (type,subtype,name1,value1,...)
systemEvent (type,subtype,table,...)
systemEvent (type,subtype,record,...)
```

Identical to calling appleEvent() with an *appID* of 0.

tableEvent

```
tableEvent (addrTable,addr,appID,type,subtype,
           name1,value1,...)
tableEvent (addrTable,addr,appID,type,subtype,table,...)
tableEvent (addrTable,addr,appID,type,subtype,record,...)
```

Identical to calling complexEvent(), but *addrTable* points to a table to be used for some or all of the input name-value pairs.

timeCreated

```
timeCreated (addrObject)
```

If the object at *addrObject* is a non-scalar, returns the date when it was created; if it is a scalar, returns false.

timeModified

```
timeModified (addrObject)
```

If the object at *addrObject* is a non-scalar, returns the date when it was last modified; if it is a scalar, returns `false`.

transactionEvent

```
transactionEvent (appID, type, subtype, transID,
                  name1, value1, ...)
transactionEvent (appID, type, subtype, transID, table, ...)
transactionEvent (appID, type, subtype, transID, record, ...)
```

Identical to calling `appleEvent()`, but the transaction ID *transID* proves to the
target process that we have permission to speak to it during this transaction.
Rarely called directly; see `setEventTransaction()`.

typeOf

```
typeOf (value)
```

Returns the string4 designator of the datatype of *value.*

unpack

```
unpack (addrSource, addrDest)
```

Coerces the binary value at *addrSource* into that binary's internal type, and stores
the result in the object at *addrDest.*

app.clearNetworkApp

```
app.clearNetworkApp ()
```

Deletes `app.idNetworkApp`, to signal that the next time `app.startWithDocu-
ment()` is called, it should launch the local copy of the application, and make the
glue table's addressee application be the local copy.

app.linkToNetworkApp

```
app.linkToNetworkApp (remoteID)
```

Sets app.idNetworkApp to *remoteID*, to signal that the next time app.startWithDocument() is called, it should make the glue table's addressee application be the remote process specified by *remoteID*.

app.start

```
app.start (glueTableName)
```

Calls app.startWithDocument() with a second parameter of nil. Normally, there is no need to call app.start() directly, but it might be necessary to call it if a glue table has no launch() script, as a way of changing the glue table's addressee application.

app.startWithDocument

```
app.startWithDocument (glueTableName, docPathname)
```

Checks app.idNetworkApp to see if we are to address a remote application. If so, copies app.idNetworkApp into the id entry of the glue table in system.verbs.apps named *glueTableName*; if not, copies the glue table's appInfo.id into its id entry. This sets the glue table's addressee application. Launches the application specified by id if it is not running, telling it to open the file at *docPathname* if it is not the empty string (or nil). Normally, should not be called directly: either a glue table's launch() script or app.start() will call it for you.

bit.clear

```
bit.clear (longInteger, whatBit)
```

Returns *longInteger* with its bit *whatBit* cleared to 0; *whatBit* is between 0 and 31, where 0 is the low-order bit.

bit.get

```
bit.get (longInteger, whatBit)
```

Returns true or false as *longInteger*'s bit *whatBit* is 1 or 0; *whatBit* is between 0 and 31, where 0 is the low-order bit.

bit.set

```
bit.set (longInteger, whatBit)
```

Returns *longInteger* with its bit *whatBit* set to 1; *whatBit* is between 0 and 31, where 0 is the low-order bit.

card.close

```
card.close ()
```

Closes the current card.

card.getObjectEnabled

```
card.getObjectEnabled (objectName)
```

Returns a boolean saying whether the object named *objectName* in the current card is enabled.

card.getObjectFlag

```
card.getObjectFlag (objectName)
```

Returns the boolean status of the object named *objectName* in the current card. The boolean status of a radio button or check box is **true** if it is highlighted or checked, **false** if not. The boolean status of an ordinary button is whether it is the default button or not.

card.getObjectText

```
card.getObjectText (objectName)
```

Returns the text of the object named *objectName* in the current card. The text of a button is its title. The text of a text field, whether user-editable or not, is the text in the field.

card.getObjectVisible

```
card.getObjectVisible (objectName)
```

Returns a boolean saying whether the object named *objectName* in the current card is visible.

card.isModal

```
card.isModal ()
```

Returns a boolean saying whether the current card is modal.

card.popup.getCheckedItem

```
card.popup.getCheckedItem (objectName)
```

Returns the number of the currently chosen menu item of the popup menu named *objectName* in the current card.

card.popup.getHasLabel

```
card.popup.getHasLabel (objectName)
```

Returns a boolean saying whether the label of the popup menu named *object-Name* in the current card is visible.

card.popup.getMenu

```
card.popup.getMenu (objectName)
```

Returns the menu item list of the popup menu named *objectName* in the current card. A menu item list is a string of menu item names separated by semicolons. A separator line is indicated by a hyphen.

card.popup.getSelectedText

```
card.popup.getSelectedText (objectName)
```

Returns the name of the currently chosen menu item of the popup menu named *objectName* in the current card.

card.popup.setCheckedItem

```
card.popup.setCheckedItem (objectName, itemNumber)
```

Causes the menu item whose number is *itemNumber* of the popup menu named *objectName* in the current card to become the currently chosen menu item.

card.popup.setHasLabel

```
card.popup.setHasLabel (objectName, visible?)
```

Shows or hides the label of the popup menu named *objectName* in the current card, depending on the boolean *visible?*.

card.popup.setMenu

```
card.popup.setMenu (objectName, string)
```

Sets the menu item list of the popup menu named *objectName* in the current card to *string*. A menu item list is a string of menu item names separated by semicolons. A separator line is indicated by a hyphen.

card.popup.setSelectedText

```
card.popup.setSelectedText (objectName, itemText)
```

Causes the menu item whose name is *itemText* of the popup menu named *object-Name* in the current card to become the currently chosen menu item.

card.run

```
card.run (addrCardObject)
```

Displays as a dialog the card object at *addrCardObject*.

card.setObjectEnabled

```
card.setObjectEnabled (objectName, enabled?)
```

Enables or disables the object named *objectName* in the current card, depending on the boolean *enabled?*.

card.setObjectFlag

```
card.setObjectFlag (objectName, boolean)
```

Sets the boolean status of the object named *objectName* in the current card to *boolean*. The boolean status of a radio button or check box is true if it is highlighted or checked, false if not. The boolean status of an ordinary button is whether it is the default button or not.

card.setObjectText

```
card.setObjectText (objectName, string)
```

Sets the text of the object named *objectName* in the current card to *string*. The text of a button is its title. The text of a text field, whether user-editable or not, is the text in the field.

card.setObjectVisible

```
card.setObjectVisible (objectName, visible?)
```

Shows or hides the object named *objectName* in the current card, depending on the boolean *visible?*.

clipboard.get

```
clipboard.get (MacOSFormat, addrObject)
```

Copies the *MacOSFormat* component of the system scrap's contents into the object at *addrObject* as a binary whose internal data type is *MacOSFormat*, which will typically be 'TEXT' or 'PICT'.

clipboard.getValue

```
clipboard.getValue (MacOSFormat)
```

Returns the *MacOSFormat* component of the system scrap's contents, unpacked so as to be of type *MacOSFormat*, which will typically be 'TEXT' or 'PICT'.

clipboard.put

```
clipboard.put (MacOSFormat, addrObject)
```

Copies the value of the object at *addrObject* onto the system scrap in the format *MacOSFormat*, which will typically be `'TEXT'` or `'PICT'`.

---

clipboard.putValue

```
clipboard.putValue (value)
```

Calls `clipboard.put()` to copy *value* onto the system scrap, supplying *MacOS-Format* from the datatype (or, if a binary, the internal datatype) of *value*.

---

clock.now

```
clock.now ()
```

Returns the date from the system clock.

---

clock.set

```
clock.set (dateTime)
```

Sets the system clock to *dateTime*.

---

clock.sleepFor

```
clock.sleepFor (seconds)
```

Within an agent script (or a script called by an agent script), tells Frontier to call the agent script again *seconds* seconds after the agent script finishes executing. An agent script that never calls this verb will be called again one second after it finishes executing. Within a modeless dialog proc, puts the dialog to sleep temporarily. It is a runtime error to use this verb outside either the agent thread or a modeless dialog proc.

clock.ticks

```
clock.ticks ()
```

Returns the number of ticks (approximately sixtieths of a second) that have elapsed since the computer was started up.

clock.waitSeconds

```
clock.waitSeconds (n)
```

Pauses for *n* seconds. *n* is coerced to an integer.

clock.waitSixtieths

```
clock.waitSixtieths (n)
```

Pauses for *n* sixtieths of a second. *n* is coerced to an integer.

date.abbrevString

```
date.abbrevString (date)
```

Returns the system's `abbrevDate` string representation of *date*.

date.dayOfWeek

```
date.dayOfWeek (date)
```

Returns the day-of-week number for *date*, where 1 represents Sunday.

date.daysInMonth

```
date.daysInMonth (date)
```

Returns the number of days in the month in which *date* falls.

date.dayString

```
date.dayString (number)
```

Returns the name of the day denoted by the day-of-week number *number,* or the empty string if *number* is not an integer from 1 to 7.

date.firstOfMonth

```
date.firstOfMonth (date)
```

Returns the date value for 12:00 midnight on the first day of the month in which *date* falls.

date.get

```
date.get (theDate, addrDay, addrMonth, addrYear,
          addrHour, addrMin, addrSec)
```

Analyzes *theDate* into numerical values (shorts) and stores them at the addresses provided in the other parameters.

date.lastOfMonth

```
date.lastOfMonth (date)
```

Returns the date value for 12:00 midnight on the last day of the month in which *date* falls.

date.longString

```
date.longString (date)
```

Returns the system's longDate string representation of *date.*

date.nextMonth

```
date.nextMonth (date)
```

Returns the date resulting from adding 1 to *date*'s month component.

date.nextWeek

```
date.nextWeek (date)
```

Returns the date resulting from adding 7 to *date*'s day component.

date.nextYear

```
date.nextYear (date)
```

Returns the date resulting from adding 1 to *date*'s year component.

date.prevMonth

```
date.prevMonth (date)
```

Returns the date resulting from subtracting 1 from *date*'s month component.

date.prevWeek

```
date.prevWeek (date)
```

Returns the date resulting from subtracting 7 from *date*'s day component.

date.prevYear

```
date.prevYear (date)
```

Returns the date resulting from subtracting 1 from *date*'s year component.

date.set

```
date.set (day, month, year, hour, min, sec)
```

Returns a date constructed from the numeric parameters; they are coerced to integers before use.

date.shortString

    date.shortString (*date*)

Returns the system's shortDate string representation of *date*.

date.tomorrow

    date.tomorrow (*date*)

Returns the date resulting from adding 1 to *date*'s day component.

date.weeksInMonth

    date.weeksInMonth (*date*)

Returns date.daysInMonth(*date*) integer-divided by 7.

date.yesterday

    date.yesterday (*date*)

Returns the date resulting from subtracting 1 from *date*'s day component.

db.close

    db.close (*pathname*)

Closes, without saving, the database at *pathname*, which must have been opened with db.open().

db.countItems

    db.countItems (*pathname, entryPathString*)

Returns the number of entries in the table object *entryPathString* in the database at *pathname*, which must have been opened with db.open(). If the object does not exist or is not a table, an error will be raised.

db.defined

> db.defined (*pathname*, *entryPathString*)

Returns a boolean reporting whether an object *entryPathString* exists in the database at *pathname*, which must have been opened with db.open().

---

db.delete

> db.delete (*pathname*, *entryPathString*)

Destroys the object *entryPathString* in the database at *pathname*, which must have been opened with db.open(). If the object does not exist, an error will be raised.

---

db.get

> db.get (*entryPathString*)

Called from an AppleScript script, returns the value of the database object named by *entryPathString*.

---

db.getNthItem

> db.getNthItem (*pathname*, *entryPathString*, *n*)

Returns the value of the *n*th entry in the table object *entryPathString* in the database at *pathname*, which must have been opened with db.open(). If the object does not exist or is not a table, an error will be raised.

---

db.getValue

> db.getValue (*pathname*, *entryPathString*)

Returns the value of the object *entryPathString* in the database at *pathname*, which must have been opened with db.open(). If the object does not exist, an error will be raised.

db.isTable

```
db.isTable (pathname, entryPathString)
```

Returns a boolean reporting whether the object *entryPathString* in the database at *pathname*, which must have been opened with db.open(), is a table. If the object does not exist, an error will be raised.

db.new

```
db.new (pathname)
```

Creates a new minimal database on disk, at *pathname*. If there is already a file or empty folder at *pathname*, it is deleted first. The new database is not opened.

db.newTable

```
db.newTable (pathname, tablePathString)
```

Creates at *tablePathString* a new table in the database at *pathname*, which must have been opened with db.open(). If an object already exists at this location, it will be deleted first. If the proposed table's parent does not exist, an error will be raised.

db.open

```
db.open (pathname, readOnly?)
```

Opens the existing database at *pathname*, ready to work with it programmatically through the other db verbs. The boolean *readOnly?* is not optional.

db.save

```
db.save (pathname)
```

Saves the database at *pathname*, which must have been opened with db.open(), returning a boolean reporting whether it did in fact save the database.

db.set

```
db.set (entryPathString, value)
```

Called from an AppleScript script, sets to *value* the database object named by *entryPathString*.

db.setValue

```
db.setValue (pathname, entryPathString, value)
```

Sets to *value* the value of the object *entryPathString* in the database at *pathname*, which must have been opened with db.open(). The object will be created if necessary. If an object already exists at this location, it will be deleted first. If the proposed table's parent does not exist, an error will be raised.

dialog.alert

```
dialog.alert (text)
```

Displays a modal dialog containing an "attention" icon, *text* coerced to a string, and an **OK** button; and sounds the system beep.

dialog.ask

```
dialog.ask (text, addrResult)
```

Displays a modal dialog containing *text* coerced to a string, a **Cancel** button and an **OK** button, which is the default. The dialog also contains an editable text field, which initially contains the value at *addrResult* coerced to a string. Returns false if the user clicks **Cancel**, true otherwise; if true, the contents of the editable text field are stored at *addrResult*.

dialog.confirm

```
dialog.confirm (text)
```

Displays a modal dialog containing *text* coerced to a string, a **Cancel** button and an **OK** button, which is the default. Returns false if the user clicks **Cancel**, true otherwise.

dialog.fileInfo

        dialog.fileInfo (*pathname*)

Displays a modal file information dialog about the file at *pathname*; if no such file exists, an error is raised.

dialog.getInt

        dialog.getInt (*text, addrResult*)

Displays a modal dialog containing *text* coerced to a string, a **Cancel** button and an **OK** button, which is the default. The dialog also contains an editable text field, which initially contains the value at *addrResult* coerced to a string. Returns `false` if the user clicks **Cancel**, `true` otherwise; if `true`, the contents of the editable text field are coerced to a short (raising an error if this is impossible) and stored at *addrResult*.

dialog.getValue

        dialog.getValue (*itemNum*)

Used in the proc of a resource-based dialog (it is a runtime error to use it in any other context), returns the value of the item of the dialog whose item number is *itemNum*. The value of a checkbox or radio button is a boolean. The value of a text item is the text, as a string. The value of a button is its title, as a string. Certain miscellaneous controls, such as a scrollbar, have a value which is a short.

dialog.hideItem

        dialog.hideItem (*itemNum*)

Used in the proc of a resource-based dialog (it is a runtime error to use it in any other context), makes invisible the item of the dialog whose item number is *itemNum*.

dialog.loadFromFile

        dialog.loadFromFile (*pathname*)

Loads from the file at *pathname* into the resource fork of the database the first `'DLOG'` and `'DITL'` whose ID is greater than or equal to 25000, plus all other

resources whose ID is greater than or equal to 25000. Returns the ID of the
`'DLOG'`.

dialog.notify

```
dialog.notify (text)
```

Displays a modal dialog containing a "chat" icon, *text* coerced to a string, and an
OK button.

dialog.run

```
dialog.run (resourceID, defaultItemNum, addrProc)
```

Displays a resource-based modal dialog. *resourceID* is the ID of a `'DLOG'`
resource already in the resource chain. *defaultItemNum* specifies the item number
of the default item in the dialog (to be "hit" automatically when the user hits Enter
or Return); typically, this is 1. *addrProc* specifies the address of the handler to be
called as the proc for the dialog.

dialog.runFromFile

```
dialog.runFromFile (pathname, defaultItemNum, addrProc)
```

Displays a modal dialog whose resources are file-based, by calling `dialog.load-FromFile()` and `dialog.run()`.

dialog.runModeless

```
dialog.runModeless (resourceID, defaultItemNum, addrProc)
```

Displays a resource-based modal dialog. *resourceID* is the ID of a `'DLOG'`
resource already in the resource chain. *defaultItemNum* specifies the item number
of the default item in the dialog (to be "hit" automatically when the user hits Enter
or Return); typically, this is 1. *addrProc* specifies the address of the handler to be
called as the proc for the dialog.

dialog.setItemEnable

```
dialog.setItemEnable (itemNum, enabled?)
```

Used in the proc of a resource-based dialog (it is a runtime error to use it in any other context), makes enabled or disabled the item of the dialog whose item number is *itemNum*, depending on whether *enabled?* is `true` or `false`.

dialog.setValue

```
dialog.setValue (itemNum, newValue)
```

Used in the proc of a resource-based dialog (it is a runtime error to use it in any other context), sets to *newValue* the value of the item of the dialog whose item number is *itemNum*. The value of a checkbox or radio button is a boolean. The value of a text item is the text, as a string. The value of a button is its title, as a string. Certain miscellaneous controls, such as a scrollbar, have a value that is a short.

dialog.showItem

```
dialog.showItem (itemNum)
```

Used in the proc of a resource-based dialog (it is a runtime error to use it in any other context), makes visible the item of the dialog whose item number is *itemNum*.

dialog.threeWay

```
dialog.threeWay (text, yesButton, noButton, cancelButton)
```

Displays a modal dialog containing *text* coerced to a string, a button that says *cancelButton*, a button that says *noButton*, and a button that says *yesButton*, which is the default. Returns 3, 2, or 1 respectively, depending on which button the user clicks.

dialog.twoWay

```
dialog.twoWay (text, OKbutton, cancelButton)
```

Displays a modal dialog containing *text* coerced to a string, a button that says *cancelButton*, and a button that says *OKbutton*, which is the default. Returns `false` if the user clicks the *cancelButton*, `true` otherwise.

dialog.yesNo

```
dialog.yesNo (text)
```

Displays a modal dialog containing *text* coerced to a string, a **No** button and a **Yes** button, which is the default. Returns `false` if the user clicks **No**, `true` otherwise.

dialog.yesNoCancel

```
dialog.yesNoCancel (text)
```

Displays a modal dialog containing *text* coerced to a string, a **Cancel** button, a **No** button, and a **Yes** button, which is the default. Returns 3, 2, or 1 respectively, depending on which button the user clicks.

editMenu.clear

```
editMenu.clear ()
```

Like choosing **Clear** from the **Edit** menu. Operates on the current target.

editMenu.copy

```
editMenu.copy ()
```

Like choosing **Copy** from the **Edit** menu. Operates on the current target.

editMenu.cut

```
editMenu.cut ()
```

Like choosing **Cut** from the **Edit** menu. Operates on the current target.

editMenu.paste

```
editMenu.paste ()
```

Like choosing **Paste** from the **Edit** menu. Operates on the current target.

editMenu.plainText

```
editMenu.plainText ()
```

Removes all text styling from the current selection in the wptext target window.

editMenu.selectAll

```
editMenu.selectAll ()
```

Like choosing **Select All** from the **Edit** menu. Operates on the current target; if a wptext or in content mode, selects the whole wptext "document," otherwise selects all lines of the outline at the same level as the current selection. In a table edit window, switches to content mode.

editMenu.setBold

```
editMenu.setBold (boolean)
```

Turns bold text styling on or off, depending on *boolean*, for the current selection in the wptext target window.

editMenu.setFont

```
editMenu.setFont (nameString)
```

Like choosing the font called *nameString* from the **Font** submenu of the **Edit** menu. Operates on the current target.

editMenu.setFontSize

```
editMenu.setFontSize (sizeInteger)
```

Like choosing the font size *sizeInteger* from the **Size** submenu of the **Edit** menu. Operates on the current target.

editMenu.setItalic

```
editMenu.setItalic (boolean)
```

Turns italic text styling on or off, depending on *boolean*, for the current selection in the wptext target window.

editMenu.setOutline

```
editMenu.setOutline (boolean)
```

Turns outline text styling on or off, depending on *boolean*, for the current selection in the wptext target window.

editMenu.setShadow

```
editMenu.setShadow (boolean)
```

Turns shadow text styling on or off, depending on *boolean*, for the current selection in the wptext target window.

editMenu.setUnderline

```
editMenu.setUnderline (boolean)
```

Turns underline text styling on or off, depending on *boolean*, for the current selection in the wptext target window.

file.bytesInFolder

```
file.bytesInFolder (folderPathname)
```

Returns the total number of bytes occupied by all files in the folder at *folderPathname.*

file.bytesOnVolume

```
file.bytesOnVolume (volumePathname)
```

Returns the number of occupied bytes in the volume at *volumePathname.*

file.close

        file.close (*pathname*)

Closes the file at *pathname*, ending a read/write access session. Returns **false** if the file is not open; raises an error if the file does not exist.

file.compare

        file.compare (*pathname1, pathname2*)

Returns a boolean reporting whether the dataforks of the files at *pathname1* and *pathname2* are identical.

file.copy

        file.copy (*source, dest*)

Copies the file or folder at *source* to a new location, maintaining creation and modification dates for copied material. If *source* is a folder, its contents, not the folder itself, are copied into *dest*, which must be a folder. If *dest* is a folder, what is being copied is copied into it; the whole *dest* pathname must denote existing folders, except that if *source* is a folder, the last element of *dest* will be created if it doesn't exist. If *dest* is a file, then *source* must be a file. Copied files replace existing files with the same name.

file.copyDataFork

        file.copyDataFork (*source, dest*)

Creates a new file at *dest* (replacing an existing file) that is identical to the file at *source* without its resource fork.

file.copyResourceFork

        file.copyResourceFork (*source, dest*)

Creates a new file at *dest* (replacing an existing file) that is identical to the file at *source* without its data fork.

file.copyToSystemFolder

```
file.copyToSystemFolder (source)
```

A utility script which simulates dropping the file or folder *source* onto the system folder's icon, copying it into its proper subfolder if it is a file of special type, and otherwise copying it into the system folder itself.

file.countLines

```
file.countLines (pathname)
```

Returns the number of end-of-lines (`cr`) in the file at *pathname*, which should not be opened beforehand. (If the file *is* opened beforehand, returns 0 if any reading has been done this session.)

file.countVolumes

```
file.countVolumes ()
```

Returns the number of volumes currently mounted.

file.created

```
file.created (pathname)
```

Returns the creation date of the file or folder at *pathname*.

file.creator

```
file.creator (pathname)
```

Returns the string4 creator code of the file at *pathname*.

file.delete

```
file.delete (pathname)
```

Deletes the file or empty folder at *pathname*. An error is raised if the item at *pathname* is a folder with contents, or is locked or "busy" (open for use by some application).

file.deleteFolder

```
file.deleteFolder (pathname)
```

Deletes the folder at *pathname* even if it has contents.

---

file.eject

```
file.eject (volumePathname)
```

Tries to eject the volume at *volumePathname*. This is not the same as unmounting the volume. Returns `false` if the volume is not ejectable.

---

file.endOfFile

```
file.endOfFile (pathname)
```

Returns `false` if on the next command it would be possible to read at least one character from the file at *pathname* using `file.read()`; `true` is returned if the file is not open or does not exist.

---

file.exists

```
file.exists (pathname)
```

Returns a boolean reporting whether the file, folder or volume at *pathname* exists.

---

file.fileFromPath

```
file.fileFromPath (pathname)
```

Returns the last element of the pathname string *pathname*.

---

file.filesInFolder

```
file.filesInFolder (folderPathname, depth)
```

Returns the total number of files and folders in the folder at *folderPathname*, to depth *depth*.

file.filesOnVolume

```
file.filesOnVolume (volumePathname)
```

Returns the number of files and folders, including invisible files and folders, in the volume at *volumePathname*.

file.filteredCopy

```
file.filteredCopy (source, dest, addrProc)
```

Selectively copies the file or folder at *source* to a new location. A utility script. The basic mechanics are identical to `file.copy()`; in fact, `file.copy()` is simply a script that calls `file.filteredCopy()` with a proc that always returns `true`. The proc, a handler pointed to by *addrProc*, must take a single parameter, a pathname, and return `true` or `false` as that file should or should not be copied. If the proc says a folder isn't to be copied, its contents are never tested.

file.findApplication

```
file.findApplication (string4)
```

Returns the pathname of the application associated with the string4 creator code *string4*. The volumes are searched beginning with the startup volume. The search ends when a volume containing the application is found. If that volume contains more than one instance of the application, the pathname of the more recent instance of the application is returned.

file.findInFile

```
file.findInFile (pathname, whatToLookFor)
```

whatToLookFor is coerced to a string, and if it appears in the file at *pathname* (which need not be opened beforehand), `true` is returned, `false` otherwise. If *pathname* doesn't exist, an error is raised.

file.findInFolder

```
file.findInFolder (folderPathname, depth, whatToLookFor)
```

whatToLookFor is coerced to a string, and if it appears in some file in the folder at *folderPathname*, searching to depth *depth*, returns the pathname of the first such file encountered; otherwise, returns the empty string.

file.folderFromPath

```
file.folderFromPath (pathname)
```

Returns all but the last element of the pathname string *pathname*.

file.foldersInFolder

```
file.foldersInFolder (folderPathname, depth)
```

Returns the total number of folders in the folder at *folderPathname*, to depth *depth*.

file.foldersOnVolume

```
file.foldersOnVolume (volumePathname)
```

Returns the number of folders, including invisible folders, in the volume at *volumePathname*.

file.followAlias

```
file.followAlias (pathname)
```

Returns the pathname of the original of the alias file at *pathname*. If what's at *pathname* is not an alias, returns the empty string. If there is nothing at *pathname*, a very weird error is raised.

file.freespaceOnVolume

```
file.freespaceOnVolume (volumePathname)
```

Returns the number of free bytes in the volume at *volumePathname*.

file.getComment

```
file.getComment (pathname)
```

Returns the comment string of the file or folder at *pathname*.

---

file.getDiskDialog

```
file.getDiskDialog (text, addrResult)
```

Displays a modal dialog permitting the user to choose a volume, including prompt *text* ("" for no prompt), a **Cancel** button, and an **Open** button. The volume initially displayed may be set by supplying a pathname at *addrResult*. Returns false if the user clicks **Cancel**, true otherwise. If true, the pathname of the chosen volume is stored at *addrResult*.

---

file.getFileDialog

```
file.getFileDialog (text, addrResult, fileTypes)
```

Displays the Mac OS StandardGetFile modal dialog, permitting the user to choose a file, including prompt *text* ("" for no prompt), a **Cancel** button, and an **Open** button. *fileTypes* specifies what types of file are listed; it may be a string4 type designator or a list of such type designators, or 0 to show all files. The directory (and file) initially displayed may be set by supplying a pathname at *addrResult*. Returns false if the user clicks **Cancel**, true otherwise. If true, the pathname of the chosen file is stored at *addrResult*.

---

file.getFolderDialog

```
file.getFolderDialog (text, addrResult)
```

Displays a modified Mac OS StandardGetFile modal dialog, permitting the user to choose a folder, including prompt *text* ("" for no prompt), a **Cancel** button, an **Open** button, and a button to choose the presently selected folder. The directory initially displayed may be set by supplying a pathname at *addrResult*. Returns false if the user clicks **Cancel**, true otherwise. If true, the pathname of the chosen folder is stored at *addrResult*.

file.getFullVersion

```
file.getFullVersion (pathname)
```

Returns the full version string of the file at *pathname.*

file.getIconPos

```
file.getIconPos (pathname, addrHoriz, addrVert)
```

Stores at *addrHoriz* and *addrVert* the components of the icon position of the file or folder at *pathname.*

file.getLabel

```
file.getLabel (pathname)
```

Returns the label of the file or folder at *pathname.*

file.getPath

```
file.getPath ()
```

Returns the current pathname prefix, which is automatically prefixed to a one-element file pathname string when it is resolved as a fileSpec.

file.getSpecialFolderPath

```
file.getSpecialFolderPath (volumePathname, folder, create?)
```

Returns the pathname of any system folder's standard subfolders. Values for *folder* are: `"Apple"`, `"Control"`, `"Desktop"`, `"Extensions"`, `"Fonts"`, `"Preferences"`, `"PrintMonitor"`, `"Startup"`, `"System"`, `"Temp"`, `"Trash"`. The folder is sought on the volume designated by *volumePathname.* If there is no system folder on that volume, an error is raised. If the requested folder doesn't exist, it is created if *create?* is `true`, otherwise an error is raised.

file.getSystemDisk

```
file.getSystemDisk ()
```

Returns the pathname of the startup volume.

file.getSystemFolderPath

```
file.getSystemFolderPath ()
```

Returns the pathname of the active system folder.

file.getVersion

```
file.getVersion (pathname)
```

Returns the version string of the file at *pathname*.

file.isAlias

```
file.isAlias (pathname)
```

Returns a boolean reporting whether the file at *pathname* is an alias. If no such file exists, an error is raised.

file.isBusy

```
file.isBusy (pathname)
```

Returns a boolean reporting whether the file at *pathname* is "busy" (open for use by an application).

file.isEjectable

```
file.isEjectable (volumePathname)
```

Returns a boolean reporting whether the volume at *volumePathname* is ejectable, like a floppy disk or a CD.

file.isFolder

```
file.isFolder (pathname)
```

Returns a boolean reporting whether the file at *pathname* is a folder (or volume). If no such file exists, an error is raised.

file.isLocked

```
file.isLocked (pathname)
```

Returns a boolean reporting whether the file or folder at *pathname* is locked.

file.isVisible

```
file.isVisible (pathname)
```

Returns a boolean reporting whether the file or folder at *pathname* is visible.

file.isVolume

```
file.isVolume (pathname)
```

Returns a boolean reporting whether *pathname* designates a volume. If no such item exists, an error is raised.

file.lock

```
file.lock (pathname)
```

Locks the file or folder at *pathname*.

file.modified

```
file.modified (pathname)
```

Returns the modification date of the file or folder at *pathname*.

file.mountServerVolume

```
file.mountServerVolume (path, userID, password)
```

Attempts to mount the remote volume *path*, which is of the form `"zone:machineID:volName"`; to signify the local zone, `"*"` may be used for the *zone*. Returns true if the volume was successfully mounted; if the server was not found, or an invalid *userID* and *password* were provided, an error is raised. For guest login, supply the empty string (`""`) as the *userID* and *password*.

file.move

```
file.move (source, destFolder)
```

Moves the file or folder at *source* into *destFolder*, which must be an existing folder on the same volume as *source*, and must not contain a file or folder with the same name as *source*.

file.new

```
file.new (pathname)
```

Creates a new file at *pathname*; if there is already a file or empty folder at *pathname* it is deleted first. The new file has type and creator codes '????', and no data. All elements of *pathname* except the last must exist.

file.newAlias

```
file.newAlias (original, dest)
```

Creates at *dest* an alias of the file, folder or volume at *original*. Everything but the last element of *dest* must exist. If there is already a file or folder at *dest*, an error is raised.

file.newFolder

```
file.newFolder (pathname)
```

Creates at *pathname* a new empty folder. Everything but the last element of *pathname* must exist. If there is already a file or folder at *pathname*, an error is raised.

file.open

```
file.open (pathname)
```

Opens the file at *pathname* for read/write access, beginning a session. An error is raised if the file couldn't be opened.

file.putFileDialog

`file.putFileDialog (`*text, addrResult*`)`

Displays the Mac OS StandardPutFile modal dialog, permitting the user to specify a folder and supply a filename, including prompt *text* (`""` for "Save as:"), an editable text field, a **New Folder** button, a **Cancel** button, and a **Save** button. Any folders the user creates with the **New Folder** button are actually created. The directory initially displayed and the initial contents of the editable text field may be set by supplying a pathname at *addrResult*. Returns `false` if the user clicks **Cancel**, `true` otherwise. If the designated file already exists, a secondary name-conflict dialog appears, and `true` is not returned unless **Replace** is clicked. If `true`, the pathname of the specified file is stored at *addrResult*; it will not be an existing folder or an illegal pathname.

file.read

`file.read (`*pathname, howManyCharacters*`)`

Reads characters from the file at *pathname*; returns a binary consisting of the characters read, or `false` if the file is not open. The read starts after the end of the previous read during the present session, or at the beginning of the file if there was no previous read. If *howManyCharacters* is too large, then if *howManyCharacters* is `infinity` the file is read to the end, but otherwise the read fails, `false` is returned, and further reading in this session becomes impossible (because the pointer is now at end-of-file).

file.readLine

`file.readLine (`*pathname*`)`

Reads characters from the file at *pathname*; returns a string (not a binary!) consisting of the characters read, or the empty string if the file is not open. The read starts after the end of the previous read during the present session, or at the beginning of the file if there was no previous read. The read continues until end-of-line is encountered (`cr`); the position pointer is moved past the end-of-line, but the end-of-line is not included in the returned string. If end-of-file is encountered, text is read successfully; but if the read starts at end-of-file, the empty string is returned.

file.rename

```
file.rename (pathname, nameString)
```

Renames the file or folder or volume at *pathname*, giving it the name *nameString*, which should be just the new name, not a pathname.

file.setComment

```
file.setComment (pathname, commentString)
```

Sets to *commentString* the comment string of the file or folder at *pathname*.

file.setCreated

```
file.setCreated (pathname, date)
```

Sets to *date* the creation date of the file or folder at *pathname*.

file.setCreator

```
file.setCreator (pathname, string4)
```

Sets to *string4* the string4 creator code of the file at *pathname*.

file.setFullVersion

```
file.setFullVersion (pathname, versionString)
```

Sets to *versionString* the full version string of the file at *pathname*.

file.setIconPos

```
file.setIconPos (pathname, horiz, vert)
```

Sets to *horiz* and *vert* the components of the icon position of the file or folder at *pathname*.

file.setLabel

```
file.setLabel (pathname, labelString)
```

Sets to *labelString* the label of the file or folder at *pathname*. *labelString* must match one of the labels registered in the *Labels* control panel, or nothing will happen.

file.setModified

```
file.setModified (pathname, date)
```

Sets to *date* the modification date of the file or folder at *pathname*.

file.setPath

```
file.setPath (pathString)
```

Sets the current pathname prefix to *pathString*, which must be the full pathname of an existing volume or folder. The change remains in force until Frontier is quit. The pathname prefix is automatically prefixed to a one-element file pathname string when it is resolved as a fileSpec.

file.setType

```
file.setType (pathname, string4)
```

Sets to *string4* the string4 type code of the file at *pathname*.

file.setVersion

```
file.setVersion (pathname, versionString)
```

Sets to *versionString* the version string of the file at *pathname*.

file.setVisible

```
file.setVisible (pathname, visible?)
```

Sets the visibility of the file or folder at *pathname* according to the boolean *visible?*.

file.size

```
file.size (pathname)
```

Returns the logical size of the file at *pathname*, the total bytes of the file's resource fork and data fork combined.

file.sureFolder

```
file.sureFolder (pathname)
```

Creates at *pathname* a new empty folder. Everything but the last element of *pathname* must exist. If there is already a file or folder at *pathname*, nothing happens.

file.type

```
file.type (pathname)
```

Returns the string4 type code of the file at *pathname*.

file.unlock

```
file.unlock (pathname)
```

Unlocks the file or folder at *pathname*.

file.unmountVolume

```
file.unmountVolume (volumePathname)
```

Tries to unmount the volume at *volumePathname*. The volume may be an ejectable medium or a remote volume. Stronger than `file.eject()`, which ejects the physical medium but retains the volume information. If the volume cannot be unmounted (because it contains files in use, for example), an error is raised.

file.visitFolder

```
file.visitFolder (folderPathString, depth, addrProc)
```

The script or handler pointed to by *addrProc* must take one parameter, which will be a reference to each file or folder in the folder denoted by *folderPathString*, recursing to the depth indicated by the integer *depth*; it must return a boolean saying whether it wants to be called again. File and folders in a folder are passed before that folder.

file.volumeBlockSize

```
file.volumeBlockSize (volumePathname)
```

Returns the block size, in bytes, of the volume at *volumePathname*.

file.volumeFromPath

```
file.volumeFromPath (pathname)
```

Returns the first element of the pathname string *pathname*.

file.volumeSize

```
file.volumeSize (volumePathname)
```

Returns the total number of bytes in the volume at *volumePathname*.

file.write

```
file.write (pathname, whatToWrite)
```

Coerces *whatToWrite* to a binary and appends it to the file at *pathname*. The file is implicitly opened and closed, but it is well to call `file.open()` and `file.close()` anyway. Raises an error if the file does not exist or is a folder; but certain conditions (for example, the file is busy) will return **false** without raising an error, so it may be wise to check the result as well.

file.writeLine

```
file.writeLine (pathname, whatToWrite)
```

Same as `file.write()` except that *whatToWrite* is first coerced to a string and cr is appended to it.

filemenu.close

```
filemenu.close ()
```

Closes the frontmost window.

filemenu.closeAll

```
filemenu.closeAll ()
```

Closes all open windows, not including the Main Window, unless the Main Window is frontmost.

filemenu.new

```
filemenu.new (pathname)
```

Creates a new minimal database at *pathname*, and opens it as a second database. If there is already a file or empty folder at *pathname*, it is deleted first.

filemenu.open

```
filemenu.open (pathname)
```

Opens the file at *pathname*. The effect is as if the file's icon had been dropped onto Frontier's icon; if the file is of a type that belongs to Frontier, it is also the same as if the file had been opened from the Finder. So, a packed object will be imported, a desktop script will be run, a second database will be opened, a HyperCard stack will have its XCMDs imported, and so on.

filemenu.save

```
filemenu.save ()
```

Saves the database.

filemenu.saveCopy

```
filemenu.saveCopy (pathname)
```

Saves the database in compact form to a file at *pathname*.

finderflags.clear

```
finderflags.clear (pathname, whichBit)
```

Clears the Finder flag *whichBit* of the file or folder at *pathname*.

finderflags.get

```
finderflags.get (pathname, whichBit)
```

Returns a boolean reporting whether the Finder flag *whichBit* of the file or folder at *pathname* is set.

finderflags.set

```
finderflags.set (pathname, whichBit)
```

Sets the Finder flag *whichBit* of the file or folder at *pathname*.

frontier.bringToFront

```
frontier.bringToFront ()
```

Brings Frontier to the front.

frontier.enableAgents

```
frontier.enableAgents (boolean)
```

Starts or stops the agent thread, depending on the value of *boolean*.

frontier.getFilePath

```
frontier.getFilePath ()
```

Returns the pathname of the database that owns the frontmost window within Frontier.

frontier.getProgramPath

```
frontier.getProgramPath ()
```

Returns the pathname of the Frontier application.

frontier.idleTime

```
frontier.idleTime ()
```

Returns the number of seconds the user has been "idle," meaning that no keys are pressed, no mouse action takes place, and no disks are inserted. Requires that the *idleTime* extension[*] be loaded at system startup.

frontier.requestToFront

```
frontier.requestToFront ([text])
```

Sounds a beep; if *text* is supplied, it appears in an alert dialog in the frontmost application. Then, pauses until the user manually brings Frontier to the front.

frontier.version

```
frontier.version ()
```

Returns the version string of the database.

kb.cmdKey

```
kb.cmdKey ()
```

Returns a boolean which is **true** if the Command key is down.

kb.controlKey

```
kb.controlKey ()
```

Returns a boolean which is **true** if the Control key is down.

---

[*] Included with the Frontier package.

kb.optionKey

```
kb.optionKey ()
```

Returns a boolean which is **true** if the Option key is down.

kb.shiftKey

```
kb.shiftKey ()
```

Returns a boolean which is **true** if the Shift key is down.

launch.anything

```
launch.anything (pathname)
```

Brings the Finder to the front, and "opens" the object designated by *pathname* just as the Finder would do. The object can be a document, an application file, a folder—anything that can be opened in this sense.

launch.appleMenu

```
launch.appleMenu (name)
```

Looks for *name* (case-sensitive) in the Apple menu, and chooses it.

launch.application

```
launch.application (pathname)
```

Starts up the application at *pathname*. Faster and more sophisticated than **launch.anything()**. Does not bring the Finder or the application to the front. Returns **true** or **false** as the application was started up successfully. If the application was already running, returns **true** immediately.

launch.appWithDocument

```
launch.appWithDocument (appPathname, docPathname)
```

Starts up the application at *appPathname* with the document at *docPathname*. Works like **launch.application()**. If the application was already running, returns true, but the document is *not* opened. If *docPathname* is the empty string, launches the application with no default document, but further programmatic interaction with the application may be impossible until a document is opened.

launch.controlPanel

    launch.controlPanel (*name*)

Looks for *name* in the *Control Panels* folder, and calls launch.anything() on that item.

launch.resource

    launch.resource (*typeString4*, *IDnumber*)

Runs the code resource of type *typeString4* and ID *IDnumber*, primarily intended to run an FKEY (*typeString4* will be 'FKEY').

launch.usingID

    launch.usingID (*string4*)

Starts up an application given its creator code, *string4*. A utility script; not recommended, because it calls launch.anything().

loginAs.loginAs

    loginAs.loginAs (*userID*, *password*)

Causes *userID* and *password* to be in force whenever a remote application is subsequently addressed through program linking. If the values are correct, prevents the Link To dialog from appearing.

mainWindow.hideButtons

    mainWindow.hideButtons ()

Hides the lower half of the Main Window.

mainWindow.hideFlag

```
mainWindow.hideFlag ()
```

Hides the flag icon in the Main Window.

---

mainWindow.hidePopup

```
mainWindow.hidePopup ()
```

Hides the agent popup menu triangle in the Main Window.

---

mainWindow.showButtons

```
mainWindow.showButtons ()
```

Shows the lower half of the Main Window.

---

mainWindow.showFlag

```
mainWindow.showFlag ()
```

Shows the flag icon in the Main Window.

---

mainWindow.showPopup

```
mainWindow.showPopup ()
```

Shows the agent popup menu triangle in the Main Window.

---

menu.addMenuCommand

```
menu.addMenuCommand (addrMenubar, menuName, itemName,
                     scriptString)
```

Modifies the menubar object at *addrMenubar* by adding an item *itemName* with a one-line script *scriptString* as the last subhead of the item *menuName*; if *menuName* doesn't exist, it will be created at summit level after everything else in the menubar. If *itemName* exists already as a subhead of *menuName*, then *scriptString* simply replaces its script.

menu.addSubMenu

```
menu.addSubMenu (addrDestMenubar, menuName,
                    addrSourceMenubar)
```

Copies the first summit-level item (call its name *firstSummit*) of the menubar object at *addrSourceMenubar* into the menubar object at *addrDestMenubar*, as a subhead of the item *menuName*, replacing an existing subhead named *first-Summit*, or, if no such subhead exists, after all subheads of *menuName*. If no item *menuName* exists, it is created at summit level after everything else in the menubar.

menu.addSuite

```
menu.addSuite (addrSuiteTable)
```

Removes all suite menus and calls the `installMenu()` script in the table at *addrSuiteTable*.

menu.clearMenubar

```
menu.clearMenubar ()
```

Removes all custom menus from the top of the screen.

menu.deleteMenuCommand

```
menu.deleteMenuCommand (addrMenubar, menuName, itemName)
```

Deletes from the menubar at *addrMenubar* the item *itemName* which is a subhead of *menuName*. If the item *itemName* has subheads, or if *itemName* or *menuName* does not exist, does nothing, returning `false`.

menu.deleteSubMenu

```
menu.deleteSubMenu (addrMenubar, menuName)
```

Deletes from the menubar at *addrMenubar* the item *menuName*, including its subheads, if it has any. If *menuName* does not exist, does nothing, returning `false`.

menu.getScript

```
menu.getScript (addrObject)
```

Sets the value of the object at *addrObject* to the script of the currently selected item of the menubar edit window which is the current target.

menu.install

```
menu.install (addrMenubar)
```

Adds the menubar at *addrMenubar* to the top of the screen after all custom menus and before the **Window** menu. If that menubar is already present at the top of the screen, does nothing.

menu.isInstalled

```
menu.isInstalled (addrMenubar)
```

Returns `true` or `false` as the menubar at *addrMenubar* is present at the top of the screen.

menu.noSuite

```
menu.noSuite ()
```

Restores the default custom menus at the top of the screen (no suite menus).

menu.remove

```
menu.remove (addrMenubar)
```

Removes the menubar at *addrMenubar* from the top of the screen. If that menubar is not already present at the top of the screen, does nothing and returns `false`.

menu.setScript

```
menu.setScript (addrScriptObject)
```

Sets the script of the currently selected item of the menubar edit window which is the current target to the value of the object at *addrScriptObject*. If that object is not a script, does nothing and returns `false`.

menu.toggle

```
menu.toggle (addrMenubar)
```

Calls `menu.install()` or `menu.remove()`, depending on the result of `menu.isInstalled()`.

menu.zoomscript

```
menu.zoomscript ()
```

Opens the script edit window of the currently selected item of the menubar edit window that is the current target.

mouse.button

```
mouse.button ()
```

Returns a boolean which is `true` if the mouse button is down.

mouse.location

```
mouse.location ()
```

Returns a pointType giving the location of the cursor measured in pixels from the top left corner of the frontmost window.

op.collapse

```
op.collapse ()
```

Collapses the subheads of the currently selected line of the outline (or script, or menubar) target window.

op.countSubs

```
op.countSubs (depth)
```

Returns the number of subheads, to a depth of *depth*, of the currently selected line of the outline (or script, or menubar) target window. For example, supplying a *depth* of 1 returns the number of immediate subheads; supplying a *depth* of `infinity` returns the total number of lines subordinate at any depth. The current state of expansion is irrelevant.

op.countSummits

```
op.countSummits ()
```

Returns the number of summit lines in the outline (or script, or menubar) target window.

op.dehoist

```
op.dehoist ()
```

In the outline (or script, or menubar) target window, undoes the effect of the most recent `op.hoist()` that has not yet been undone.

op.dehoistAll

```
op.dehoistAll ()
```

In the outline (or script, or menubar) target window, undoes the effect of every `op.hoist()`.

op.deleteLine

```
op.deleteLine ()
```

Deletes the currently selected line of the outline (or script, or menubar) target window. If the selection was the *n*th line of its bundle, the selection afterwards is the *n*th line of its bundle, unless there is no *n*th line anymore, in which case it is the (*n*–1)th line of its bundle, unless there are no lines in the bundle anymore, in which case it is the line to which the bundle was subordinate. The mode is unchanged; if it was content mode, the insertion point is at the end of the selected line.

op.deleteSubs

```
op.deleteSubs ()
```

Deletes the subheads of the currently selected line of the outline (or script, or menubar) target window.

op.demote

```
op.demote ()
```

Moves to the right all lines below and in the same bundle as the currently selected line of the outline (or script, or menubar) target window.

op.expand

```
op.expand (depth)
```

Expands the subheads of the currently selected line of the outline (or script, or menubar) target window, revealing at most *depth* new levels. Returns **true** if any new levels were revealed.

op.firstSummit

```
op.firstSummit ()
```

Like saying `op.go(flatup,infinity)`.

op.flatCursorKeys

```
op.flatCursorKeys (boolean)
```

Sets whether up arrow and down arrow keys in outline (or script, or menubar) edit windows should cause movement flatup and flatdown, or up and down among siblings, according to whether *boolean* is **true** or **false**; changes certain other navigation key combinations as well. Reinitialized to **false** at startup.

op.fullCollapse

```
op.fullCollapse ()
```

Collapses all non-summit lines of the outline (or script, or menubar) target window.

op.fullExpand

```
op.fullExpand ()
```

Expands all lines of the outline (or script, or menubar) target window.

op.getCursor

```
op.getCursor ()
```

Returns a number which can be used later to select again the currently selected line of the outline (or script, or menubar) target window.

op.getDisplay

```
op.getDisplay ()
```

Returns a boolean which is **false** if display updating in the target outline (or script, or menubar) edit window is currently frozen.

op.getLineText

```
op.getLineText ()
```

Returns the text of the current line of the outline (or script, or menubar) target window.

op.getRefCon

```
op.getRefCon ()
```

Returns the "secret" scalar value of the currently selected line of the outline (or script) target window.

op.go

```
op.go (dir, count)
```

Navigates the selection relative to the currently selected line of the outline (or script, or menubar) target window, *count* times in the direction *dir*, which can be up, down, left, right, flatup, flatdown, or nodirection. Movement is with regard to the current state of expansion. Returns false only if it is impossible to navigate in the specified direction *at all*; otherwise, returns true and navigates either *count* times or, if that's impossible, as many times as possible. Puts the window in selection mode.

op.hoist

```
op.hoist ()
```

Causes the subheads of the currently selected line of the outline (or script, or menubar) target window to appear as the outline's summits. The effect lasts until undone with op.dehoist() or op.dehoistAll(), or until the database is saved with the window closed, or until Frontier is quit.

op.insert

```
op.insert (string, dir)
```

Creates a new line whose text is *string*, adjacent to the currently selected line of the outline (or script, or menubar) target window in the direction *dir*, which can be up, down, left, or right. If that's impossible (for example, left is specified when a summit-level line is selected), false is returned and nothing happens. The mode is unchanged: if selection mode, the new line is selected; if content mode, the insertion point is at the end of the new line.

op.insertAtEndOfList

```
op.insertAtEndOfList (string)
```

Creates a new line whose text is *string*, as the last subhead of the currently selected line of the outline (or script, or menubar) target window, selecting the new line in selection mode.

op.level

```
op.level ()
```

Returns an integer representing the level of the currently selected line of the outline (or script, or menubar) target window. Summit lines are at level 1.

op.promote

```
op.promote ()
```

Moves to the left all immediate subheads of the currently selected line of the outline (or script, or menubar) target window.

op.reorg

```
op.reorg (dir, count)
```

Moves the currently selected line of the outline (or script, or menubar) target window *count* times in the direction *dir*, which can be up, down, left, or right (like hitting Command-u, Command-d, Command-l, or Command-r, respectively, *count* times). If *dir* is right, subheads will be expanded if necessary. Returns false only if the line cannot be moved in the specified direction *at all*; otherwise, returns true and moves the line either *count* times or, if that's impossible, as many times as possible. Does not alter the mode or (if content mode) the selection within the line.

op.setCursor

```
op.setCursor (bookmarkNum)
```

When *bookmarkNum* is the result of an earlier call to op.getCursor(), selects (in selection mode) the line of the outline (or script, or menubar) target window that was selected when op.getCursor() was called, regardless of how that line has been collapsed, moved, or cut and pasted, within the same outline. If the line has been deleted, returns false.

op.setDisplay

```
op.setDisplay (boolean)
```

If *boolean* is `false`, freezes display updating in the target outline (or script, or menubar) edit window until the window is closed or until reset with *boolean* being `true`.

op.setLineText

```
op.setLineText (string)
```

Sets the text of the current line of the outline (or script, or menubar) target window to *string*, selecting the line in selection mode.

op.setRefCon

```
op.setRefCon (scalarValue)
```

Sets to *scalarValue* the "secret" value of the currently selected line of the outline (or script) target window.

op.sort

```
op.sort ()
```

ASCII-sorts the lines in the same bundle as the currently selected line of the outline (or script, or menubar) target window. In selection mode, the selection travels with its line. (In content mode, it is unpredictable what line will contain the selection afterwards.)

op.subsExpanded

```
op.subsExpanded ()
```

Returns a boolean reporting whether or not the immediate subheads of the currently selected line of the outline (or script, or menubar) target window are expanded.

op.tabKeyReorg

> `op.tabKeyReorg (boolean)`

Sets whether Tab and Shift-Tab in outline edit windows move a line rightward and leftward, or type text, according as *boolean* is `true` or `false`. Reinitialized to `true` at startup.

op.visit

> `op.visit (addrProc)`

An outline (or script, or menubar) must already be the target. The subheads of the currently selected line, each followed by its subheads (and so on), will be selected in turn, and the script or handler pointed to by *addrProc* will be called; it must take no parameters, and must return a boolean saying whether it wants to be called again.

op.wipe

> `op.wipe ()`

Deletes all lines of the outline (or script, or menubar) target window, leaving a single empty summit line.

passwordDialog.run

> `passwordDialog.run (addrName, addrPassword)`

Displays a modal dialog containing prompt text, a **Cancel** button and an **OK** button, which is the default. The dialog also contains two labelled editable text fields; the first initially contains the value at *addrName* coerced to a string. Characters typed into the second text field are displayed as bullets. Returns `false` if the user clicks **Cancel**, `true` otherwise; if `true`, the contents of the editable text fields are stored at *addrName* and *addrPassword*, respectively (unencrypted). The prompt and the text field labels can be customized by calling a modified version of the script.

pbs.getLinks

    pbs.getLinks (*string*)

Returns a list of all the href parameter values of <a> tags in *string*.

pbs.stripHTML

    pbs.stripHTML (*string*)

Returns *string* stripped of all HTML tags.

pbs.utilities.parseRelativeURL

    pbs.utilities.parseRelativeURL (*baseURL*, *relativeURL*)

Returns the absolute URL of the entity denoted by *relativeURL* relative to *baseURL*, in accordance with RFC 1808.

pict.expressions

    pict.expressions (*boolean*)

Sets whether or not the Frontier picture whose edit window is currently the target should have its text expressions evaluated. If the target is not a Frontier picture edit window, an error is raised.

pict.getPicture

    pict.getPicture (*addrBinaryPICT*)

Stores at *addrBinaryPICT* a binary of internal datatype 'PICT' converted from the Frontier picture whose edit window is presently the target. If the target is not a Frontier picture edit window, an error is raised.

pict.PICTtoPicture

    pict.PICTtoPicture (*addrBinaryPICT*, *addrPicture*)

Stores at *addrPicture* a Frontier picture object converted from the binary of internal datatype 'PICT' at *addrBinaryPICT.*

pict.pictureToPICT

```
pict.pictureToPICT (addPicture, addrBinaryPICT)
```

Stores at *addrBinaryPICT* a binary of internal datatype `'PICT'` converted from the Frontier picture object at *addrPicture*.

pict.scheduleUpdate

```
pict.scheduleUpdate (seconds)
```

Sets the interval at which the Frontier picture object whose edit window is currently the target should have its text expressions evaluated. If the target is not a Frontier picture edit window, an error is raised.

pict.setPicture

```
pict.setPicture (binaryPICTvalue)
```

Stores in the object whose edit window is presently the target a Frontier picture converted from the binary of internal datatype `'PICT'` *binaryPICTvalue*. If the target is not a Frontier picture edit window, an error is raised.

point.get

```
point.get (pt, addrX, addrY)
```

Separates the point *pt* into its *x*- and *y*-components and places them in the objects at *addrX* and *addrY*, respectively.

point.set

```
point.set (x, y)
```

Returns the point whose *x*- and *y*-components are the shorts *x* and *y*.

rectangle.get

```
rectangle.get (rect, addrTop, addrLeft, addrBottom,
               addrRight)
```

Separates the rect *rect* into its top, left, bottom, and right components and places them in the objects at *addrTop*, *addrLeft*, *addrBottom*, and *addrRight*, respectively.

rectangle.inset

```
rectangle.inset (rect, widthDelta, heightDelta)
```

Returns the rect whose left and right are each *widthDelta* closer to the center of *rect* and whose top and bottom are each *heightDelta* closer to the center of *rect*.

rectangle.outset

```
rectangle.outset (rect, widthDelta, heightDelta)
```

Returns the rect whose left and right are each *widthDelta* farther from the center of *rect* and whose top and bottom are each *heightDelta* farther from the center of *rect*.

rectangle.random

```
rectangle.random (top, left, bottom, right)
```

Returns a pseudo-random rect falling within *top*, *left*, *bottom*, and *right*.

rectangle.set

```
rectangle.set (top, left, bottom, right)
```

Returns the rect whose top, left, bottom, and right components are the shorts *top*, *left*, *bottom*, and *right*.

regex.extract

```
regex.extract (searchFor, addrString, addrList, [groups,
               caseSensitive?])
```

Creates at *addrList* a list of the substrings in the string at *addrString* matching the grep expression *searchFor*. The optional parameters default to {0} and `false` respectively; the use of *groups* is too complicated to detail here (see the regex documentation).

regex.join

```
regex.join (delimString, addrList)
```

Returns a string made up of the strings in the list of strings at *addrList* interspersed with *delimString*.

regex.split

```
regex.split (searchFor, addrString, [caseSensitive?])
```

Returns a list of strings which are the pieces of the string at *addrString* when all substrings matched by the grep expression *searchFor* have been removed; the list also includes the strings matching any group expressions in *searchFor*. *caseSensitive?* defaults to `false`.

regex.subst

```
regex.subst (searchFor, replaceWith, addrString,
             [caseSensitive?, maxSubstitutions])
```

Returns a string which is the result of searching in the string at *addrString* for the grep expression *searchFor* and replacing it with *replaceWith*. Or, *replaceWith* can be the address of a script or handler which generates the replacement string. The optional parameters default to `false` and `infinity`, respectively.

rez.countResources

```
rez.countResources (pathname, type)
```

Returns the number of resources of type *type* in the file at *pathname*. If there is no file at *pathname*, an error is raised.

rez.countResTypes

```
rez.countResTypes (pathname)
```

Returns the number of different resource types in the file at *pathname*. If there is no file at *pathname*, an error is raised.

rez.deleteNamedResource

```
rez.deleteNamedResource (pathname, type, name)
```

Attempts to delete the resource in the file at *pathname* of type *type* and name *name*, returning a boolean reporting whether the action succeeded. If there is no file at *pathname*, an error is raised.

rez.deleteResource

```
rez.deleteResource (pathname, type, ID)
```

Attempts to delete the resource in the file at *pathname* of type *type* and ID *ID*, returning a boolean reporting whether the action succeeded. If there is no file at *pathname*, an error is raised.

rez.getNamedResource

```
rez.getNamedResource (pathname, type, name,
                      addrDestination)
```

Returns true or false, as the file at *pathname* does indeed have a resource of type *type* and name *name*; if true, stores that resource's data as a binary at *addrDestination*. If there is no file at *pathname*, an error is raised.

rez.getNthResInfo

```
rez.getNthResInfo (pathname, type, n, addrID, addrName)
```

Returns true or false, as the file at *pathname* does indeed have an *n*th resource of type *type*. If true, stores that resource's ID number at *addrID*, and stores either the empty string or that resource's name if it has one at *addrName*. If there is no file at *pathname*, an error is raised.

rez.getNthResource

```
rez.getNthResource (pathname, type, n, addrName,
                    addrDestination)
```

Returns true or false, as the file at *pathname* does indeed have an *n*th resource of type *type*. If true, stores that resource's data as a binary at *addrDestination*, and stores either the empty string or that resource's name if it has one at *addrName*. If there is no file at *pathname*, an error is raised.

rez.getNthResType

```
rez.getNthResType (pathname, n, addrDestination)
```

Returns true or false, as the file at *pathname* does indeed have an *n*th resource type; if true, stores the string4 designator for that type at *addrDestination*. If there is no file at *pathname*, an error is raised.

rez.getResource

```
rez.getResource (pathname, type, ID, addrDestination)
```

Returns true or false, as the file at *pathname* does indeed have a resource of type *type* and ID *ID*; if true, stores that resource's data as a binary at *addrDestination*. If there is no file at *pathname*, an error is raised.

rez.getResourceAttributes

```
rez.getResourceAttributes (pathname, type, ID)
```

Returns the short representing the attributes of the resource of type *type* and ID *ID* in the file at *pathname*. If the file or the resource doesn't exist, an error is raised.

rez.getStringResource

        rez.getStringResource (*pathname*, *ID*, *addrDestination*)

Returns true or false, as the file at *pathname* does indeed have a resource of type 'STR ' and ID *ID*; if true, converts that resource's data to a string at *addrDestination*. If there is no file at *pathname*, an error is raised.

rez.namedResourceExists

        rez.namedResourceExists (*pathname*, *type*, *name*)

Returns a boolean reporting whether the file at *pathname* contains a resource of type *type* and name *name*. If there is no file at *pathname*, an error is raised.

rez.putNamedResource

        rez.putNamedResource (*pathname*, *type*, *name*, *addrSource*)

Stores the binary at *addrSource* into the file at *pathname* as a resource of type *type* and name *name*, replacing an existing resource if necessary and supplying a unique ID otherwise. If there is no file at *pathname*, an error is raised.

rez.putResource

        rez.putResource (*pathname*, *type*, *ID*, *addrSource*)

Stores the binary at *addrSource* into the file at *pathname* as a resource of type *type* and ID *ID*, replacing an existing resource if necessary. If there is no file at *pathname*, an error is raised.

rez.putStringResource

        rez.putStringResource (*pathname*, *ID*, *theString*)

Converts the string *theString* (not an address!) to a Pascal string and stores it into the file at *pathName* as a resource of type 'STR ' and ID *ID*, replacing an existing resource if necessary. If there is no file at *pathname*, an error is raised.

rez.resourceExists

`rez.resourceExists (`*`pathname,`* *`type,`* *`ID`*`)`

Returns a boolean depending on whether the file at *pathname* contains a resource of type *type* and ID *ID*. If there is no file at *pathname*, an error is raised.

rez.setResourceAttributes

`rez.setResourceAttributes (`*`pathname,`* *`type,`* *`ID,`* *`theShort`*`)`

Sets the short representing the attributes of the resource of type *type* and ID *ID* in the file at *pathname* to *theShort*. If the file or the resource doesn't exist, an error is raised.

rgb.get

`rgb.get (`*`rgb,`* *`addrRed,`* *`addrGreen,`* *`addrBlue`*`)`

Separates the rgb *rgb* into its red, green, and blue components and places them in the objects at *addrRed*, *addrGreen*, and *addrBlue*, respectively.

rgb.set

`rgb.set (`*`red,`* *`green,`* *`blue`*`)`

Returns the rgb whose red, green, and blue components are the shorts *red, green,* and *blue.*

script.clearBreakpoint

`script.clearBreakpoint ()`

Clears a breakpoint at the currently selected line of the script (or outline, or menubar) target window.

script.compile

`script.compile (`*`addrScriptObject`*`)`

Compiles the script object at *addrScriptObject.*

script.getBreakpoint

```
script.getBreakpoint ()
```

Returns a boolean reporting whether the currently selected line of the script (or outline, or menubar) target window is a breakpoint.

script.getLanguage

```
script.getLanguage (addrScriptObject)
```

Returns the OSA dialect of the script object at *addrScriptObject*, typically "User-Talk" or "AppleScript".

script.isComment

```
script.isComment ()
```

Returns a boolean reporting whether the currently selected line of the script (or outline, or menubar) target window is a comment.

script.makeComment

```
script.makeComment ()
```

Turns the currently selected line of the script (or outline, or menubar) target window into a comment.

script.removeSource

```
script.removeSource (addrScriptObject)
```

Clears the source code from the script object at *addrScriptObject*, converting it to a compiled code object.

script.setBreakpoint

```
script.setBreakpoint ()
```

Sets a breakpoint at the currently selected line of the script (or outline, or menubar) target window.

script.setLanguage

```
script.setLanguage (addrScriptObject, languageString)
```

Sets the OSA dialect of the script object at *addrScriptObject* to *languageString*, which is typically `"UserTalk"` or `"AppleScript"`.

---

script.unComment

```
script.unComment ()
```

Turns the currently selected line of the script (or outline, or menubar) target window into a non-comment.

---

script.uncompile

```
script.uncompile (addrScriptObject)
```

Clears from memory the explicitly compiled version of the script object at *addrScriptObject*.

---

semaphores.lock

```
semaphores.lock (name, timeOutTicks)
```

Checks to see whether the semaphore *name* is registered as locked; if so, waits until it is unlocked (or, if *timeOutTicks* elapses first, throws an error). If and when *name* is not locked, locks it.

---

semaphores.unlock

```
semaphores.unlock (name)
```

Unlocks the semaphore *name*. It is not an error if no such semaphore is locked.

---

speaker.beep

```
speaker.beep ()
```

Plays the system alert sound.

speaker.ouch

```
speaker.ouch ()
```

Plays a brief, high-pitched "blip" sound.

speaker.playNamedSound

```
speaker.playNamedSound (name)
```

Plays a 'snd ' resource called *name*, which must be in the resource chain already.

speaker.sound

```
speaker.sound (duration, amplitude, frequency)
```

Plays a customized square-wave sound.

stack.create

```
stack.create ()
```

Creates an empty stack and returns its address.

stack.dispose

```
stack.dispose (addrStack)
```

Releases the database storage occupied by the stack pointed to by *addrStack*.

stack.pop

```
stack.pop (addrStack)
```

Pops the topmost value off the stack pointed to by *addrStack*, and returns it. If the stack is empty, a runtime error is generated.

stack.push

    stack.push (addrStack, value)

Pushes *value* onto the stack pointed to by *addrStack*.

stack.visit

    stack.visit (addrStack, addrProc)

The script or handler pointed to by *addrProc* must take one parameter, which will be the value of each item on the stack at *addrStack* in turn, and must return a boolean saying whether it wants to be called again.

string.addCommas

    string.addCommas (number)

Returns a string representation of *number* with commas appropriately inserted.

string.commentDelete

    string.commentDelete (string)

Assuming that *string* is a line of a script, returns everything in it except the trailing comment, if any.

string.countFields

    string.countFields (string, delimiterChar)

Returns the number of fields in *string* when the delimiter character is *delimiterChar.*

string.countWords

    string.countWords (string)

Returns the number of words in *string*, given the current word-delimiter.

string.dateString

```
string.dateString ()
```

Returns the system's `abbrevDate` string representation of the current date.

---

string.delete

```
string.delete (string, startIndex, count)
```

Returns a string consisting of *string* with *count* characters removed starting at index *startIndex* of *string*. It is not an error for *count* to be too large: all of *string* starting at index *startIndex* is removed. It is not an error for *startIndex* to be too large: none of *string* is removed. If *count* is 0 or *startIndex* is 0, none of *string* is removed.

---

string.filledString

```
string.filledString (character, length)
```

Returns a string made up of *length* occurrences of *character*.

---

string.firstSentence

```
string.firstSentence (string)
```

Returns everything in *string* up to and including the first full stop which is followed by a space, or *string* itself if this is not found.

---

string.firstWord

```
string.firstWord (string)
```

Returns the first word of *string*, given the current word-delimiter.

---

string.getWordChar

```
string.getWordChar ()
```

Returns the current word-delimiter.

string.hex

```
string.hex (integer)
```

Returns a hexadecimal string representation of *integer*, in four or eight digits. *integer* is coerced to an integer if necessary.

string.insert

```
string.insert (substring, originalString, startIndex)
```

Returns a string consisting of *originalString* with *substring* inserted into it starting at index *startIndex* of the result.

string.isAlpha

```
string.isAlpha (character)
```

On Mac OS, returns true if *character* is in the ASCII ranges 65–90, 97–122, 128–159, 174–175, 190–191, 203–207, 216–217, 229–239, or 241–245.

string.isNumeric

```
string.isNumeric (character)
```

Returns true if *character* is a digit.

string.iso8859encode

```
string.iso8859encode (string)
```

Returns a string made up of the characters of *string*, each character, if high-ASCII, having been passed through a translate table—the table at user.html.prefs.iso8859map if there is one, a default hard-coded table otherwise.

string.isPunctuation

```
string.isPunctuation (character)
```

Returns true if *character* is in the ASCII ranges 33–47 or 58–63.

string.kBytes

```
string.kBytes (integer)
```

Returns a string ending in K representing the smallest integral number of kilobytes larger than *integer* bytes. (That is, *integer* is divided by 1024 and rounded up.)

string.lastWord

```
string.lastWord (string)
```

Returns the last word of *string*, given the current word-delimiter.

string.lower

```
string.lower (string)
```

Returns *string* with every character rendered to its lowercase equivalent in accordance with the system's definitions. Bug: if *string* is longer than about 32000 characters, *string* is returned unchanged.

string.memAvailString

```
string.memAvailString ()
```

Returns memAvail() integer-divided by 1024, coerced to a string ending with "K"; generally differs from string.kBytes(memAvail()) by 1.

string.mid

```
string.mid (string, startIndex, count)
```

Returns a string consisting of *count* characters of *string* starting at index *startIndex* of *string*. It is not an error for *count* to be too large: all of *string* starting at index *startIndex* is returned. It is not an error for *startIndex* to be too large: the empty string is returned. If *count* is 0 or negative, the empty string is returned. If *startIndex* is 0, it is translated to 1.

string.nthChar

```
string.nthChar (string, index)
```

Returns the char at index *index* of *string*; or, if *index* is 0 or too large, returns the empty string.

string.nthField

```
string.nthField (string, delimiterChar, n)
```

Returns the *n*th field of *string* when the delimiter character is *delimiterChar*; if *n* is too large or 0, returns the empty string.

string.nthWord

```
string.nthWord (string, n)
```

Returns the *n*th word of *string*, given the current word-delimiter.

string.parseHTTPargs

```
string.parseHTTPargs (string)
```

Assuming that *string* is of the form `"name1=value1&name2=value2..."`, returns a list of strings of the form `{"name1", "value1", "name2", "value2"...}`, where all strings are URL-decoded.

string.patternMatch

```
string.patternMatch (substring, string)
```

Returns the index of the first occurrence of *substring* in *string*, or 0 if it doesn't occur.

string.popLeading

```
string.popLeading (string, character)
```

Returns a string made from *string* with any occurrences of *character* removed from its start.

string.popTrailing

```
string.popTrailing (string, character)
```

Returns a string made from *string* with any occurrences of *character* removed from its end.

string.replace

```
string.replace (string, oldSubstring, newSubstring)
```

Returns a string consisting of *string* with the first occurrence of *oldSubstring* replaced by *newSubstring*. If *newSubstring* is `""`, the first occurrence of *oldSubstring* is removed.

string.replaceAll

```
string.replaceAll (string, oldSubstring, newSubstring)
```

Returns a string consisting of *string* with all occurrences of *oldSubstring* replaced by *newSubstring*. If *newSubstring* is `""`, all occurrences of *oldSubstring* are removed.

string.setWordChar

```
string.setWordChar (character)
```

Set the word-delimiter to *character.*

string.timeString

```
string.timeString ()
```

Returns the system's string representation of the current time.

string.upper

```
string.upper (string)
```

Returns *string* with every character rendered to its uppercase equivalent in accordance with the system's definitions. Bug: if *string* is longer than about 32000 characters, *string* is returned unchanged.

string.URLdecode

```
string.URLdecode (string)
```

Returns *string* with %-octet codes and + replaced by the characters they represent.

string.URLencode

```
string.URLencode (string)
```

Returns *string* with URL-illegal characters replaced by their %-encoded form.

sys.appIsRunning

```
sys.appIsRunning (processIdentifier)
```

Returns a boolean reporting whether the application *processIdentifier* is running.

sys.bringAppToFront

```
sys.bringAppToFront (processIdentifier)
```

Brings the application *processIdentifier* to the front. If *processIdentifier* is not a running process, returns false and does nothing. If *processIdentifier* is a faceless background application, an error is raised.

sys.browseNetwork

```
sys.browseNetwork (text, creatorCode, addrResult)
```

Displays the Mac OS PPCBrowser modal dialog, permitting the user to choose a process running on this computer or any computer on the AppleTalk network with program linking turned on. The prompt is *text*, or, if " " is passed, "Choose a program to link to:"; the dialog has a **Cancel** button and an **OK** button, which is the default. *creatorCode* can be a string4 creator designator to specify one application, or 0 to allow all applications to appear. Returns false if the user clicks **Cancel**, true otherwise. If true, a 'targ' binary specifying the chosen process is stored at *addrResult*; this binary may be used as an application ID in an appleEvent() call.

sys.countApps

```
sys.countApps ()
```

Returns the number of running processes, including faceless background applications that do not appear in the system's Application menu.

sys.frontmostApp

```
sys.frontmostApp ()
```

Returns the name of the frontmost application.

sys.getAppPath

```
sys.getAppPath (processIdentifier)
```

Returns the pathname of the application file running as *processIdentifier*.

sys.getAppSize

```
sys.getAppSize (pathname)
```

Returns the preferred RAM allocation for the application at *pathname*, as shown in the Finder's **Get Info** window.

sys.getMinAppSize

```
sys.getMinAppSize (pathname)
```

Returns the minimum RAM allocation for the application at *pathname*, as shown in the Finder's **Get Info** window.

sys.getNthApp

```
sys.getNthApp (n)
```

Returns the name of the *n*th running process, including faceless background applications that do not appear in the system's Application menu; the order is the order in which the processes were started up.

sys.hasBundle

```
sys.hasBundle (pathname)
```

Returns a boolean reporting whether the file at *pathname* has its bundle bit set.

sys.memAvail

```
sys.memAvail ()
```

Returns the amount of free RAM memory.

sys.OSVersion

```
sys.OSVersion ()
```

Returns the system's version string.

sys.setAppSize

```
sys.setAppSize (pathname, bytes)
```

Sets to *bytes* the preferred RAM allocation for the application at *pathname*, as shown in the Finder's **Get Info** window. If the application is running, an error is raised. If the file at *pathname* is not an application, returns **false** and does nothing.

sys.setBundle

```
sys.setBundle (pathname, set?)
```

Sets the bundle bit of the file at *pathname* according to the boolean *set?*.

sys.setMinAppSize

```
sys.setMinAppSize (pathname, bytes)
```

Sets to *bytes* the minimum RAM allocation for the application at *pathname*, as shown in the Finder's **Get Info** window. If the application is running, an error is raised. If the file at *pathname* is not an application, returns **false** and does nothing.

sys.systemTask

```
sys.systemTask ()
```

Yields time to other processes.

table.assign

```
table.assign (addrObject, value)
```

Sets the value of the object at *addrObject* to *value*, possibly creating the object if it doesn't exist.

table.copy

```
table.copy (addrObject, addrTable)
```

Copies the object at *addrObject* into the table at *addrTable*. The new object's name and value are the same as those of the object at *addrObject*. If there is not a table at *addrTable*, an error is raised.

table.copyContents

```
table.copyContents (addrSourceTable, addrDestTable)
```

Calls `table.copy()` on each entry of the table at `addrSourceTable()`.

table.emptyTable

```
table.emptyTable (addrTable)
```

Destroys all the entries in the table at *addrTable*. Returns the number of objects deleted. If *addrTable* does not point to a table, a runtime error is raised.

table.getCursor

```
table.getCursor ()
```

Returns the address of the currently selected entry of the target table window.

table.go

```
table.go (dir, count)
```

Navigates the selection within the target table window relative to the current selection, *count* times in the direction *dir*, which can be **up** or **down** (or **nodirection**). If *count* is **infinity**, moves to the start or end of the table and returns **true**; otherwise makes no move unless the move requested is possible, and returns **true** or **false** reporting whether it was possible.

table.goto

```
table.goto (n)
```

Selects the *n*th entry of the target table window. If there is no *n*th entry, no move is made and **false** is returned.

table.gotoAddress

```
table.gotoAddress (addrObject)
```

Calls **edit()** on the parent table of the object at *addrObject*, then calls **table.gotoName()** so that object's entry is selected in the edit window.

table.gotoName

```
table.gotoName (nameString)
```

Selects the entry named *nameString* in the target table window. If there is no such entry, no move is made and **false** is returned.

table.move

```
table.move (addrObject, addrTable)
```

Copies the object at *addrObject* into the table at *addrTable*, and deletes the object at *addrObject*. The new object's name and value are the same as those of the object at *addrObject*. If there is not a table at *addrTable*, an error is raised.

table.moveAndRename

```
table.moveAndRename (addrSourceObject, addrDestObject)
```

Assigns to the object at *addrDestObject* (creating it if necessary) the value of the object at *addrSourceObject*, and destroys the object at *addrSourceObject.*

table.moveContents

```
table.moveContents (addrSourceTable, addrDestTable)
```

Calls `table.move()` on each entry of the table at `addrSourceTable()`.

table.promptNewItem

```
table.promptNewItem (newWhat, dataType)
```

A table edit window must be the target, or an error will result. Displays a modal dialog with prompt text asking for the name of a new cell of type *newWhat* with `"new"` removed from it, an editable text field initially containing *newWhat*, a **Cancel** button and an **OK** button, which is the default. Returns `false` if the user clicks **Cancel**, `true` otherwise; if `true`, an object of type *dataType* whose name is the contents of the editable text field is created in the target table.

table.rename

```
table.rename (addrObject, nameString)
```

Changes the name of the object at *addrObject* to *nameString.** If *addrObject* is a legal address but there is no such object, returns `false` and does nothing.

table.sortBy

```
table.sortBy (byWhat)
```

Sorts the entries in the target table window according to *byWhat*, which must be `"name"`, `"value"`, or `"kind"`.

---

* A bug allows you to rename a table entry so that its name is the same as that of another entry in the same table, so your script should avoid this by checking explicitly first.

table.surePath

```
table.surePath (string)
```

Creates in the database all nonexistent tables in the object path *string*, which must not be a partial reference (only "root" should be omitted). Returns the number of tables created.

table.uniqueName

```
table.uniqueName (startOfName, addrTable)
```

Returns the address of a nonexistent entry of the table at *addrTable* whose name starts with *startOfName* and ends with a number.

table.validate

```
table.validate (addrTable)
```

Checks the internal consistency of the table at *addrTable*.

table.visit

```
table.visit (addrTable, addrProc)
```

The script or handler pointed to by *addrProc* must take one parameter, which will be the address of the table at *addrTable* and then the address of each of its entries, a subtable being passed before its entries, to infinite depth; it must return a boolean saying whether it wants to be called again.

target.clear

```
target.clear ()
```

Explicitly returns to the default target situation; if the target's edit window was invisible, also closes it.

target.get

```
target.get ()
```

Returns the address of the object which is the target (contrast `window.front-most()` which returns a string), or the empty string if there is no target.

target.set

```
target.set (addrObject)
```

Explicitly declares the non-scalar object at *addrObject* to be the target; if the object's edit window is not open, opens it invisibly. (If the object's edit window is already open, neither its visibility nor its position in the layering order are changed.) If the object at *addrObject* is a scalar, returns `false` and does nothing.

thread.evaluate

```
thread.evaluate (string)
```

Initiates evaluation of *string* as a UserTalk expression in a new thread and returns the ID of that thread.

thread.evaluateTo

```
thread.evaluateTo (string, addrDatabaseEntry)
```

Initiates evaluation of *string* as a UserTalk expression in a new thread and returns the ID of that thread; when the thread concludes, its result will be stored in the database at *addrDatabaseEntry*.

thread.exists

```
thread.exists (ID)
```

Returns a boolean according as a thread whose ID is *ID* exists or not.

thread.getCurrentID

```
thread.getCurrentID ()
```

Returns the ID number of the thread within which the call is made.

thread.getNthID

```
thread.getNthID (n)
```

Returns the ID number of the *n*th thread, or 0 if *n* is greater than the number of threads.

thread.isSleeping

```
thread.isSleeping (ID)
```

Returns a boolean reporting whether the thread whose ID is *ID* is sleeping or not. It is an error if no such thread exists.

thread.kill

```
thread.kill (ID)
```

Aborts the thread whose ID number is *ID*. It is an error if there is no such thread.

thread.sleep

```
thread.sleep (ID)
```

Puts to sleep the thread whose ID number is *ID*; the thread pauses until it is told to resume with thread.wake(). It is an error if there is no such thread.

thread.sleepFor

```
thread.sleepFor (seconds)
```

Puts to sleep the thread within which the verb is called; the thread pauses for *seconds* seconds or until it is told to resume with thread.wake(), whichever comes first.

thread.wake

```
thread.wake (ID)
```

Awakens the thread whose ID is *ID*. It is an error if there is no such thread. It is not an error if the thread was not asleep.

toys.alphaChar

```
toys.alphaChar (character)
```

Returns true if *character* is a non-diacritic letter, a digit, or underscore.

toys.commentDelete

```
toys.commentDelete (string)
```

Returns the first field of *string* where « is the delimiter character.

toys.dropNonAlphas

```
toys.dropNonAlphas (string)
```

Returns *string* stripped of any character which is not a non-diacritic letter or digit.

toys.emptyFolder

```
toys.emptyFolder (pathname)
```

Deletes the contents of the folder at *pathname*. A utility script which calls file.delete() and file.deleteFolder(). Does not raise an error or return false if *pathname* doesn't exist.

toys.getCursorAddress

```
toys.getCursorAddress ()
```

If window.frontmost() is not an edit window, raises an error. Otherwise, if it is *not* a table, returns the address of the object whose edit window it is; if it *is* a table, returns the address of the entry selected within it.*

---

* This verb is a utility script; because of a bug in the way it is coded, it will break if the target is not window.frontmost().

toys.innerCaseName

```
toys.innerCaseName (string)
```

Returns *string* processed as follows: it is translated to lowercase, and then stripped of spaces, with the character after each removal translated to uppercase.

---

toys.listToOutline

```
toys.listToOutline (list, addrOutline)
```

Converts the list *list* to an outline and inserts it into the outline at *addrOutline* starting down from the currently selected line.

---

toys.outlineToList

```
toys.outlineToList (addrOutline)
```

Returns a list of strings (and lists), reflecting the currently selected line of the outline at *addrOutline*, and all lines below it in the same bundle, and all subheads of these to an infinite depth.

---

toys.padWithZeros

```
toys.padWithZeros (theNumber, length)
```

Returns a string of length *length* made up of zeros followed by *theNumber* coerced to a string; or, if *theNumber* is already longer than *length*, returns *theNumber.*

---

toys.popStringSuffix

```
toys.popStringSuffix (string, [character])
```

Returns a string consisting of *string* with the rightmost instance of *character* and everything after it removed. The default for *character* is " . ".

toys.puncChar

```
toys.puncChar (character)
```

Returns true on Mac OS if *character* is in the ASCII ranges 33–47, 58–64, 91–96, 123–126, 160–173, 176–189, 192–202, 208–215, 218–228, 240, or 246–254.

toys.readFileIntoTextObject

```
toys.readFileIntoTextObject (pathname, addrDestination)
```

Creates or replaces an object at *addrDestination*, to be a wptext consisting of the data of the file at *pathname*.

toys.readWholeFile

```
toys.readWholeFile (pathname)
```

Returns a binary consisting of the data of the file at *pathname*.

toys.sureFilePath

```
toys.sureFilePath (pathname)
```

Creates all nonexistent folders in the pathname *pathname*.

toys.threadCall

```
toys.threadCall (verbNameString, paramList)
```

Calls the verb named in *verbNameString*, handing it as parameters the values in the list *paramList*, in a new thread and returns the ID of that thread.

toys.touchPath

```
toys.touchPath (pathname)
```

Sets to clock.now() the modification date of the file or folder at *pathname*, and that of all its enclosing folders.

toys.uniqueTableName

```
toys.uniqueTableName (startOfName, addrTable,
                      numberOfDigits)
```

Just like `table.uniqueName()`, but the number appended to *startOfName* is *numberOfDigits* long, so the resulting names will sort properly.

toys.URLsplit

```
toys.URLsplit (string)
```

Assuming string is a URL, returns a list of three strings: the URL scheme name prefix, which must be present; the domain (*not* ending in a slash); and the path within the domain.

toys.writeWholeFile

```
toys.writeWholeFile (pathname, data, type, creator, date)
```

Coerces *data* to a binary and writes it out as a file at *pathname*,* giving the file type and creator codes *type* and *creator* (which are string4s) and a creation date of *date*.

trigCmd

```
trigCmd.abs (number)
trigCmd.acos (number)
trigCmd.asin (number)
trigCmd.atan (number)
trigCmd.ceil (number)
trigCmd.cos (number)
trigCmd.cosh (number)
trigCmd.exp (number)
trigCmd.floor (number)
trigCmd.log (number)
trigCmd.log10 (number)
trigCmd.sin (number)
```

---

* See `file.new()`.

```
trigCmd.sinh (number)
trigCmd.sqrt (number)
trigCmd.tan (number)
trigCmd.tanh (number)
```

Standard mathematical functions.

window.bringToFront

```
window.bringToFront (windowReference)
```

Brings to the front and makes visible the window referred to by *windowReference*. Returns `false` if *windowReference* does not resolve to an open window.

window.close

```
window.close (windowReference)
```

Closes the window referred to by *windowReference*. Returns `false` if *windowReference* does not resolve to an open window, or if the window is the Main Window.

window.dbStats

```
window.dbStats ()
```

Puts up a windoid reporting some statistics relevant to the state of the database.

window.frontmost

```
window.frontmost ()
```

Returns a string reference to the frontmost visible window; or, if called from a script initiated with its edit window's **Run** or **Debug** button, the first visible window behind that window. Returns the empty string if there is no such window.

window.getPosition

```
window.getPosition (windowReference, addrX, addrY)
```

Puts the screen coordinates of the top-left of the content area of the window referred to by *windowReference* into the objects at *addrX* and *addrY.* Returns `false` if *windowReference* does not resolve to an open window.

window.getSize

```
window.getSize (windowReference, addrWidth, addrHeight)
```

Puts the width and height of the content area of the window referred to by *windowReference* into the objects at *addrWidth* and *addrHeight.* Returns `false` if *windowReference* does not resolve to an open window.

window.getTitle

```
window.getTitle (windowReference)
```

Returns the title of the window referred to by *windowReference,* or the empty string if *windowReference* does not resolve to an open window.

window.hide

```
window.hide (windowReference)
```

Makes the window referred to by *windowReference* invisible. Returns `false` if *windowReference* does not resolve to an open window.

window.isFront

```
window.isFront (windowReference)
```

Returns `true` or `false` as `window.frontmost()` returns a reference to the window referred to by *windowReference.*

window.isHidden

```
window.isHidden (windowReference)
```

Returns `false` or `true` as *windowReference* refers to an open visible window.

window.isMenuScript

```
window.isMenuScript (windowReference)
```

Returns `true` or `false` as *windowReference* refers to an open menu item script window.

---

window.isModified

```
window.isModified (addrObject)
```

Returns a boolean reporting whether there is an object at *addrObject* which is a non-scalar that is dirty. Supplying a parameter of `"root"` reveals whether the database is dirty.

---

window.isOpen

```
window.isOpen (windowReference)
```

Returns `true` or `false` reporting whether *windowReference* refers to an open window.

---

window.isVisible

```
window.isVisible (windowReference)
```

Returns `true` or `false` reporting whether *windowReference* refers to an open visible window.

---

window.msg

```
window.msg (string)
```

Posts *string* (coerced to a string) to the status message area of the window that is actually frontmost, if it has one. The message remains until overwritten by another message, or erased with `window.msg("")`, or the window is closed.

window.next

```
window.next (windowReference)
```

Returns a string reference to the window (visible or invisible) behind the window referred to by *windowReference*, or the empty string if *windowReference* does not resolve to an open window or has no window behind it.

window.open

```
window.open (addrObject)
```

Opens, and makes visible and frontmost, the edit window of the non-scalar object at *addrObject*. Returns `false` if this object is a scalar.

window.quickScript

```
window.quickScript ()
```

Opens and brings to the front the Quick Script window.

window.scroll

```
window.scroll (direction, count)
```

Scrolls the target window as if by clicking, *count* times, the scroll arrow that moves the window's content in the direction *direction*. Returns `true` if the window was scrolled at all.

window.sendToBack

```
window.sendToBack (windowReference)
```

Sends to the back the window referred to by *windowReference*. Returns `false` if *windowReference* does not resolve to an open window.

window.setModified

```
window.setModified (addrObject, dirty?)
```

If the object at *addrObject* is a non-scalar, sets the object's dirtiness to the boolean *dirty?*. Supplying a parameter of `"root"` sets the dirtiness of the database.

window.setPosition

```
window.setPosition (windowReference, X, Y)
```

Positions the window referred to by *windowReference* so that the screen coordinates of the top-left of its content area are *X* and *Y*, curtailing the values of *X* and *Y* if necessary just enough to prevent any part of the window from being offscreen. Returns `false` if *windowReference* does not resolve to an open window.

window.setSize

```
window.setSize (windowReference, width, height)
```

Sets the width and height of the content area of the window referred to by *windowReference* to *width* and *height*, curtailing the values of *width* and *height* if necessary just enough to prevent either from being larger than the visible dimension of the screen, or smaller than a reasonable minimum. Returns `false` if *windowReference* does not resolve to an open window.

window.setTitle

```
window.setTitle (windowReference, newTitle)
```

Sets the title of the window referred to by *windowReference* to *newTitle*. Returns `false` if *windowReference* does not resolve to an open window.

window.show

```
window.show (windowReference)
```

Makes the window referred to by *windowReference* visible. Returns `false` if *windowReference* does not resolve to an open window.

window.visit

```
window.visit (addrProc)
```

The script or handler pointed to by *addrProc* must take one parameter, which will be a reference to each open window in front-to-back order, and must return a boolean saying whether it wants to be called again.

window.zoom

```
window.zoom (windowReference)
```

Equivalent to clicking the zoom button in the window referred to by *windowReference.*

---

wp.clearTabs

```
wp.clearTabs ()
```

Clears all tabs in all paragraphs containing any of the current selection in the wptext target window.

---

wp.getDisplay

```
wp.getDisplay ()
```

Returns a boolean which is **false** if display updating in the target wptext (or content mode) edit window is currently frozen.

---

wp.getIndent

```
wp.getIndent ()
```

Returns the first-line indent, measured in pixels from the left edge, of the current selection in the wptext target window; or returns **-1** if the selection includes text with different indents.

---

wp.getLeftMargin

```
wp.getLeftMargin ()
```

Returns the left margin, measured in pixels from the left edge, of the current selection in the wptext target window; or returns **-1** if the selection includes text with different left margins.

wp.getRightMargin

    wp.getRightMargin ()

Returns the right margin, measured in pixels from the right edge (the end of the ruler), of the current selection in the wptext target window; or, returns −1 if the selection includes text with different right margins.

wp.getRuler

    wp.getRuler ()

Returns a boolean reporting whether the ruler is visible in the wptext target window.

wp.getSelect

    wp.getSelect (*addrStart, addrEnd*)

Stores at *addrStart* and *addrEnd* the positions of the start and end, respectively, of the target window's current selection. These positions are integers measured by counting characters from the start of the "document": the start is 0; between the first and second character is 1; and so forth. The target window must be a wptext or in content mode.

wp.getSelText

    wp.getSelText ()

Returns the text (as a string) of the current selection in the target window. The target window must be a wptext or in content mode.

wp.getText

    wp.getText ()

Returns the text of the entire "document" in the target window. If the target is not a wptext, this means the current line or cell. The target window must be a wptext or in content mode.

wp.go

```
wp.go (dir, count)
```

Moves the insertion point in the target window relative to its present position (and, in a wptext, scrolls to reveal the insertion point). *dir* can be `left`, `right`, `up` or `down` (or `nodirection`). Movement left or right starts at the left or right end of the current selection, respectively. If *count* is too large, movement is to the start or end. If the target is not a wptext, movement up or down goes to the start or end. Returns `true` if *any* movement was possible. The target window must be a wptext or in content mode.

wp.insert

```
wp.insert (string)
```

Inserts *string* into the target window according to the same rules as typing or pasting: if there is a selection, it is replaced; otherwise, the text appears after the insertion point. Afterwards, the insertion point is after what was just inserted. The target window must be a wptext or in content mode.

wp.inTextMode

```
wp.inTextMode ()
```

Returns a boolean reporting whether the target outline, script, table, or menubar window is in selection mode (returns `false`) or content mode (returns `true`).

wp.rulerLength

```
wp.rulerLength ()
```

Returns the length of the ruler of the wptext target window.

wp.selectLine

```
wp.selectLine ()
```

Selects the line containing the end of the current selection in the target window; if the target is not a wptext, this means the whole "document." The target window must be a wptext or in content mode.

wp.selectParagraph

```
wp.selectParagraph ()
```

Selects the paragraph containing the end of the current selection in the target window; if the target is not a wptext, this means the whole "document." The target window must be a wptext or in content mode.

wp.selectWord

```
wp.selectWord ()
```

Selects the word containing the end of the current selection in the target window. The target window must be a wptext or in content mode.

wp.setDisplay

```
wp.setDisplay (boolean)
```

If *boolean* is `false`, freezes display updating in the target wptext (or content mode) edit window until the window is closed or until reset with *boolean* being `true`.

wp.setIndent

```
wp.setIndent (pixels)
```

Sets the first-line indent of all paragraphs containing any of the current selection in the wptext target window, to *pixels*, measuring in pixels from the left edge. Because of a bug, you should subtract 3 from *pixels* beforehand.

wp.setJustification

```
wp.setJustification (justification)
```

For all paragraphs containing any of the current selection in the wptext target window, sets the justification to *justification*, which can be 1–4 for left, center, right, and decimal.

wp.setLeftMargin

`wp.setLeftMargin (`*pixels*`)`

Sets the left margin of all paragraphs containing any of the current selection in the wptext target window, to *pixels*, measuring in pixels from the left edge. Because of a bug, you should subtract 3 from *pixels* beforehand.

wp.setRightMargin

`wp.setRightMargin (`*pixels*`)`

Sets the right margin of all paragraphs containing any of the current selection in the wptext target window, to *pixels*, measuring in pixels from the right edge. Because of a bug, you should add 3 to *pixels* beforehand.

wp.setRuler

`wp.setRuler (`*visible?*`)`

Makes the ruler visible or invisible in the wptext target window, depending on the boolean *visible?*.

wp.setSelect

`wp.setSelect (`*start, end*`)`

Selects text in the target window starting at position *start* and ending at position *end*. These positions are integers measured by counting characters from the start of the "document": the start is 0; between the first and second character is 1; and so forth. If *end* is too large, the selection will reach to the end. The target window must be a wptext or in content mode.

wp.setSpacing

`wp.setSpacing (`*verticalPixels*`)`

For all paragraphs containing any of the current selection in the wptext target window, sets line spacing. There are 12 vertical pixels to the (nominal) inch; *verticalPixels* is added to a default value based on the font and size (so supplying 0 restores the default, and supplying a negative value is legal).

wp.setTab

```
wp.setTab (position, justification, leaderChar)
```

For all paragraphs containing any of the current selection in the wptext target window, creates a tab at *position* pixels from the left, whose justification is *justification*, which can be 1–4 for left, center, right, and decimal, and whose leader character is *leaderChar* (supply ' ' for no leader character). Because of a bug, you should subtract 3 from *position* beforehand.

wp.setText

```
wp.setText (string)
```

Replaces the text of the entire "document" in the target window with *string*. If the target is not a wptext, this means the current line or cell. The target window must be a wptext or in content mode.

wp.setTextMode

```
wp.setTextMode (boolean)
```

Causes the target outline, script, table, or menubar window to be in selection mode (if *boolean* is `false`) or content mode (if *boolean* is `true`).

wordInfo.getNthWordOffset

```
wordInfo.getNthWordOffset (string, n)
```

Returns the index of the start of the *n*th word of *string*.

47

Apple Event Suites

This chapter provides reference information for the verbs described under "Apple Event Suites" in Chapter 32, *Driving Other Applications.*

core.close

```
core.close (appID, whatObject, [saving?, savingIn])
```

Closes *whatObject*. The values for *saving?* are **yes**, **no**, and **ask**; the default is **no**.

```
theApp = "Scriptable Text Editor"
with objectModel
    core.close(theApp, window[1])
```

core.count

```
core.count (appID, container, whatToCount)
```

Returns the number of *whatToCount*s (some type of element) in *container*. If there is no container, pass **null**.

```
theApp = "Finder"
with objectModel, finder
    msg(core.count(theApp, null, disk))
```

core.create

```
core.create (appID, whatClass, [data, propertyList, where])
```

Makes a new *whatClass* (some type of element). This concept is sometimes called "make" in other literature.

core.dataSize

```
core.dataSize (appID, whatObject, [whatType])
```

Returns the size of the data of *whatObject*; *whatType* says what type of object *whatObject* should be considered as, because this can make a difference to the size of its data.

```
theApp = "Scriptable Text Editor"
with objectModel, window[1]
    msg(core.dataSize(theApp, word[4], text))
        « 4, because that's how many letters it has
    msg(core.dataSize(theApp, word[4]))
        « 58, because it includes style information
```

core.delete

```
core.delete (appID, whatObject)
```

Deletes the element *whatObject*.

core.duplicate

```
core.duplicate (appID, whatObject, toWhere)
```

Makes a copy of the element *whatObject* at *toWhere*. This concept is sometimes called "clone" in other literature.

```
theApp = "Scriptable Text Editor"
with objectModel, document[1]
    core.duplicate(theApp, word[2], after(word[3]))
```

core.exists

```
core.exists (appID, whatObject)
```

Returns a boolean reporting whether *whatObject* exists.

core.get

```
core.get (appID, whatObject, [whatType])
```

Returns the data of *whatObject*; *whatType* says what type of object *whatObject* should be considered as.

```
theApp = "Scriptable Text Editor"
with objectModel, window[1]
    msg(core.get(theApp, word[4], text))
        « "time"
        « if we omit the third parameter, we get a mess...
        « because of the style information
```

core.getAs

```
core.getAs (appID, whatObject, whatType)
```

Identical to core.get() except that *whatType* isn't optional.

core.move

```
core.move (appID, whatObject, toWhere)
```

Moves the element *whatObject* to *toWhere*.

core.open

```
core.open (appID, whatObject)
```

Opens *whatObject*.

core.print

```
core.print (appID, whatObject)
```

Prints *whatObject*.

core.quit

```
core.quit (appID, [saving?])
```

Quits the application; possible values for *saving?* are **yes**, **no**, and **ask**, and the default is **ask**.

core.save

```
core.save (appID, whatObject, [filePathname, fileType])
```

Saves to disk. *fileType* is a string4 type code. There is a bug in the glue; the *file-Pathname* is coerced to an alias, when it should be coerced to a fileSpec. This is true for the save() and saveAs() of just about all application glue as well.

core.saveAs

```
core.saveAs (appID, whatObject, filePathname, [fileType])
```

Identical to core.save() except that *filePathname* isn't optional.

core.set

```
core.set (appID, whatObject, toWhat)
```

Sets the value of *whatObject* to *toWhat*. A powerful command, the commonest way of getting real work done. This example boldifies the selected stretch of text.

```
local (theApp = "Scriptable Text Editor")
with objectModel, STE « Scriptable Text Editor glue
    stylRec = {onStyles:{bold}, offStyles:{}}
    core.set(theApp, selection.style, stylRec)
```

misc.beginTransaction

```
misc.beginTransaction (appID)
```

Asks the application to start a transaction and to return a *transactionID*.

misc.copy

```
misc.copy (appID)
```

Like choosing **Copy** from the application's **Edit** menu. Most applications must be frontmost for this command to work.

misc.createPublisher

    misc.createPublisher (*appID*, *whatObject*, *whatFile*)

Causes the application to make a publish-and-subscribe publisher of *whatObject* in the edition file *whatFile*.

misc.cut

    misc.cut (*appID*)

Like choosing **Cut** from the application's **Edit** menu. Most applications must be frontmost for this command to work.

misc.doMenu

    misc.doMenu (*appID*, *whatMenuItem*)

Like choosing *whatMenuItem*. In this example, we ask FileMaker Pro to boldify the currently selected text.

```
theApp = "FileMaker Pro"
with objectModel
    misc.doMenu(theApp, menu["Style"].menuitem["Bold"])
```

misc.doScript

    misc.doScript (*appID*, *string*)

Causes the application to execute a script in its internal scripting language.

misc.editGraphic

    misc.editGraphic (*appID*, *graphicArea*)

Tells the application to open *graphicArea* for editing, so that the user can edit it; when the user has finished the editing session, returns the edited graphic.

misc.endTransaction

```
misc.endTransaction (appID, transactionID)
```

Asks the application to end a transaction. This is the only verb in the database that calls `transactionEvent()`.

misc.imageGraphic

```
misc.imageGraphic (appID, graphic, format, antialiasing?,
                   dithering?, rotationRec, scale,
                   translationPoint, flipHorizontal?,
                   flipVertical?, quality,
                   structuredGraphic?)
```

Performs a graphic image conversion. All but the first three parameters are optional. The *format* is a string4 type code such as `'EPS '`, `'PICT'`, or `'TIFF'`. The *quality* can be `draft`, `high`, or `regular`. To avoid passing large amounts of data, both the *graphic* and the result can be an alias to a file.

misc.isUniform

```
misc.isUniform (appID, whatObject, whatProperty)
```

Returns a boolean telling whether the object(s) *whatObject* all have the same value for the property *whatProperty*.

misc.paste

```
misc.paste (appID)
```

Like choosing **Paste** from the application's **Edit** menu. Most applications must be frontmost for this command to work.

misc.redo

```
misc.redo (appID)
```

Like choosing **Redo** from the application's **Edit** menu.

misc.revert

```
misc.revert (appID, whatObject)
```

Reverts the object *whatObject* to its most recently saved state.

misc.select

```
misc.select (appID, whatObject)
```

Causes *whatObject* to become the current selection. A shorthand for using `core.set()` with direct object `selection`.

```
theApp = "Scriptable Text Editor"
with objectModel
    misc.select(theApp, window[1].word[2])
    « same as saying:
    « core.set(theApp, selection, window[1].word[2])
```

misc.show

```
misc.show (appID, whatObject, [whatWindow, whatPoint])
```

Asks that the object *whatObject* be brought into view.

misc.undo

```
misc.undo (appID)
```

Like choosing **Undo** from the application's **Edit** menu.

required.openApplication

```
required.openApplication (appID)
```

Sent to an application immediately after it is started up. Tells the application to perform its initialization tasks, such as opening an empty document. Apple Computer says that you should never call this verb (only the Finder may do so).

required.openDocument

```
required.openDocument (appID, documentPathname)
```

Sent to an application when the user opens one of its documents in the Finder; if the application wasn't running, sent instead of `required.openApplication()`. Tells the application to open the document.

required.printDocument

```
required.printDocument (appID, documentPathname)
```

Sent to an application when the user selects one of its documents in the Finder and chooses **Print** from the Finder's **File** menu. Tells the application to print the document.

required.quitApplication

```
required.quitApplication (appID)
```

Sent to an application when the Finder wants it to quit (for example, because the user has chosen **Shut Down** from the **Special** menu). Tells the application to perform any shutdown tasks, then terminate.

Index

Symbols

& (ampersand) in CGI arguments, 159
&& (ampersands) for logical and, 451
* (asterisk) for multiplication, 106, 161, 450
@ (at sign) for addresses, 82–85, 450
\ (backslash), 454
   to escape special characters, 109
   in grep expressions, 154
   for line continuation, 40
   in Web site, 413
! (bang) for logical not, 451
!= (bang equal) for logical inequality, 450
[ ] (brackets), 454
   in array notation, 112
   in object model, 315
   to reference database objects, 60
^ (caret) to dereference addresses, 83–85,
     450
: (colon), 454
   for application-specific indexing, 317
   in on statements, 56
   as pathname delimiter, 157, 291
   as record item delimiter, 99
, (comma), 454
   adding into numbers, 158
   to separate multiple parameters, 56
{ } (curly braces), 454
   for bundling code, 41–43, 407
   delimiting list and record literals, 99
$ (dollar sign) in URLs, 383
. (dot), 315, 454
   dot-notation, 28, 37–38

= (equal sign)
   == (equality) operator, 115, 117, 450
   to calculate menu items, 243
   for evaluating text in pictures, 262
   for object assignment, 35, 170, 449
> (greater than) operator, 451
>= (greater than or equals) operator, 451
(hash mark) for directives, 393
– (hyphen)
   decrement (--) operator, 161, 450
   for item deletion, 116, 449
   as menu item, 243
   for subtraction, 161, 449
---- direct parameter, 304
« (left guillemot) symbol, 45–46, 454
< (less than) operator, 451
<= (less than or equals) operator, 451
((left parenthesis) for menu items, 243
( ) (parentheses), 454
% (percent sign)
   in CGI arguments, 159
   for modulus, 161, 450
| (pipe)
   in glossPatch expressions, 417
   in Script Debugger, 349
| | (pipes) for or operator, 451
+ (plus sign)
   addition operator, 106, 161, 449
   in CGI arguments, 159
   concatenation operator, 115, 449
   increment (++) operator, 161, 450
? (question mark) in URLs, 383

' (quote) to delimit chars/string4s, 109, 454
" (quote) to delimit strings, 109, 454
; (semicolon) ending script lines, 41–43,
        454
/ (slash) for division, 106, 161, 450
≠ (not equal) operator, 450

Numbers

2-button modal dialogs, 225
3-button modal dialogs, 225

A

aborting running programs, 144
aborting threads, 201
abs( ), 161, 455
absolute URLs, converting from
        relative, 159
absolute values, 161
accessing suite menus, 250
ACGI applications, 375–388
    security and, 386
action scripts, 233
#activeURLs directive, 399
add( ), 377
addition, 106, 161, 449
address of (@) operator, 82–85, 450
address parameters, 85–90
address( ), 97
addresses
    for applications, 324–326
    coercing, 105
    constructing object references with, 85
    datatype for, 98
    dereferenced, 83
    non-scalars, obtaining, 172
    syntax of, 82–85
    of table entries, 172
    for target objects, obtaining, 175
    of verbs, 88–89
    as Web page object datatype, 406
    window references, 217
addressType datatype, 98
#adrPageData directive, 398
after( ), 318
age, object, 173
agents, 140, 255–257
    agent messages, 256–257
    debugging agent scripts, 256

polling, 248
    system.agents entry, 255
alert dialogs, 224
alias( ), 97
aliases, 297
    coercing, 105
    pathnames of aliased items, 298
aliasType datatype, 99
#alink directive, 397
allocating memory for Frontier, 27
alphabetic/alphanumeric, testing for, 151
ampersand (&) in CGI arguments, 159
ampersands (&&) for logical and, 451
and (&&) operator, 451
app verbs, 314, 325, 462–463
app.clearNetworkApp( ), 325, 462
app.linkToNetworkApp( ), 325, 463
app.start( ), 314, 463
app.startWithDocument( ), 325, 463
appending comments to commands, 45
appInfo subtable, 323
Apple events, 302–311, 340–344, 446–448
    Apple Event Registry, 311
    custom, 344–346
    suites, 311–314
    timeouts, 310
    transactions and, 309–310
Apple menu items, launching, 291
appleEvent( ), 305, 307–308, 455
AppleScript, 239, 332–335
applications
    addressing, 324–326
    associating with creator codes, 295
    dialogs in, 238–239
    driving Frontier from, 336–346
    droplets, 265, 270–272
    glue (see glue)
    launching, 292
    non-scriptable, 301
    processes, verbs for, 288
    RAM partitions, setting, 295
    shared menus, 246–249
    switching to/from, hook at, 258
    transactions between, 309–310
arithmetic operations, 161–163
arrays, 112–117
    iterating through with for...in, 120
    syntax of, 112–114
    verbs for, 114

ASCII characters, encoding for Web, 158
assignment (=) operator, 35, 170, 449
assignment commands, 35
asterisk (*) for multiplication, 106, 161, 450
asynchronous CGI applications, 375–388
asynchronous events, 308
at sign (@) for addresses, 82–85, 450
attributes, resource, 197
audio, verbs for, 289
#autoParagraphs directive, 398
awakening threads, 201

B

#background directive, 397, 426
backing up database, 26, 260, 273–277
backslash (\), 454
    to escape special characters, 109
    in grep expressions, 154
    for line continuation, 40
    in Web sites, 413
Backup (Main menu item), 260
backups.backuproot( ), 132, 273
bang (!) for logical not, 451
bang-equal (!=) operator, 450
Baron, Doug, xx, 142, 282
basicStuff (see samples.basicStuff entries)
batch export, 275–277
Batch Import (Batch menu item), 276
Baxter, John, xxi, 165
BBEdit editor, 347–352, 428–430
beeping, 289
before( ), 318
beginningOf( ), 318
beginsWith operator, 116, 451
#bgcolor directive, 397
binaries, 98, 100–101
    operations on, 117
    testing equality of, 117
binary( ), 96, 100
binaryType datatype, 98, 100
bit verbs, 162–163, 197, 463–464
bit.clear( ), 163, 463
bit.get( ), 163, 463
bit.set( ), 163, 464
bitmap patterns, datatype for, 98
body directives (in <body> tags), 397
<bodytext> pseudo-tags, 405
boldface, 187
bookmarking position in outlines, 182

Bookmarks submenu, 28
boolean indexing, 318
boolean( ), 96
booleans, 98
    coercing, 104
booleanType datatype, 98
braces { }, 454
    for bundling code, 41–43, 407
    delimiting list and record literals, 99
brackets ([ ]), 454
    in array notation, 112
    in object model, 315
    to reference database objects, 60
break statement, 119
breaking looping constructs, 119
breakpoints, 141
    operations on, 183
bringToFront( ), 323
Brown, Roger, 442
browsers (see suites.webBrowser)
bugs
    closing edit windows in content
        mode, 178
    coercing lists to strings, 104
    commenting case constructs, 130
    crashing system to find, 133–134
    declaring local variables, 72
    mouse.location( ), 286
    system vs. private scraps, 287
    try...else construct, 132
    in window close hooks, 215
buildTableOutline( ) utility, 182
built-in modal dialogs, 224–226
built-in verbs, 38, 137
bundle keyword, 41, 73
bundles, 16
    breakpoints and, 141
    managing file bundle bits, 296
    sorting lines of, 182
busy, checking files if, 298
buttons
    custom two- and three-button
        dialogs, 225
    hiding/showing in Main Window, 222
bytes, converting, 162

C

cadillac renderer, 403
calculating menu items, 243

Calhoun, Kevin, 442
call result values (verbs), 52–53
callback (see proc handlers)
callbacks, XCMD, 445
callXCMD( ), 456
canEdit( ) (see suites.odbEditor.editors entry)
Cantor, Georg, 106
Capture AE control panel, 302
card verbs (MacBird), 234–235, 464–467
card.close( ), 234, 464
card.getCheckedItem( ), 235
card.getObjectEnabled( ), 235, 464
card.getObjectFlag( ), 235, 464
card.getObjectText( ), 235, 464
card.getObjectVisible( ), 235, 465
card.getSelectedText( ), 235
card.isModal( ), 235, 465
card.popup.getCheckedItem( ), 465
card.popup.getHasLabel( ), 235, 465
card.popup.getMenu( ), 235, 465
card.popup.getSelectedText( ), 465
card.popup.setCheckedItem( ), 466
card.popup.setHasLabel( ), 235, 466
card.popup.setMenu( ), 235, 466
card.popup.setSelectedText( ), 466
card.run( ), 234, 466
card.setCardAttributes( ), 308
card.setCheckedItem( ), 235
card.setObjectEnabled( ), 235, 466
card.setObjectFlag( ), 235, 467
card.setObjectText( ), 235, 467
card.setObjectVisible( ), 467
card.setSelectedText( ), 235
cards (MacBird), 232–238
caret (^) to dereference addresses, 83–85, 450
cascadeWindows( ) utility, 126
case (capitalization), 453
    first letter of words, 157
    operations on, 151
case construct, 129–130
    breakpoints and, 141
cell object type, 341
CGI applications, 375–388
    security and, 386
CGI arguments, parsing, 159
changed Web sites, releasing, 419
char( ), 96

characters (chars), 98, 109–111
    ASCII, encoding for Web, 158
    coercing to/from strings, 104
    escaping special characters, 109
    generating string of repeating, 155
    operations on, 151
    (see also strings)
charType datatype, 98
checkmarks in menu items, 243
    Window menu, 214
chevron («) symbol, 45–46, 454
classAds suite, 433–437
classes, verb, 177–178
classified ads (example), 433–437
#clayCompatibility directive, 399
clean root, 277–278
clients, 368–370
Clip2GIF utility, 263, 321
clipboard verbs, 286–288, 467–468
clipboard, executing UserTalk scripts on, 337
clipboard.get( ), 288, 467
clipboard.getValue( ), 288, 467
clipboard.put( ), 287, 468
clipboard.putValue( ), 287, 468
clipToGifInDatabase( ) utility, 322
clock.now( ), 165, 290, 468
clock.set( ), 290, 468
clock.sleepFor( ), 227, 255, 468
clock.ticks( ), 290, 469
clock.waitSeconds( ), 198, 469
clock.waitSixtieths( ), 198, 469
close( ), 176, 456
closeWindow.addToChangedPages( ), 419
closing
    card-based dialogs, 234
    database, hooks at, 258
    edit windows, 176
    files for access, 190
    windows, 23, 214–215, 218–219
code (see scripts)
codeType datatype, 99
coerceTo( ) utility, 108
coerceValue( ), 108
coercion, 21, 100–108
    dates and, 164
    implicit, 100
    list of possibilities for, 102–105
    mixed-operand operations and, 107

coercion (*continued*)
  moving data to clipboard and, 287
  at runtime, 108
  verbs for, 96–97
collapsing outline subheads, 16, 182
colon (:), 454
  for application-specific indexing, 317
  in on statements, 56
  as pathname delimiter, 157, 291
  as record item delimiter, 99
colors
  converting hex triples to rgb, 102
  directives for, 397
  Mac OS colors, 102
  RGB (see rgbs)
  system.verbs.colors entry, 102
columns in tables, size of, 24
comma (,), 454
  adding into numbers, 158
  to separate multiple parameters, 56
command key (see modifier keys)
Command key shortcuts, 244
Command-period, 144
commands, 35
  appending comments to, 45
  making into comments, 44
comments
  deleting from lines, 156
  for files/folders, 295
  operations on, 183
  in scripts, 44–46
Commercial Developers (Suites menu
      item), 320
Common Styles submenu (Main
      menu), 15, 216
compacting copy of database, 274
comparing
  files, 298
  strings, 453
Compile button (edit window), 135
compiled code, datatype for, 99
compiler entries (see entries at
      system.compiler)
compiling scripts, 135–138
  errors when, 137
  menu item scripts, 245
complexEvent( ), 308, 456
compressing packed object files, 265

concatenating
  joining outline lines, 185
  records and lists, 115
  strings, 115, 151
conditional constructs, 128–130
configuring Frontier for CGI, 376
confirm dialogs, 225
  stopDialog example, 227–229
  stopDialog2 example, 235–236
constants, 93
  direction constants, 99
  for glue, 324
  for months, 168
containment, object, 315–317
contains operator, 116, 451
content mode
  closing edit windows while in, 178
  outline edit windows, 15–16
  table edit windows, 21–22
Content Server, 438
continue statement, 119
Control key (see modifier keys)
control panels, launching, 291
control structures, 118–134
converting
  case (capitalization), 151
  decimal and hexadecimal numbers, 162
  HTML-related string
      conversions, 158–160
  table entry types (see coercion)
convertToGMT( ) utility, 166
coordinates, 219
copying
  data and resource forks, 296
  database, 274
  files and folders, 296
  objects, 172
  text, verbs for, 179
  (see also cutting/copying and pasting)
core event suite, 312
core verbs, 312, 554–557
'core'/'setd' events, 303, 340–342
corruption of database, 273
counting, 554
  fields in strings, 156, 524
  lines in files, 192, 483
  mounted volumes, 292, 483
  outline summits and subheads, 183, 506
  processes currently running, 288, 531

counting (*continued*)
    resource types, 193
    resources, 193
    table entries, 281, 472
    threads, 199
    words, 157, 524
countInString( ) utility, 153
cr constant, 65
crashing system on purpose, 133–134
creating objects, 170
curly braces { }, 454
    for bundling code, 41–43, 407
    delimiting list and record literals, 99
cursors
    location of, 286
    "rolling beachball" shape, 289
custom
    Apple Events, 344–346
    directives, 408–409
    menus, 240, 254
    user-based pseudo-hooks, 260
customBackup( ), 260
customStartup( ), 260
cutting/copying and pasting
    in table edit windows, 21
    verbs for, 179
    in wptext edit windows, 19
cycleWindowsBackwards( ) utility, 221

D

daemon table (inetd), 363–365
data
    batch exports, 275–277
    multiple databases, 279–282
data forks, copying, 296
data parameter (core/setd verbs), 304
data resources, 193–197
data stacks, 143, 205–206
database (Frontier.root), 25–31
    AppleScript and (see AppleScript)
    backing up, 26, 260, 273–277
    clean root, 277–278
    editing objects in other
        applications, 347–353
    editing risks, 26, 59–60
    elements of, 29–31
    exporting objects from, 264–273
    hooks at opening/closing, 258
    importing into (see importing)

importing XCMDs/UCMDs, 196, 209
    multiple databases, 279–282
    navigating in, 27–29
    searching, 29
    verbs for managing, 280–282
    version of, 285
    window for (see Main Window)
database entries
    creating tables for, 170
    dialogs for creating
        (examples), 229–231, 237–238
    exploring for ideas, 145
    exporting to packed object
        files, 265–268
    modal dialog for entering, 225
    names of, 28
        calculating at runtime, 60–62, 85
    navigating to, 27–29
    operations on, 169–173
    references to (see object references)
    referring to, 58–68
    scripts as, 37–39
    testing if objects are, 172
    variables vs., 58–59
    verbs for managing, 281
datafiles, 188–191
datatypes, 37, 95–111
    arrays (see arrays)
    converting (see coercion)
    HyperTalk and, 339
    obtaining (see typeOf( ))
    operations on (see operations)
    precedence of, 107
    type-dependent operations, 106
    for Web page objects, 406
Date & Time control panel, 105
date and time, 164–168
    coercing, 105
    creation date for files/folders, 295
    dateType datatype, 98
    event timeouts, 310
    LongDateTime datatype, 165
    minutesSinceShip agent, 257
    modification date for files/folders, 295
    operations on, 166–168
    parsing and manipulating, 166–167
    scheduling tasks, 358
    setting clock, 290
    since computer started up, 290

date and time (*continued*)
 yielding for fixed intervals, 198
 (see also entries under clock)
date verbs, 166–167, 469–472
date( ), 96
date.abbrevString( ), 167, 469
date.dayOfWeek( ), 167, 469
date.daysInMonth( ), 167, 469
date.dayString( ), 167, 470
date.firstOfMonth( ), 167, 470
date.get( ), 166, 470
date.lastOfMonth( ), 167, 470
date.longString( ), 167, 470
date.nextMonth( ), 167, 470
date.nextWeek( ), 167, 471
date.nextYear( ), 167, 471
date.prevMonth( ), 167, 471
date.prevWeek( ), 167, 471
date.prevYear( ), 167, 471
date.set( ), 166, 471
date.shortString( ), 167, 472
date.tomorrow( ), 167, 472
date.weeksInMonth( ), 167, 472
date.yesterday( ), 167, 472
dateToLDT( ) utility, 165
daveNetOutline renderer, 402
db verbs, 281, 472–475
db.close( ), 281, 472
db.countItems( ), 281, 472
db.defined( ), 281, 473
db.delete( ), 281, 473
db.get( ), 334, 473
db.getNthItem( ), 281, 473
db.getValue( ), 281, 473
db.isTable( ), 281, 474
db.new( ), 281, 474
db.newTable( ), 281, 474
db.open( ), 281, 474
db.save( ), 281, 474
db.set( ), 334, 475
db.setValue( ), 281, 475
Debug button (edit window), 140
debugging
 agent scripts, 256
 compile errors, 137
 crashing system on purpose, 133–134
 Debug mode, 140–143
 menu item scripts, 245

decimal, converting to/from
 hexadecimal, 162
declaring local variables, 70–72
decrement (--) operator, 161, 450
default
 parameter values, 57
 timeouts, 310
#defaultFileName directive, 401
#defaultTemplate directive, 400
#define directive, 409
defined( ), 66, 91–92, 171, 456
#defineScript directive, 409
delete( ), 114, 171, 457
deleting
 aborting threads, 201
 array items, 114
 clearing tabs, 187
 database entries, 281
 files, 297
 folders, 297
 menu items and submenus, 253
 object destruction, 171
 outline contents, 180
 from records and lists, 116
 references, 195
 script/menubar/outline lines, 180
 stacks and stack values, 205–206
 from strings, 116, 150, 153, 155
 suite menus, 250
 table entries, 171
 trailing comments, 156
demoting outline lines, 182
dereference (^) operator, 83–85, 450
dereferenced addresses, 83
desktop scripts, 140, 264, 268–269
destroying objects, 171
dialog verbs, 224–227, 475–479
dialog.alert( ), 224, 475
dialog.ask( ), 225, 475
dialog.confirm( ), 225, 475
dialog.fileInfo( ), 226, 476
dialog.getInt( ), 225, 476
dialog.getValue( ), 227, 476
dialog.hideItem( ), 227, 476
dialog.loadFromFile( ), 232, 476
dialog.notify( ), 224, 477
dialog.run( ), 226–227, 477
dialog.runFromFile( ), 477

dialog.runModeless( ), 226–227, 477
dialog.setItemEnable( ), 227, 478
dialog.setValue( ), 227, 478
dialog.showItem( ), 227, 478
dialog.threeWay( ), 225, 478
dialog.twoWay( ), 225, 479
dialog.yesNo( ), 225, 479
dialog.yesNoCancel( ), 225, 479
dialogs
    card-based, 232–238
    dialog-based droplets, 271–272
    manipulating items of, 227
    in other applications, 238–239
    pre-configured (verbs for), 224–226
    resource-based, 226–232
    shared menus and, 249
    titles of, 214
    types of, 223
diamond-mark in Window menu, 214
direct object, Apple event, 304
direction constants, 99, 105
direction( ), 97
directionType datatype, 99
directives (Web site), 393–401
    custom, 408–409
    special outline directives, 409
#directivesOnlyAtBeginning directive, 399
dirtyness, object, 173
disks (see volumes)
displayString( ), 110, 457
division (/) operator, 106, 161, 450
doAsAppleScript( ) utility, 334
DocServer (Main menu item), 145
DocServer help program, 144–145
documentation
    for glue tables, 324
    on UserTalk, 144–145
    on writing XCMDs and UCMDs, 441
dollar sign ($) in URLs, 383
domain for with statement (see with
            (keyword))
dot (.), 315, 454
dot-notation, 28, 37–38
double quote (") to delimit strings, 109, 454
double( ), 96
double-clicking text, hook for, 259
doubleToDate( ) utility, 164
doubleType datatype, 97
Droplet Developer (Suites menu item), 270

droplets, 265, 270–272
    (see also entries under system.droplet)
#dropNonAlphas directive, 401, 420
dynamic scoping, 78–79
dynamic Web sites, 431–438

E

edit windows, 13–24
    choosing type of, 14
    closing, 176
    Compile button, 135
    contents (see non-scalars)
    Debug button, 140
    manipulating (see windows)
    outline edit windows, 14–18
    Run button, 138
    table edit windows, 19–24
    verbs for manipulating, 216–222
    verbs operating in, 178–187
    wptext edit windows, 19
Edit with App (Main menu item), 233
edit( ), 175, 457
editing
    database, danger of, 26, 59–60
    MacBird cards, 233
    menus, 241
    non-scalars, 174–187
    objects dirtyness and, 173
    objects in other applications, 347–353
    pictures, 261–262
    scripts while debugging, 140
    strings (see strings, operations on)
    table entries, 22
    ToDo List, 357
editMenu verbs, 179, 187, 479–481
editMenu.clear( ), 179, 479
editMenu.copy( ), 179, 479
editMenu.cut( ), 179, 479
editMenu.paste( ), 179, 480
editMenu.plainText( ), 187, 480
editMenu.selectAll( ), 480
editMenu.setBold( ), 187, 480
editMenu.setFont( ), 179, 480
editMenu.setFontSize( ), 179, 480
editMenu.setItalic( ), 187, 481
editMenu.setOutline( ), 187, 481
editMenu.setShadow( ), 187, 481
editMenu.setUnderline( ), 187, 481
ejectable volumes, testing for, 292

element objects, 315–318
else statement
    breakpoints and, 141
    case (see case construct)
    if...else (see if...else construct)
    try...else (see try...else construct)
empty strings/lists/records, 117
encoding strings for Web, 158–159
endOf( ), 318
endsWith operator, 116, 452
entries, database (see database entries)
entries, table (see table entries)
entryMaker dialog (example), 229–231
entryMaker2 dialog (example), 237–238
enum( ), 97
enumeratorType datatype, 99
eponymous handlers, 47–54
    Handler Rule, 48–50
equal sign (=)
    == (equality) operator, 115, 117, 450
    to calculate menu items, 243
    for evaluating text in pictures, 262
    for object assignment, 35, 170, 449
equality, testing for, 115–117
erasing (see deleting)
errors
    application error messages, 330
    causing on purpose, 132–133
    compile errors, 137
    crashing system on purpose, 133–134
    runtime errors and error dialogs, 143
    trapping, 131–132
    webserver suite and, 378
escape characters in Web sites, 413
escaping special characters, 109
Eudora, 67–68, 319, 329
eudora.get( ), 324
eudora.saveAs( ), 320
eudora.visitMessages( ), 320
evaluate( ), 93–94, 457
evaluateThread( ), 247
evaluating (see testing)
evaluation, special cases for, 91–94
events
    Apple (see Apple events)
    asynchronous, 308
    suites, 311–314
    timeouts, 310
    transactions and, 309–310
    user, verbs for, 286
execution path, 141
existence, testing for, 555
    files, 297, 484
    resources, 195
    threads, 201, 537
#expandGlossaryItems directive, 400, 412
expanding outline subheads, 16, 182
Export (Main menu item), 265
Export dialog, 265
export.callImporter( ), 272
export.callImportSubmenu( ), 272
export.importFolder( ), 268
export.importSubmenu( ), 267, 272
exporting, 264–273
    AppleScript scripts, 335
    batch exports, 275–277

F

faceless droplets, 271
.fcgi filename extension, 376
fields, parsing strings for, 155–156
file verbs, 190–192, 293–297, 481–496
file.bytesInFolder( ), 293, 481
file.bytesOnVolume( ), 293, 481
file.close( ), 190, 482
file.compare( ), 298, 482
file.copy( ), 296, 482
file.copyDataFork( ), 296, 482
file.copyResourceFork( ), 296, 482
file.copyToSystemFolder( ), 297, 483
file.countLines( ), 192, 483
file.countVolumes( ), 292, 483
file.created( ), 295, 483
file.creator( ), 295, 483
file.delete( ), 297, 483
file.deleteFolder( ), 297, 484
file.eject( ), 293, 484
file.endOfFile( ), 191, 484
file.exists( ), 297, 484
file.fileFromPath( ), 157, 484
file.filesInFolder( ), 293, 484
file.filesOnVolume( ), 293, 485
file.filteredCopy( ), 296, 485
file.findApplication( ), 295, 485
file.findInFile( ), 191, 485
file.findInFolder( ), 192, 486

file.folderFromPath( ), 157, 486
file.foldersInFolder( ), 293, 486
file.foldersOnVolume( ), 293, 486
file.followAlias( ), 298, 486
file.freespaceOnVolume( ), 293, 486
file.getComment( ), 295, 487
file.getDiskDialog( ), 225, 487
file.getFileDialog( ), 225, 487
file.getFolderDialog( ), 225, 487
file.getFullVersion( ), 295, 488
file.getIconPos( ), 296, 488
file.getLabel( ), 296, 488
file.getPath( ), 291, 488
file.getSpecialFolderPath( ), 294, 488
file.getSystemDisk( ), 294, 488
file.getSystemFolderPath( ), 294, 489
file.getVersion( ), 285, 295, 489
file.isAlias( ), 297, 489
file.isBusy( ), 298, 489
file.isEjectable( ), 292, 489
file.isFolder( ), 297, 489
file.isLocked( ), 296, 490
file.isVisible( ), 296, 490
file.isVolume( ), 297, 490
file.lock( ), 296, 490
file.modified( ), 295, 490
file.mountServerVolume( ), 292, 490
file.move( ), 296, 491
file.new( ), 190, 297, 491
file.newAlias( ), 297, 491
file.newFolder( ), 297, 491
file.open( ), 190, 491
file.putFileDialog( ), 225, 492
file.read( ), 191, 492
file.readLine( ), 191, 492
file.rename( ), 297, 493
file.setComment( ), 295, 493
file.setCreated( ), 295, 493
file.setCreator( ), 295, 493
file.setFullVersion( ), 295, 493
file.setIconPos( ), 296, 493
file.setLabel( ), 296, 494
file.setModified( ), 295, 494
file.setPath( ), 291, 494
file.setType( ), 295, 494
file.setVersion( ), 295, 494
file.setVisible( ), 296, 494
file.size( ), 190, 294, 495
file.sureFolder( ), 297, 495

file.type( ), 295, 495
file.unlock( ), 296, 495
file.unmountVolume( ), 293, 495
file.visitFolder( ), 128, 495
file.volumeBlockSize( ), 293, 496
file.volumeFromPath( ), 157, 496
file.volumeSize( ), 293, 496
file.write( ), 191, 496
file.writeLine( ), 191, 496
#fileExtension directive, 401, 420
fileloop( ) looping construct, 122–125
FileMaker Pro, 309, 317, 342–344
filemenu verbs, 218, 497
filemenu.close( ), 218, 497
filemenu.closeAll( ), 218, 497
filemenu.new( ), 497
filemenu.open( ), 497
filemenu.save( ), 497
filemenu.saveCopy( ), 274, 497
files
    comparing, 298
    copying, 296
    counting in folders/volumes, 293
    creation/modification dates for, 295
    datafiles, 188–191
    directives for, 400
    file-based resources, 231–232
    icon position, 296
    information on, 226, 294–296
    iterating through, 122–125, 128
    managing file system (verbs
        for), 290–299
    manipulating, verbs for, 296–298
    operations on
        pathname-related, 157
    packed object files, 265–268
    pathnames of (see paths and
        pathnames)
    resources of, 193–197
    selection dialogs (verbs for), 225–226
    size of, 294
filesInFolder( ) utility, 294
filespec( ), 97
filespecType datatype, 99, 105, 406
filtering Web data, 414–416
#filters directive, 414–416
finalFilter( ), 414–416, 429
Find & Replace Dialog, 29
finderEvent( ), 309, 457

finderflags.clear( ), 296, 498
finderflags.get( ), 296, 498
finderflags.set( ), 296, 498
finding (see searching)
finding and replacing strings, 152–155
findNth( ) utility, 153, 156
fixed( ), 96
fixedType datatype, 97
FKEY, launching, 292
flags, hiding/showing (Main Window), 222
floating-point numbers, 97
focus of outline, zooming, 182
folders
 copying, 296
 counting on volumes, 293
 creation/modification dates for, 295
 dialog for selecting, 225
 icon position, 296
 importing, 268
 iterating through files in, 122–125, 128
 manipulating, verbs for, 296–298
 obtaining name of, 157
 searching within, 192
 system folder, 294, 297
 tables as exported packed objects, 266
 verbs for, 293
folderSize( ) utility, 294
Follow button (Debug mode), 142
fonts
 in windows, 216
 setting, 179
 setting styles, 187
 in windows, 15
footer directives, 398
for looping construct, 119–120
for...in looping construct, 120
formatting windows, 215–216
forms, HTML, 380–382
free space
 in Frontier's heap, 285
 RAM, 290
 of volumes, 293
Frontier
 as ACGI application, 376
 allocating memory for, 27
 Apple events, responding to, 340–341
 bugs (see bugs)
 database of (see database)
 driving from Web pages, 345

edit windows, 13–24
how to use (examples), 10–12
pictures (see pictures)
querying, verbs for, 285
remote communications with, 371–374
running Web server with, 364, 386–387
switching to/from, hook for, 258
versions of, xix, 285
Frontier Do Script command (Nisus
 Writer), 337
Frontier SDK, 441
Frontier Text Files folder, 349–350
frontier.bringToFront( ), 498
frontier.clickers.type2CLK( ), 272
frontier.enableAgents( ), 498
frontier.finder2click( ), 272
frontier.getFilePath( ), 285, 498
frontier.getProgramPath( ), 285, 499
frontier.idleTime( ), 499
frontier.requestToFront( ), 238–239, 499
Frontier.root (see database)
frontier.version( ), 285, 499
frontierPath agent, 257
FTP, Frontier for (example), 369–370
#ftpSite directive, 419
#fullTimeNetConnection directive, 401

G

generating HTML, directives for, 398
Gestalt Manager, 290
gestalt( ), 290, 458
getAs( ), 329
getBinaryType( ), 101, 458
GIF images, 263, 321
global
 database entries as, 59
 variables, 79–80
glossaries, 412–413
glossPatch mechanism, 416–418, 425, 429
glossSub( ) utility, 424
glue, 207–209, 319–331
 anatomy of glue tables, 323–324
 challenges, tips, and pitfalls, 326–331
 Open Glue Script (Frontier menu
  item), 145
 performance tuning, 321–322
Go button (Debug mode), 141
Go To button (error dialog), 137, 143
goto (see looping constructs, interrupting)

graphics (see pictures)
greater than (>) operator, 451
greater than or equal (>=) operator, 451
grep search and replace, 153–155

H

Handler Rule, 48–50
handlers
    for dialogs, 226–232
    global variables and, 79–80
    parameters (see parameters)
    passed in visit constructs, 125–128
    scope of, 74–76, 78–79
        (see also scope)
hash mark (#) for directives, 393
header directives (in <head> tags), 396
height (see size)
help
    comments in scripts, 44–46
    DocServer program, 144–145
    exporing database for ideas, 145
hexadecimal numbers, 162
hiding
    card-based dialog items, 235
    dialog items, 227
    files and folders, 296
    Main Window elements, 222
    windows, 218–219
hierarchy
    of directives, 395
    menubar, 242–243
    in outline edit windows, 16
hoisting, 182
hooks, 140, 258–260
    for closing windows, 214–215
    as interrupts, 259
    user-based psuedo-hooks, 260
HTML
    directives (see directives)
    forms, 380–382
    generating, directives for, 398
    hyperlinks, 416–418, 425, 427
    string.processHtmlMacros( ), 422
    suites.html entry, 421–423
    #template.html files, 428
    user.html (see entries under user.html)
    (see also Web)
html suite (see entries under user.html)

html.addToChangedPages( ), 215
html.buildFromOutline( ), 419, 423
html.buildObject( ), 421
html.buildPageTable( ), 421
html.data.page entry, 407
html.data.page.bodyText entry, 414
html.data.page.renderedText entry, 414
html.data.standardMacros entry, 407,
        409–412
    (see also entries under standardMacros)
html.ftpText( ), 422
html.getFileName( ), 423
html.getOneDirective( ), 423
html.getPref( ), 423
html.getSiteTable( ), 423
html.loadImageFile( ), 423
html.processMacros( ), 422, 429
html.runDirectives( ), 422
html.runOutlineDirectives( ), 422
html.traversalSkip( ), 423
html.ucmds.getOutlineHtml( ), 403
HTML-related string conversions, 158–160
HyperCard, 338–340
hyperlinks
    for navigation, 427
    relative, 416–418, 425
HyperTalk messages, 338
hyphen (-)
    decrement (--) operator, 161, 450
    for item deletion, 116, 449, 116
    as menu item, 243
    for subtraction, 161, 449

I

IAC... verbs, 446–448
icons for files/folders, 296
if...else construct, 128–129
    case construct vs., 130
#imgFileCreator directive, 401
implicit coercion, 100, 107
implicit local scope, 77–78
implicit script compilation, 135–136
importing
    AppleScript files, 335
    from batch exports, 276–277
    folders of packed objects, 268
    menu item scripts, 267
    packed objects, 266–268

importing (*continued*)
    PICTs, 261–262
    suites, 252, 267
    XCMD/UCMD extensions, 196, 209
importSuite( ), 267
importXCMD( ) utility, 196
In button (Debug mode), 142
#includeMetaCharset directive, 396
increment (++) operator, 161, 450
indentation, 40–41, 187
index, array, 112
indexing, 317–318, 330
inetd suite, 363–365
inetd.start( ), 364
inetd.startOne( ), 364
inetd.supervisor( ), 364
infinite looping constructs, 50, 121
infinity constant, 92
information
    on files, 226, 294–296
    modal dialogs for, 224
    system, querying for, 285–299
input
    file-selection dialogs, 225–226
    modal dialogs for, 225
inserting substrings into strings, 149
inserting text, 179
integers
    datatypes for, 97
    integer input dialogs, 225
    operations on, 106
interaction levels, 311
internal types, 100
interrupts, hooks as, 259
intervals
    for running agent scripts, 255
    updating pictures, 263
intType datatype, 97
isDatabaseEntry( ) utility, 172
isLower( ) utility, 151
#isoFilter directive, 399
isRunning( ), 323
isUpper( ) utility, 151
italics, 187
iterating (see looping constructs)

J

#javascript directive, 397
joining (see concatenating)

Jump (Open menu item), 28
Jump dialog, 28
jumping to database entries, 28
justifying paragraphs, 187

K

kb.cmdKey( ), 286, 499
kb.controlKey( ), 286, 499
kb.optionKey( ), 286, 500
kb.shiftKey( ), 286, 500
kernel, 136
kernel( ), 137
keyboard
    modifier keys (see modifier keys)
    shortcuts, 244
keywords, UserTalk, 35, 92
Kill button (Debug mode), 142

L

labels for files/folders, 296
Latin-1 encoding, 158
launch verbs, 291–292, 500–501
launch( ), 324, 325
launch.anything( ), 291, 500
launch.appleMenu( ), 291, 500
launch.application( ), 292, 500
launch.appWithDocument( ), 292, 500
launch.controlPanel( ), 291, 501
launch.resource( ), 292, 501
launch.usingID( ), 292, 501
Lawton, Scott, xxi, 77
Lawton's Law, 77
layers, window, 220–221
ldtToDate( ) utility, 165
left guillemot («) character, 45–46, 454
left parenthesis (for menu items, 243
length (see size)
less than (<) operator, 451
less than or equal (<=) operator, 451
lettering (see fonts)
line continuation character (\), 40
line spacing, 187
linefeed character (\n), 109
lines of script, evaluating, 53–54
lines of text
    counting within file, 192
    selecting, 179
#link directive, 397

links, Web site
    for navigation, 427
    relative, 416–418, 425
linksToOtherPages( ) utility, 427
list( ), 97
listeners, 363
listFolderToDisk( ) utility, 189
lists
    adding new items to, 115
    array notation with, 113
    coercing, 104
    coercing to strings (bug), 104
    converting to/from outlines, 185
    empty, 117
    iterating through with for...in, 120
    listType datatype, 99
    operations on, 115–116
    (see also records)
lists (list datatype), 57
listWindows( ) utility, 221
Load Existing Site (Web menu item), 420
loading existing Web sites, 420
local (keyword), 70, 72
local processes, 308
local scope, 69–72
    handlers, 75–76
    implicit, 77–78
local variables, 70–72, 142
location parameters, 318, 330
locking and unlocking
    files and folders, 296
    semaphores, 203–204
logging scheduled tasks, 360
logical equality (==) operator, 115, 117
logical operators, 450–452
loginAs.loginAs( ), 501
long( ), 96
LongDateTime datatype, 165
longType datatype, 97
Lookup button (Debug mode), 142
loop( ), 121
looping constructs, 118–125
    errors and, 131
    infinite, 50, 121
    interrupting, 119
lowercase (see case)
#lowerCaseFileNames directive, 401, 420

M

Mac OS colors, 102
MacBird application, 232–238
    dialog-based droplets, 271–272
macintosh entries (see entries under
        system.macintosh)
macros, 379, 407–412
macros.imageTag( ), 429
Main Window, 27, 213
    hiding/showing elements of, 222
    (see also windows)
mainWindow.hideButtons( ), 222, 501
mainWindow.hideFlag( ), 222, 502
mainWindow.hidePopup( ), 222, 502
mainWindow.showButtons( ), 222, 502
mainWindow.showFlag( ), 222, 502
mainWindow.showPopup( ), 222, 502
Make New Suite (Suites menu item), 251
margins, managing, 187
math operations, 161–163
#maxFileNameLength directive, 401, 420
memAvail( ), 285, 458
memory
    allocating for Frontier, 27
    determining how much free, 285, 290
    shared menus and, 248
    used for folder contents, 293
    volume free space, 293
Menu Bar button (Main Window), 242
Menu Editor (Open menu item), 242
menu item scripts, 244–245, 253
    importing, 267
    managing and coding, 243
    testing, 222
menu items
    Apple menu items, 291
    separator lines, 243
menu verbs, 252–254, 502–505
menu.addMenuCommand( ), 252, 502
menu.addSubMenu( ), 253, 503
menu.addSuite( ), 254, 503
menu.clearMenubar( ), 254, 503
menu.deleteMenuCommand( ), 253, 503
menu.deleteSubMenu( ), 253, 503
menu.getScript( ), 253, 504
menu.importSuite( ), 252
menu.install( ), 254, 504
menu.isInstalled( ), 254, 504

menu.noSuite( ), 254, 504
menu.remove( ), 254, 504
menu.setScript( ), 253, 505
menu.toggle( ), 254, 505
menu.zoomscript( ), 253, 505
menubars, 240, 242–245
    coercing to strings, 103
    exporting as packed objects, 266
    manipulating, verbs for, 252–254
    menuBarType datatype for, 97
    opening, 242
    relationship with menus, 243–244
    system.menubars entry, 31
    system.misc.menubar entry, 31, 242
    wptext verbs for, 178–186
menus, 240–242
    calculated names for, 244
    importing suites, 252, 267
    keyboard shortcuts, 244
    manipulating, verbs for, 252–254
    modal menus, 241, 250
    modifying, 241
    relationship with menubars, 243–244
    scripts as menu items, 139
    shared menus, 241, 246–249
        keyboard shortcuts, 244
messages
    agent messages, 256–257
    error messages from applications, 330
    HyperTalk messages, 338
    receiving from other applications, 344
    status (see status messages)
    system level (see Apple events)
#meta directive, 396
Microsoft Internet Explorer (see
        suites.webBrowser)
Microsoft Word, Apple events
        and, 341–342
migrate( ) utility, 282
migrating to clean root, 277–278
MIME types, 382
minus sign (see hyphen)
minutesSinceShip agent, 257
misc verbs, 313, 557–560
misc.beginTransaction( ), 313, 557
misc.copy( ), 313, 557
misc.createPublisher( ), 313, 558
misc.cut( ), 313, 558
misc.doMenu( ), 313, 558

misc.doScript( ), 313, 558
misc.editGraphic( ), 313, 558
misc.endTransaction( ), 313, 559
misc.imageGraphic( ), 313, 559
misc.isUniform( ), 313, 559
misc.paste( ), 313, 559
misc.redo( ), 313, 559
misc.revert( ), 313, 560
misc.select( ), 313, 560
misc.show( ), 313, 560
misc.undo( ), 313, 560
'misc'/'dosc' events, 341–344
missed scheduled tasks, 360
mod( ), 161, 458
modal dialogs, 223, 224–226
    card-based (example), 235–236
    resource-based (example), 227–229
modal menus, 241, 250
modeless dialogs, 223
    card-based (example), 237–238
    resource-based (example), 229–231
modes, 178
    of outline edit windows, 15–16
    of table edit windows, 21–22
    toggling, 16
modes.monitor( ), 250, 257
modifier keys
    hooks for modifier-double-clicking, 259
    verbs to detect events with, 286
modifying (see editing)
modulus (%) operator, 161, 450
months, 166–167
mounting/unmounting volumes, 292
mouse events, verbs for, 286
mouse.button( ), 286, 505
mouse.location( ), 286, 505
movable modal dialogs, 223
moving
    files and folders, 296
    object values onto clipboard, 287
    objects, 172
    table entries, 172
msg( ), 73, 224, 458
    agents and, 257
multidimensional arrays, 113
multiple databases, 279–282
multiple parameters, 56
multiplication, 107
multiplication (*) operator, 106, 161, 450

multithreading, 198–204
    agent scripts and, 255
    managing threads, 200–202
    semaphores, 202–204
    shared menus and, 247

N

nameOf( ), 91–92, 114, 171–172, 458
names
    addresses vs., 83
    array items, 114
    calculating at runtime, 60–62, 85
    capitalization of, 298
    database entries, 28
    days of week, 167
    droplets, 270
    evaluating, special cases for, 91–94
    folders, 157
    guaranteeing uniqueness of, 170
    for menu items, calculating, 243
    paths (see paths and pathnames)
    referencing resources with, 194
    renaming files, 297
    suite titles, 251
    table entries, 170–172
    verbs, 37–38
    volumes, 157
    window titles, 214, 221
name-value pairs (in tables), 19
navigation
    within database, 27–29
    within scripts, menubars,
        outlines, 17–18, 180–182
    within tables, 23–24, 180
    among window layers, 220–221
    in wptext situations, 179
navigation links, 427
NetEvents application, 362
netEvents.listenStream( ), 364
netEvents.nameToAddress( ), 364
netEvents.readStream( ), 364
netEvents.writeStream( ), 364
'netf'/'valu' events, 373
NetFrontier suite, 371–374
netFrontier.ATclient.get( ), 373
netFrontier.remoteDelete( ), 372
netFrontier.remoteGet( ), 372
netFrontier.remotePut( ), 372
netFrontier.remoteQuit( ), 372

netFrontier.remoteRun( ), 372
Netscape Navigator, 345
    (see also suites.webBrowser)
networking, 362–370
    CGI applications, 375–388
    remote communications, 371–374
New Card (Table menu item), 233
New menu items (Table menu), 22
New Script (Main menu item), 256, 377
New Site (Web menu item), 391
new( ), 117, 170, 459
newCulture renderer, 402
newsPage suite, 431–433
nil constant, 93
Nisus Writer, 337–338, 353
#noHintsInHeader directive, 398
non-evaluation, 91–92
non-scalars, 97
    age and dirtyness of, 173
    coercion of, 103
    menubars (see menubars)
    obtaining addresses of, 172
    pictures (see pictures)
    verbs for editing, 174–187
non-scriptable applications, 301
not (!) operator, 451
not equal (!=) operator, 450
Notepad (Open menu item), 30
notify dialogs, 224
null container, 316
number( ), 96
numbers, 97
    adding commas into, 158
    attaching zeros to, 155
    coercing, 104
    counting (see counting)
    datatype-dependent operations on, 106
    hexadecimal, 162
    integer input dialogs, 225
    math, 161–163
    parsing dates into, 166
    testing for numeric characters, 151

O

Object Database Map, 29
Object DB button, 27
object model syntax, 306–307
object pointers (see addresses)

object references
    calculating at runtime, 60–62, 85
    constructing with addresses, 85
    partial, 62–66
    with statement for, 66–68
object specifiers, 304
object types (see datatypes)
objectHierarchy outline, 324
objects
    age and dirtiness of, 173
    card objects (see card-based dialogs)
    containment, 315–317
    direct object, Apple event, 304
    edit windows and, 14
    editing in other applications, 347–353
    exporting from database, 264–273
    packed objects, 264–268
    reading data into, 188
    stacks of, 143, 205–206
    target, 174–177
    UserTalk object model, 315–319
    writing as datafiles, 189
objects, operations on, 169–173
objSpec( ), 97, 306
objspecType datatype, 99
odbEditor suite (see suites.odbEditor
        entries)
odbEditor.edit( ), 348
odbServer suite, 365–367
odbServer.commandDecode( ), 366
odbServer.commandEncode( ), 366
odbServer.commandSend( ), 366
odbServer.serverStart( ), 366
on statement, 47
    breakpoints and, 141
    codeType and, 99
    (see also eponymous handlers)
op verbs, 180–183, 505–512
op.collapse( ), 182, 505
op.countSubs( ), 183, 506
op.countSummits( ), 183, 506
op.dehoist( ), 182, 506
op.dehoistAll( ), 182, 506
op.deleteLine( ), 180, 506
op.deleteSubs( ), 180, 507
op.demote( ), 182, 507
op.expand( ), 182, 507
op.firstSummit( ), 180, 507
op.flatCursorKeys( ), 507

op.fullCollapse( ), 182, 508
op.fullExpand( ), 182, 508
op.getCursor( ), 182, 508
op.getDisplay( ), 176, 508
op.getLineText( ), 180, 508
op.getRefCon( ), 186, 508
op.go( ), 181, 509
op.hoist( ), 182, 509
op.insert( ), 180, 509
op.insertAtEndOfList( ), 180, 509
op.level( ), 183, 510
op.promote( ), 182, 510
op.reorg( ), 180, 510
op.setCursor( ), 182, 510
op.setDisplay( ), 176, 511
op.setLineText( ), 180, 511
op.setRefCon( ), 186, 511
op.sort( ), 182, 511
op.subsExpanded( ), 183, 511
op.tabKeyReorg( ), 512
op.visit( ), 126–127, 512
op.wipe( ), 180, 512
Open Glue Script (Frontier menu
        item), 145
Open Scripting Architecture
        (OSA), 337–340
openDocument( ), 323
opening
    database, hooks at, 258
    files for access, 190
    packed objects, 266–268
    second database, 280
    verbs for, 291
    windows, 23, 214, 218–219
operating system
    querying, verbs for, 285–299
    version of, 290
operations
    datatype-dependent, 106
    with mixed operands, 107–108
operators, reference on, 449–452
opJoin( ) utility, 185
opSplit( ) utility, 184
Option key (see modifier keys)
or (||) operator, 451
order
    of arithmetic operation, 161
    datatype precedence, 107
    outline organization, 17–18

order (*continued*)
    parameters, 56
    resolving object references, 62–63
    shortcut key assignment, 244
    sorting lines in outline bundles, 182
    sorting table entries, 180
    table entries (sort order), 24
    window layers, 220–221
OSA (Open Scripting
            Architecture), 337–340
    dialects, verbs for, 138
    OSA Menu extension, 246, 269
OSAX scripting extension, 207, 334–335
Out button (Debug mode), 142
outdated database data, 273
Outgoing Frontier Objects folder, 265
outline edit windows, 14–18
outliner suite, 186
outlines
    coercing to strings, 103
    converting to/from lists, 185
    defining directives in, 393
    directives for, 409
    menubars as, 242–243
    navigating in and reorganizing, 17–18
    outlineType datatype, 97
    rendering, 401–404
    scripts as, 39–44
    secret line values, 186
    splitting and joining lines of, 184
    templates, 404–406
    verbs for (see op verbs)
    as Web page object datatypes, 406

P

pack( ), 101, 459
packed objects, 264–268
    batch imports of, 276
padding strings, 155
pageFilter( ), 414–417
PageSpinner editor, 347–350
paragraphs
    counting fields in, 156
    justifying, 187
    selecting, 179
parameters, 35, 54–57
    addresses as, 85–90
    default values for, 57

handlers and, 48
    multiple, 56
parentheses ( ), 454
parentOf( ), 91–92, 171, 459
parsing
    CGI arguments, 159
    dates, 166–167
    strings, 155–158
    URLs, 159
partial object references, 62–66
partitions, RAM, 295
passing by value, 55
    circumventing, 86
passwordDialog.run( ), 225, 512
passwords
    dialog for inputting, 225
Paste Into Frontier (Frontier menu), 145
pasting (see cutting/copying and pasting)
paths and pathnames, 28, 291
    of aliased items, 298
    for applications, 289
    coercing fileSpecs and aliases to, 105
    creating all folders along, 297
    desktop scripts, 269
    execution paths, 141
    Frontier application file, 285
    frontierPath agent, 257
    iterating through with
            fileloop( ), 122–125
    parsing, 157–158
    people table and, 30
    prefix for, 291
    for resolving data entry
            references, 62–63, 65–66
    of system folder, 294
patternMatchAfter( ) utility, 152
patternType datatype, 98
pbs.deleteListItem( ), 446
pbs.getLinks( ), 160, 513
pbs.stripHTML( ), 160, 513
pbs.utilities.parseRelativeURL( ), 159, 513
people table, 30
people.[user.initials] table, 60–66
percent sign (%)
    in CGI arguments, 159
    for modulus, 161, 450
performance
    advice for running Web server, 386–387
    backing up database, 273

performance (*continued*)
    memory allocation and, 27
    OSAX scripting extension, 207, 334–335
    scheduling (see Scheduler suite)
    script extensions, 441–448
    tuning glue, 321–322
    XCMD and UCMD scripting
        extensions, 207–209
persistence of database entries, 59
physicalSize( ) utility, 295
pict.expressions( ), 263, 513
pict.getPicture( ), 262, 513
pict.PICTtoPicture( ), 261, 513
pict.pictureToPICT( ), 261, 514
pict.scheduleUpdate( ), 263, 514
pict.setPicture( ), 262, 514
PICTs, 261–262
pictures, 261–263
pictureType datatype, 97
pipe (|)
    in glossPatch expressions, 417
    in Script Debugger, 349
pipes (||) as or operator, 451
plus sign (+)
    in CGI arguments, 159
    addition operator, 106, 161, 449
    concatenation operator, 115, 449
    increment (++) operator, 161, 450
point( ), 97
point.get( ), 102, 514
point.set( ), 102, 514
pointers to objects (see addresses)
points, 99, 101–102, 104, 514
pointType datatype, 99, 101–102
popAllTrailing( ) utility, 155
popup menus, 222, 235, 465–466, 502
ports, 362
position
    icons for files/folders, 296
    mouse cursor, detecting, 286
    windows, 215, 219
post method, 380
pow( ) utility, 162
PPCBrowser modal dialog, 225
precedence
    of arithmetic operations, 161
    datatype, 107
pre-evaluation, 92
prefix, pathname, 291

preflightSite( ) utility, 425
prettyOutline renderer, 402
previewing Web sites, 418
printDocument( ), 323
private scrap, 287
proc handlers
    for dialogs, 226–232
    in visit constructs, 125–128
processes
    local vs. remote, 308
    selecting (dialog for), 226
    verbs for, 288–289
processes, simultaneous (see threads)
#processMacros directive, 400
promoting outline lines, 182
properties, object, 315
property lists, 319
protocols, networking, 368
pseudo-hooks, 260
pseudo-random numbers, 162
punctuation, 151, 453–454
putting threads to sleep, 201

Q

QuarkXPress, 327, 330
querying operating system, 285–299
question mark (?) in URLs, 383
Quick Script window, 138
quit( ), 323
quote to delimit characters/strings, 109, 454

R

radio buttons
    in MacBird cards, 233
RAM
    determining how much free, 290
    setting partitions for applications, 295
random( ), 162, 459
rDelete( ) utility, 150
reading from files, 188–189, 191
readyMainWindow( ) utility, 224
recalculation scripts, 233
record( ), 97
records
    array notation with, 113
    coercing, 104
    empty, 117
    iterating through with for...in, 120

records (*continued*)
  operations on, 115–116
  .recordType datatype, 99
  (see also lists)
rect( ), 97
rectangle.get( ), 102, 515
rectangle.inset( ), 163, 515
rectangle.outset( ), 163, 515
rectangle.random( ), 163, 515
rectangle.set( ), 102, 515
rectangles (rects), 99, 101–102, 515
  coercing, 104
rectType datatype, 99, 101–102
recursion, 81
redirection (CGI), 382
refCon, 186
references to database objects (see object
          references)
regex suite, 153–155, 516
regex.extract( ), 154, 516
regex.join( ), 154, 516
regex.split( ), 154, 516
regex.subst( ), 154, 516
relative hyperlinks, 416–418, 425
relative URLs, making absolute, 159
releasing Web sites, 418–420
remainder (%) operator, 161, 450
remote communications with
          Frontier, 371–374
remote processes, 308
remote server volumes, mounting, 292
remote sites, releasing Web pages to, 419
remoteGet( ), odbServer for
          (example), 366–367
removing (see deleting)
rendering engine, 390
  directives for, 399
  outline renderers, 401–404
rendering Web sites, 390
  previewing and releasing, 418–420
#renderOutlineWith directive, 400, 401–402
reorganizing (see order)
replace( ), 318
replacing (see finding and replacing
          strings)
required event suite, 312, 560–561
required.openApplication( ), 312, 560
required.openDocument( ), 312, 561
required.printDocument( ), 312, 561

required.quitApplication( ), 312, 561
reserved words (UserTalk), 35, 92
resolving object references (see object
          references)
resource-based dialogs, 226–232
resources, 193–197
  attributes of, 197
  copying resource forks, 296
  file-based, 231–232
  launching, 292
  resource chain, 196
  verbs for managing, 193–197
resources, sound, 289
result values (verbs), 52–53
return character (\r), 109
return keyword, 53
rez verbs, 193–195, 517–520
rez.countResources( ), 193, 517
rez.countResTypes( ), 193, 517
rez.deleteNamedResource( ), 195, 517
rez.deleteResource( ), 195, 517
rez.getNamedResource( ), 194, 517
rez.getNthResInfo( ), 194, 518
rez.getNthResource( ), 194, 518
rez.getNthResType( ), 193, 518
rez.getResource( ), 194, 518
rez.getResourceAttributes( ), 197, 518
rez.getStringResource( ), 195, 519
rez.namedResourceExists( ), 195, 519
rez.putNamedResource( ), 194, 519
rez.putResource( ), 194, 519
rez.putStringResource( ), 195, 519
rez.resourceExists( ), 195, 520
rez.setResourceAttributes( ), 197, 520
rgb( ), 97
rgb.get( ), 102, 520
rgb.set( ), 102, 520
rgbs (RGB color sets), 99, 101–102
  coercing, 104
  converting hex triples, 102
rgbType datatype, 99, 101–102
rightN( ) utility, 150
rIndex( ) utility, 114
rollBeachball( ), 289, 459
root database, resolving references in, 62
Rosenthol, Leonard, 246
rulers in wptexts, 187
Run button (edit window), 138
Run Selection (Main menu item), 53, 139

running scripts, 138–140
runtime coercion, 108
runtime errors (see errors)

S

samples.basicStuff entries
    buildFolder( ), 371
    buildFolderOutline( ), 124
    buildTableOutline( ), 181
    createNewSuite( ), 251
    megabyte( ), 162
    nightlyBackup( ), 298
    quarkDocServer entry, 327
    reallyEmptyTrash( ), 131
    resourceMap( ), 194
    resourceStealer( ), 196
    rimshot( ), 289
    runClipboard( ), 94
    windowFormatter entry, 216
    windowFormatter( ), 138, 222
samples.NewIn96.add2lines( ), 191
saving
    copy of database, 274
    databases, 281
scalars, 97
    coercing, 104
    editing in external editors, 350–352
Scheduler suite, 358–361
scheduler.addTask( ), 361
scheduler.monitor( ), 359
schedulerMonitor agent, 255, 257
scheduling, 358–361
    missed tasks, 360
    user.scheduler entries, 358–361
scope
    dynamic, 78–79
    of handlers, 74–76
    implicit local, 77–78
    local, 69–72
    of variables, 73–74
scratchpad entry, 30
screenSize( ) utility, 220
script-based droplets, 270–271
Script Debugger, 348–349
Script Editor format, 269
script verbs, 183, 520–522
script.clearBreakpoint( ), 183, 520
script.compile( ), 138, 520
script.getBreakpoint( ), 183, 521

script.getLanguage( ), 138, 521
script.isComment( ), 183, 521
script.makeComment( ), 183, 521
script.removeSource( ), 138, 521
script.setBreakpoint( ), 183, 521
script.setLanguage( ), 138, 522
script.unComment( ), 183, 522
script.uncompile( ), 138, 522
scriptError( ), 132–133, 378, 460
scripts, 11–12
    aborting with Command-period, 144
    agent (see agents)
    AppleScript (see AppleScript)
    breakpoint and comment verbs, 183
    CGI, 377–378, 386
    coercing to strings, 103
    comments in (see comments)
    compiling, 135–138
    as database entries, 37–39
    debugging (see debugging)
    desktop scripts, 140, 264, 268–269
    errors (see errors)
    extensions for, 441–448
    hooks (see hooks)
    HyperCard, 338–340
    indentation (subordination) in, 40–41
    interrupting execution with hooks, 259
    as menu items, 244–245, 253
    non-outline rendering of, 41–43
    OSA, 337–340
    OSAX extension, 207, 334–335
    as outlines, 39–44
    running, 138–140
    scheduling (see Scheduler suite)
    scriptType datatype, 97
    suites, 250–252
    testing, 51–52, 256
    testing if menu item scripts, 222
    verbs for script objects, 138
    as Web page object datatype, 406
    whitespace in, 453
    wptext verbs for, 178–186
    XCMD and UCMD extensions, 207–209
scrolling windows, 221
SDK, Frontier, 441
search paths, 62–63, 65–66
searching
    CGI script for (example), 383–385
    database, 29

searching (*continued*)
  within files, 191
  finding/replacing strings, 152–155
  grep search and replace, 153–155
  to resolve object references, 62–66
seconds (see time)
secret outline line values, 186
security
  CGIs and, 386
  imported packed objects, 267
selection in wptext situations, 179
selection mode
  outline edit windows, 15–16, 17–18
  table edit windows, 21–22
semaphores, 202–204, 522
semaphores.lock( ), 203–204, 522
semaphores.unlock( ), 203–204, 522
semicolon (;) ending script lines, 41–43,
    454
sentences, verb for, 157
separator lines (in menus), 243
servers, 362–367
  advice for running, 386–387
Set Command Key (edit window), 244
setBinaryType( ), 101, 460
setEventInteraction( ), 311, 460
setEventTimeout( ), 310, 460
setEventTransactionID( ), 309, 460
setObj( ), 305
sets of shorts, 99, 101–102
  coercing, 104
shared menus, 241, 246–249
  keyboard shortcuts, 244
Shift key (see modifier keys)
shortcuts, keyboard, 244
showing (see hiding)
siblings (outline edit window), 16
Simmons, Brent, 327, 441, 446
single quote (') to delimit
    chars/string4s, 109, 454
single( ), 96
singleType datatype, 97
#siteDefaultName directive, 411
#siteOutlineHeadFont directive, 411
siteOutliner renderer, 402
#siteOutlineSubtextFont directive, 411
size
  arrays (number of items in), 114
  file size, 190, 294

ruler length, 187
setting application RAM partitions, 295
string length, 150
table columns, 24
text (see fonts)
volumes, 293
windows, 215, 219
sizeOf( ), 91–92, 114, 295, 461
slash (/) for division, 106, 161, 450
sleeping threads, 201
smart quotes (", "), 109, 454
Software Developer's Kit, Frontier, 441
sorting (see order)
sortList( ) utility, 186
sound, verbs for, 289
space in UserTalk code, 453
spacing of text lines, 187
speaker.beep( ), 289, 522
speaker.ouch( ), 289, 523
speaker.playNamedSound( ), 289, 523
speaker.sound( ), 289, 523
special characters, escaping, 109
special forms, 91
specifiers, array, 112
splitting outline lines, 184
square brackets [ ], 454
  in array notation, 112
  to reference database objects, 60
square-wave sounds, 289
stable menus, 241
stack.create( ), 205, 523
stack.dispose( ), 206, 523
stack.pop( ), 205, 523
stack.push( ), 205, 524
stack.visit( ), 126, 524
stacks, 143, 205–206
  suites.stacks entry, 205
  system.compiler.stack entry, 143
standardMacros.embeddedUserTalk( ), 412
standardMacros.glossaryPatcher( ), 416
standardMacros.imageRef( ), 410
standardMacros.linkPrev( ), 411
standardMacros.outlineSite( ), 411
standardMacros.pageFooter( ), 398, 405
standardMacros.pageHeader( ), 396, 405
standardMacros.renderObject( ), 410
standardMacros.spacePixels( ), 411
startup volume, 294

status messages, 27, 199, 223–224, 250
    msg( ), 73, 224, 257, 458
    StatusMessage agent, 257
    window.msg( ), 224, 545
Step button (Debug mode), 142
"stop" icon (see breakpoints)
stopDialog dialog (example), 227–229
stopDialog2 dialog (example), 235–236
stopping agents, 256
'STR ' resources, 195
'STR#' resources, 195
streams, 362
string verbs, 149–153, 155–159, 162–167,
    524–530
string( ), 96
string.addCommas( ), 158, 524
string.commentDelete( ), 156, 524
string.countFields( ), 156, 524
string.countWords( ), 157, 524
string.dateString( ), 167, 525
string.delete( ), 150, 525
string.filledString( ), 155, 525
string.firstSentence( ), 157, 525
string.firstWord( ), 157, 525
string.getWordChar( ), 156, 525
string.hex( ), 162, 526
string.insert( ), 149, 526
string.isAlpha( ), 526
string.isNumeric( ), 151, 526
string.iso8859encode( ), 158, 526
string.isPunctuation( ), 151, 526
string.kBytes( ), 162, 527
string.lastWord( ), 157, 527
string.length( ), 150
string.lower( ), 151, 527
string.megabyteString( ), 162
string.memAvailString( ), 285, 527
string.mid( ), 150, 527
string.nthChar( ), 150, 528
string.nthField( ), 156, 528
string.nthWord( ), 157, 528
string.parseHTTPargs( ), 159, 385, 528
string.patternMatch( ), 152, 528
string.popLeading( ), 155, 528
string.popTrailing( ), 155, 529
string.processHtmlMacros( ), 422
string.replace( ), 529
string.replaceAll( ), 529
string.setWordChar( ), 156, 529

string.timeString( ), 167, 529
string.upper( ), 151, 529
string.URLdecode( ), 159, 530
string.URLencode( ), 159, 530
string4( ), 96
string4Type datatype, 98, 104
strings, 109–111
    array notation with, 113
    case (capitalization) of, 151, 157
    chars (see characters)
    coercing, 104
    coercing lists to (bug), 104
    coercing non-scalars to, 103
    comparing, 453
    dates as (see date and time)
    directives as, 394
    empty, 117
    escaping special characters, 109
    hexadecimal numbers, 162
    operations on, 115–116, 149–160
        HTML-related, 158–160
        parsing, 155–158
        working with substrings, 116,
            149–150
    searching for (see searching)
    stringType datatype, 98
    as Web page object datatype, 406
stripTestFailers( ) utility, 152
strListToTable( ) utility, 195
stubs, 51–52
subheads, outline, 16, 182
submenus (see menus)
subordination in UserTalk scripts, 40–41
substrings (see strings)
subtables, 20
subtraction (–) operator, 161, 449
suites, 241, 250–252, 311–314
suites entry, 30, 250
suites.backups.lastfile entry, 273
suites.batchExporter.batchImport( ), 276
suites.classAds entry, 433–437
suites.docs.verbList entry, 145
suites.export entry, 267–268, 272
suites.fileMakerLib entry, 327
suites.html entry, 421–423
suites.inetd entry, 363–365
suites.newsPage entry, 431–433
suites.odbEditor entries, 348–352
suites.odbServer entry, 365–367

suites.outliner entry, 186
suites.samples.basicStuff (see
        samples.basicStuff entries)
suites.stacks entry, 205
suites.todo entries, 357
suites.webBrowser entry, 257, 345
suites.webserverScripts.samples.tellParams
        entry, 378
summit-level lines, 16
Swift, Tom, 109
switchTextEditor( ) utility, 349
sys verbs, 288–290, 295–296, 530–533
sys.appIsRunning( ), 288, 323, 530
sys.bringAppToFront( ), 289, 323, 530
sys.browseNetwork( ), 226, 530
sys.countApps( ), 288, 531
sys.frontmostApp( ), 288, 531
sys.getAppPath( ), 289, 531
sys.getAppSize( ), 295, 531
sys.getMinAppSize( ), 295, 531
sys.getNthApp( ), 289, 531
sys.hasBundle( ), 296, 532
sys.memAvail( ), 290, 532
sys.OSVersion( ), 290, 532
sys.setAppSize( ), 295, 532
sys.setBundle( ), 532
sys.setBundles( ), 296
sys.setMinAppSize( ), 295, 532
sys.systemTask( ), 198, 533
syscrash( ), 133–134, 461
system
    clock (see date and time)
    querying, verbs for, 285–299
    scrap, 287
    version of, 290
system entry, 31
system folder, 294, 297
system.agents entry, 31, 255
system.compiler entry, 31, 135
system.compiler.kernel entry, 136
system.compiler.language entry, 136
system.compiler.semaphores entry, 204
system.compiler.stack entry, 143
system.deskscripts entry, 268
system.droplet.closedown entry, 270
system.droplet.fileList entry, 271
system.droplet.path entry, 270
system.droplet.startup entry, 270
system.extensions entry, 31, 207

system.macintosh entry, 311
system.macintosh.constants entry, 311
system.macintosh.core entry, 312
system.macintosh.misc entry, 313
system.macintosh.objectModel entry, 311
system.menubars entry, 31, 246
system.misc.closeWindow( ), 215
system.misc.cmd2click entry, 259
system.misc.control2click entry, 259
system.misc.menubar entry, 31, 242
system.misc.option2click entry, 259
system.resume entry, 31
system.shutdown entry, 31
system.startup entry, 31
system.suspend entry, 31
system.verbs.apps entry, 31, 314, 319
system.verbs.apps.netEvents entry, 362
system.verbs.builtins.app entry, 314
system.verbs.colors entry, 102
system.verbs.constants.cr entry, 65
system.verbs.traps entries, 340–341
system.verbs.traps.WWWΩ.sdoc entry, 376
systemEvent( ), 308, 461
system-level messages (see Apple events)

T

tab character (\t), 109
table edit windows, 19–24
table entries, 19–21
    creating new, 22–23, 170
    cutting, copying, pasting, 21
    destroying all, 171
    exporting as packed objects, 266
    names for, 28, 170–172
    operations on, 170–172
    order of, 24
    sorting, 180
    wptext verbs for, 180
table verbs, 170–172, 180, 533–536
table.assign( ), 170, 533
table.copy( ), 172, 533
table.copyContents( ), 172, 533
table.emptyTable( ), 171, 533
table.getCursor( ), 180, 533
table.go( ), 180, 534
table.goto( ), 180, 534
table.gotoAddress( ), 175, 534
table.gotoName( ), 180, 534
table.move( ), 172, 534

table.moveAndRename( ), 172, 535
table.moveContents( ), 172, 535
table.promptNewItem( ), 225, 535
table.rename( ), 172, 535
table.sortBy( ), 180, 535
table.surePath( ), 170, 536
table.uniqueName( ), 170, 536
table.validate( ), 274, 536
table.visit( ), 127, 536
tableEvent( ), 308, 461
tableOutliner renderer, 403
tables
    checking consistency of, 274
    coercing to strings, 103
    coercion (see coercion)
    creating for database entry, 170
    deleting all entries of, 171
    dialog for database entry creation, 225
    droplet tables (see droplets)
    exporting as packed objects, 266
    glue tables (see glue)
    iterating through with table.visit( ), 127
    in MacBird card data, 234
    navigating within, 23–24
    sorting entries in, 180
    subtables, 20
    tableType datatype, 97
    as Web page object datatype, 406
    wptext verbs for, 178–180
tabs, verbs for, 187
#tagSubstitution directive, 400
target objects, 174–177
target.clear( ), 176, 536
target.get( ), 175, 537
target.set( ), 175, 218, 537
TCP/IP, 362–370
tcpCmd entries, 368–370
tcpCmd.upload( ), 369
TCPserver.listener( ), 373
#template directive, 404
#template.html files, 428
templates, 404–406
testEachChar( ) utility, 152
testing
    CGI scripts, 378
    script lines, 53–54
    scripts, 51–52, 256
    (see also debugging)
#text directive, 397

text
    double-clicking, hook for, 259
    manipulating in wptext situations, 179
    in pictures, 262–263
    processing (see wptext edit windows)
    size and style (see fonts)
    (see also characters; strings)
text editors for editing database
        objects, 347–353
#textFileCreator directive, 400
TextMachine application, 154
thread verbs, 199–201, 537–538
thread.evaluate( ), 199, 247, 537
thread.evaluateTo( ), 537
thread.exists( ), 201, 537
thread.getCount( ), 199
thread.getCurrentID( ), 201, 537
thread.getNthID( ), 201, 538
thread.isSleeping( ), 201, 538
thread.kill( ), 201, 538
thread.sleep( ), 201, 538
thread.sleepFor( ), 201, 538
thread.wake( ), 201, 538
#threadedRendering directives, 400
threads, 198–204
    agent scripts and, 255
    number of presently running, 199
    semaphores, 202–204
    shared menus and, 247
three-button modal dialogs, 225
time (see date and time)
timeCreated( ), 173, 461
timeModified( ), 173, 462
timeouts, Apple event, 310
titleCase( ) utility, 157
titles
    Main Window, 214
    for suites, 251
    #title directive, 396
    <title> pseudo-tags, 405
    window, 214, 221
ToDo List tool, 357
Toggle Breakpoint (Script menu item), 141
Toggle Comment (Script menu item), 44
tokenType datatype, 99
toys.alphaChar( ), 151, 539
toys.commentDelete( ), 156, 539
toys.dropNonAlphas( ), 151, 539
toys.emptyFolder( ), 297, 539

toys.getCursorAddress( ), 172, 539
toys.innerCaseName( ), 151, 540
toys.iso8859filter( ), 158
toys.listToOutline( ), 185, 540
toys.outlineToList( ), 185, 540
toys.padWithZeros( ), 155, 540
toys.popStringSuffix( ), 150, 540
toys.puncChar( ), 151, 541
toys.readFileIntoTextObject( ), 189, 541
toys.readWholeFile( ), 188, 541
toys.sureFilePath( ), 297, 541
toys.threadCall( ), 200, 541
toys.touchPath( ), 541
toys.uniqueTableName( ), 170, 542
toys.URLsplit( ), 159, 542
toys.writeWholeFile( ), 189, 542
transactionEvent( ), 309, 462
transactions, 309–310
trapping errors, 131–132
traverse( ) utility, 124, 181, 184
triangles in outline edit windows, 15–17
trigCmd verbs, 162, 542
trigonometric operations, 162
trimming strings, 155
tripletToRGB( ) utility, 102
troubleshooting
    address parameters, 89
    crashing system on purpose, 133–134
    imported packed objects, 267
    indefinite number of parameters, 57
    infinite looping constructs, 50
    shared menus, 249
    shortcut key assignment, 244
    textfiles of database objects, 349–350
true, false constants (see booleans)
try...else construct, 131–132
tryError variable, 133
tweaking glue, 321–322
two-button modal dialogs, 225
twoLevelOutline renderer, 402
typeOf( ), 95, 172, 462
types (see datatypes)

U

UCMD scripting extension, 207–209, 441,
    446–448
UCMDmain( ), 446
ucmds.getOutlineHtml( ), 403

unary increment/decrement
    operators, 161, 450
underlines, 187
unlocking (see locking and unlocking)
unmounting volumes, 293
unpack( ), 101, 462
updating pictures, interval for, 263
uploadAndRename( ) utility, 369
uppercase (see case)
#url directive, 408
URLs, 159–160
    (see also Web)
#useGlossPatcher directive, 400, 416
user entry, 30
user events, verbs for, 286
user interaction levels, 311
user.classAds.prefs.hoursTillArchive
    entry, 435
user.classAds.templates.catPage entry, 436
user.classAds.templates.newPageReturn
    entry, 436
user.classAds.templates.previewPage
    entry, 436
user.classAds.templates.singleAd entry, 436
user.classAds.website.#template entry, 436
user.droplets entries (see droplets)
user.hooks.closeWindow table, 214
user.html.glossary entry, 413, 429
user.html.lists.changedPages entry, 419
user.html.macros entry, 407
user.html.prefs.iso8859map entry, 158
user.html.renderers entry, 402–403
user.html.templates entry, 405
user.inetd.shutdown entry, 364
user.newsPage entry, 431–433
user.preferences.outlineFont entry, 216
user.preferences.outlineFontSize entry, 216
user.preferences.scriptFont entry, 216
user.preferences.scriptFontSize entries, 216
user.preferences.tableFont entry, 216
user.preferences.tableFontSize entry, 216
user.scheduler entries, 358–361
user.switchUser( ), 260
user.webserver.errorPage entry, 378
user.webserver.fileNotFoundPage
    entry, 378
user.webserver.mimeTypes entry, 382

user.webserver.utilities entry, 379, 381
    processMacros( ), 379
    showMatching( ), 384
    showThisRec( ), 385
user.websites entry, 391–393
user.websites.#filters entry, 392
user.websites.#ftpSite entry, 391
user.websites.glossary entry, 392
user.websites.images entry, 392
user.websites.#template entry, 392
user.websites.tools entry, 392
user-based pseudo-hooks, 260
users, switching, 260
UserTalk
    AppleScript and (see AppleScript)
    basic syntax of, 35–37
    constants, 93
    control structures, 118–134
    datatypes (see datatypes)
    documentation on, 144–145
    keywords (reserved words), 35, 92
    object model, 315–319
    scripting extensions for, 441–448
    scripts (see scripts)
    XCMD and UCMD scripting
        extensions, 207–209

V

variables, 35
    calculating names of, 60–62, 85
    consulting in Debug mode, 142
    converting datatypes of (see coercion)
    database entries vs., 58–59
    global, 79–80
    for iterations (see looping constructs)
    operations on, 169–173
    scope of, 73–74
    (see also datatypes; scope)
verbs, 35, 37–39
    address parameters to, 85–90
    addresses of, 88–89
    Apple events (see Apple events)
    call result values, 52–53
    classes of, 177–178
    names for, 37–38
    parameters of (see parameters)
    suites.docs.verbList entry, 145
    (see also under specific verb name)

version
    database, 285
    file, 295
    Frontier application, xix, 285
    operating system, 290
vertical bar (|)
    in glossPatch expression, 417
    in Script Debugger, 349
vertical bars (| |) as or operator, 451
visibility (see hiding)
visit constructs, 125–128
#vlink directive, 397
volumes
    counting files and folders on, 293
    dialog for selecting, 225
    iterating though files on, 122–125
    obtaining names of, 157
    startup, 294
    verbs for managing, 292

W

waiting for fixed intervals, 198, 469
Web
    BBEdit as front-end editor, 428–430
    browsers (see suites.webBrowser)
    CGI applications, 375–388
    datatypes for Web page objects, 406
    directives, 393–401, 408–409
    dynamic Web sites, 431–438
    filtering Web data, 414–416
    glossaries, 412–413
    hyperlinks, 416–418, 425, 427
    loading existing sites, 420
    macros for, 407–412
    previewing and releasing sites, 418–420
    running Web server with
        Frontier, 386–387
    site management, 389–391
    site tables (user.websites), 391–393
    suites.webserverScripts entry, 377
    templates, 404–406
    user.websites (see entries under
        user.websites)
    utility scripts for, 424–428
Web pages, driving Frontier from, 345
Web servers, 364
webBrowser (suites.webBrowser), 257, 345
WebServer submenu (Suites menu), 376

webServer.data.agentScript( ), 387
webServer.httpHeader( ), 377, 382
Websites folder, 418
while looping construct, 121
whitespace, 453
willBeImplicitLocal( ) utility, 171
Window menu, 214
window verbs, 216–222
window.bringToFront( ), 221, 543
window.close( ), 176, 218, 543
window.dbStats( ), 274, 543
window.frontmost( ), 175, 220, 543
window.getPosition( ), 219, 544
window.getSize( ), 219, 544
window.getTitle( ), 221, 544
window.hide( ), 218, 544
window.isFront( ), 221, 544
window.isHidden( ), 219, 544
window.isMenuScript( ), 222, 545
window.isModified( ), 173, 545
window.isOpen( ), 218, 545
window.isVisible( ), 219, 545
window.msg( ), 224, 545
window.next( ), 220, 546
window.open( ), 175, 216, 546
window.quickScript( ), 546
window.scroll( ), 221, 546
window.sendToBack( ), 221, 546
window.setModified( ), 173, 546
window.setPosition( ), 219, 547
window.setSize( ), 219, 547
window.setTitle( ), 221, 547
window.show( ), 218, 547
window.update( ), 222
window.visit( ), 125–126, 547
window.zoom( ), 220, 548
windows
    contents (see non-scalars)
    dialog (see dialogs)
    fonts in, 15, 216
    formatting, 215–216
    layering, 220–221
    manipulating, 213–216
    opening, closing, hiding, 23, 214–215,
        218–219
    samples.basicStuff.windowFormatter
        entry, 216
    status (see status messages)
    titles of, 214, 221

windoids vs., 223
window references, 217
    (see also edit windows)
Winer, Dave, xx
with statement, 66–68, 76–77, 331
    object containment and, 316–317
wordInfo.getNthWordOffset( ), 553
word-processing text (see wptext edit
        windows)
words, 156–157, 179
workspace entry, 29
workspace.notepad entry, 30
wp verbs, 179, 187, 548–553
wp.clearTabs( ), 187, 548
wp.getDisplay( ), 176, 548
wp.getIndent( ), 187, 548
wp.getLeftMargin( ), 187, 548
wp.getRightMargin( ), 187, 549
wp.getRuler( ), 187, 549
wp.getSelect( ), 179, 549
wp.getSelText( ), 179, 549
wp.getText( ), 179, 549
wp.go( ), 179, 550
wp.insert( ), 179, 550
wp.inTextMode( ), 178, 550
wp.rulerLength( ), 187, 550
wp.selectAll( ), 179
wp.selectLine( ), 179, 550
wp.selectParagraph( ), 179, 551
wp.selectWord( ), 179, 551
wp.setDisplay( ), 176, 551
wp.setIndent( ), 187, 551
wp.setJustification( ), 187, 551
wp.setLeftMargin( ), 187, 552
wp.setRightMargin( ), 187, 552
wp.setRuler( ), 187, 552
wp.setSelect( ), 179, 552
wp.setSpacing( ), 187, 552
wp.setTab( ), 187, 553
wp.setText( ), 179, 553
wp.setTextMode( ), 178, 553
wptext edit windows, 19
wptext objects
    defining directives in, 393
    editing in Nisus Writer, 353
    reading data into, 189
    templates, 404–406
    verbs for (see wp verbs)
    as Web page object datatype, 406

wpTextType datatype, 97
wpToNisus( ) utility, 353
writeBackgroundImageFile( ) utility, 426
writing scripts, 51–52
writing to files, 189, 191
'WWWΩ'/'sdoc' events, 376

X

XCMD scripting extension, 207–209,
    441–445
  importing, 196, 209
XCmdBlock structure, 442
XCmdPtr pointer, 442

Y

years, 166
yes-no, yes-no-cancel modal dialogs, 225
yielding to processes, 198

Z

zeros, attaching to numbers, 155
zooming outline focus, 182
zooming windows, 215, 220

About the Author

Matt Neuburg started programming computers in 1968, when he was 14 years old, as a member of a literally underground high school club, which met once a week to do timesharing on a bank of PDP-10s by way of primitive teletype machines. He also occasionally used Princeton University's IBM-360/67, but gave it up in frustration when one day he dropped his punch cards. He majored in Greek at Swarthmore College, and received his Ph.D. from Cornell University in 1981, writing his doctoral dissertation (about Aeschylus) on a mainframe. He proceeded to teach Classical languages, literature, and culture at many well-known institutions of higher learning, most of which now disavow knowledge of his existence, and to publish numerous scholarly articles unlikely to interest anyone. Meanwhile he obtained an Apple IIc and became hopelessly hooked on computers again, migrating to a Macintosh in 1990. He wrote some educational and utility freeware, became an early regular contributor to the online journal *TidBITS*, and in 1995 left academe to edit *MacTech Magazine*. In August 1996 he became a freelancer, which means he has been looking for work ever since.

Colophon

Our look is the result of reader comments, our own experimentation, and feedback from distribution channels. Distinctive covers complement our distinctive approach to technical topics, breathing personality and life into potentially dry subjects.

The animal appearing on the cover of *Frontier: The Definitive Guide* is popularly known as an American buffalo, correctly called a bison. Bison are the largest mammals in North America: mature males stand about 6.5 feet tall at the shoulder, and weight almost a ton. They generally live in herds, which vary in size and movement. Bison are usually dark brown, and their front half is overdeveloped, with especially pronounced withers and a thick growth of long, dark hair. Bulls fight over cows during mating season (late summer), and have a generally unpredictable temperament, ranging from quiet to fierce. Born after a gestation period of nine months, the young nurse for about a year, and mature after two or three years. A bison's lifespan is typically 20–25 years, with some living 10 or 15 years longer than average.

Bison were revered by the Plains tribe of Native Americans, who put every piece of them to practical and/or ceremonial use, including hide, hair, flesh, bones, organs, horns, hooves, excrement, and fat. Held especially sacred was the extremely rare white bison, appearing only once in several million. The Plains society was intertwined with and entirely dependent upon the bison.

By most estimates, there were approximately 60 million bison in North America in the early eighteenth century. Intent on settling the American West and overtaking the Native American civilization, white settlers very nearly brought about the complete extinction of bison by the late nineteenth century through methodical hunting for food and sport. Some of the most extreme incidents were linked with the construction and journeys of the Union Pacific Railroad in the 1860s, first through the killing of huge numbers of bison to supply workers with food, and later through passengers' sport of shooting as many bison as they pleased from the train and taking only the tongue as a delicacy, leaving piles of carcasses along the tracks.

By the mid-1880s, white settlers had effectively wiped out the bison population, reduced from their former millions to well under a thousand; only the efforts of a few cattlemen and conservationists in the early twentieth century saved them from extinction. Today, bison herds in wildlife preserves and other ranges are sufficient to ensure survival of the species.

Edie Freedman designed the cover of this book, using a 19th-century engraving from the Dover Pictorial Archive. The cover layout was produced with Quark XPress 3.32 using the ITC Garamond font. Whenever possible, our books use RepKover™, a durable and flexible lay-flat binding. If the page count exceeds RepKover's limit, perfect binding is used.

The inside layout was designed by Nancy Priest and implemented in FrameMaker 5.0 by Mike Sierra. The text and heading fonts are ITC Garamond Light and Garamond Book. The illustrations that appear in the book were created in Adobe Photoshop 4 by Robert Romano. This colophon was written by Nancy Wolfe Kotary.

More Titles from O'Reilly

Developing Web Content

WebMaster in a Nutshell, Deluxe Edition

By O'Reilly & Associates, Inc.
1st Edition September 1997
374 pages, includes CD-ROM & book
ISBN 1-56592-305-7

The Deluxe Edition of *WebMaster in a Nutshell* is a complete library for web programmers. The main resource is the Web Developer's Library, a CD-ROM containing the electronic text of five popular O'Reilly titles: *HTML: The Definitive Guide, 2nd Edition*; *JavaScript: The Definitive Guide, 2nd Edition*; *CGI Programming on the World Wide Web*; *Programming Perl, 2nd Edition*—the classic "camel book," written by Larry Wall (the inventor of Perl) with Tom Christiansen and Randal Schwartz; and *WebMaster in a Nutshell*. The Deluxe Edition also includes a printed copy of *WebMaster in a Nutshell*.

WebMaster in a Nutshell, Deluxe Edition, makes it easy to find the information you need with all of the convenience you'd expect from the Web. You'll have access to information webmasters and programmers use most for development—complete with global searching and a master index to all five volumes—all on a single CD-ROM. It's incredibly portable. Just slip it into your laptop case as you commute or take off on your next trip and you'll find everything at your fingertips with no books to carry.

System requirements: The CD-ROM is readable on all Windows and UNIX platforms. Current implementations of the Java Virtual Machine for the Mac platform do not support the Java search applet in this CD-ROM. Mac users can purchase the World Wide Web version (see www.oreilly.com/books/javaref/ for more information). A web browser that supports HTML 3.2, Java, and Javascript, such as Netscape 3.0 or Internet Explorer 3.0, is required.

HTML: The Definitive Guide, 2nd Edition

By Chuck Musciano & Bill Kennedy
2nd Edition May 1997
552 pages, ISBN 1-56592-235-2

This complete guide is chock full of examples, sample code, and practical, hands-on advice to help you create truly effective web pages and master advanced features. Learn how to insert images and other multimedia elements, create useful links and searchable documents, use Netscape extensions, design great forms, and lots more. The second edition covers the most up-to-date version of the HTML standard (HTML version 3.2), Netscape 4.0 and Internet Explorer 3.0, plus all the common extensions.

Frontier: The Definitive Guide

By Matt Neuburg
1st Edition February 1998
616 pages, 1-56592-383-9

This definitive guide is the first book devoted exclusively to teaching and documenting Userland Frontier, a powerful scripting environment for web site management and system level scripting. Packed with examples, advice, tricks, and tips, *Frontier: The Definitive Guide* teaches you Frontier from the ground up. Learn how to automate repetitive processes, control remote computers across a network, beef up your web site by generating hundreds of related web pages automatically, and more.

Whether you're a complete Frontier beginner or an experienced hand in need of a clear, ordered reference, you'll find what you need to make the most of Frontier. Even if you've never programmed before, you'll be cranking out web pages or writing your own custom scripts in no time, joining the ranks of thousands of Macintosh power users who make Frontier their "home base." (Covers Frontier 4.2.3 for the Macintosh.)

This book covers Frontier's:

- Simple but sophisticated multithreaded scripting language (UserTalk) and its elegant debugging environment
- Integrated database with instant access to data, text, outlines, and tables
- Totally automated environment, where scripts can create dialogs, open windows, edit text, alter menus
- Hook to the System and ability to read and write files, open documents, read the clipboard
- Base for sending and receiving Apple events, so it can drive any scriptable application and can be integrated into otherscripting applications
- Network application and ability to function as a client or server over AppleTalk or the Internet

WebMaster in a Nutshell

By Stephen Spainhour & Valerie Quercia
1st Edition October 1996
374 pages, ISBN 1-56592-229-8

Web content providers and administrators have many sources for information, both in print and online. *WebMaster in a Nutshell* puts it all together in one slim volume for easy desktop access. This quick reference covers HTML, CGI, JavaScript, Perl, HTTP, and server configuration.

O'REILLY™

TO ORDER: **800-998-9938** • **order@oreilly.com** • **http://www.oreilly.com/**
OUR PRODUCTS ARE AVAILABLE AT A BOOKSTORE OR SOFTWARE STORE NEAR YOU.
FOR INFORMATION: **800-998-9938** • **707-829-0515** • **info@oreilly.com**

Developing Web Content

CGI Programming on the World Wide Web

By Shishir Gundavaram
1st Edition March 1996
450 pages, ISBN 1-56592-168-2

This book offers a comprehensive explanation of CGI and related techniques for people who hold on to the dream of providing their own information servers on the Web. It starts at the beginning, explaining the value of CGI and how it works, then moves swiftly into the subtle details of programming.

Information Architecture for the World Wide Web

By Louis Rosenfeld & Peter Morville
1st Edition January 1998
224 pages, ISBN 1-56592-282-4

Learn how to merge aesthetics and mechanics to design web sites that "work." This book shows how to apply principles of architecture and library science to design cohesive web sites and intranets that are easy to use, manage, and expand. Covers building complex sites, hierarchy design and organization, and techniques to make your site easier to search. For webmasters, designers, and administrators.

Learning VBScript

By Paul Lomax
1st Edition July 1997
616 pages, includes CD-ROM
ISBN 1-56592-247-6

This definitive guide shows web developers how to take full advantage of client-side scripting with the VBScript language. In addition to basic language features, it covers the Internet Explorer object model and discusses techniques for client-side scripting, like adding ActiveX controls to a web page or validating data before sending it to the server. Includes CD-ROM with over 170 code samples.

Web Client Programming with Perl

By Clinton Wong
1st Edition March 1997
228 pages, ISBN 1-56592-214-X

Web Client Programming with Perl shows you how to extend scripting skills to the Web. This book teaches you the basics of how browsers communicate with servers and how to write your own customized web clients to automate common tasks. It is intended for those who are motivated to develop software that offers a more flexible and dynamic response than a standard web browser.

JavaScript: The Definitive Guide, 3rd Edition

By David Flanagan & Dan Shafer
3rd Edition June 1998
728 pages, ISBN 1-56592-392-8

This third edition of the definitive reference to JavaScript covers the latest version of the language, JavaScript 1.2, as supported by Netscape Navigator 4.0. JavaScript, which is being standardized under the name ECMAScript, is a scripting language that can be embedded directly in HTML to give web pages programming-language capabilities.

Designing for the Web: Getting Started in a New Medium

By Jennifer Niederst
with Edie Freedman
1st Edition April 1996
180 pages, ISBN 1-56592-165-8

Designing for the Web gives you the basics you need to hit the ground running. Although geared toward designers, it covers information and techniques useful to anyone who wants to put graphics online. It explains how to work with HTML documents from a designer's point of view, outlines special problems with presenting information online, and walks through incorporating images into web pages, with emphasis on resolution and improving efficiency.

O'REILLY™

TO ORDER: **800-998-9938** • **order@oreilly.com** • **http://www.oreilly.com/**

OUR PRODUCTS ARE AVAILABLE AT A BOOKSTORE OR SOFTWARE STORE NEAR YOU.

FOR INFORMATION: **800-998-9938** • **707-829-0515** • **info@oreilly.com**

Web Review Studio Series

Designing with JavaScript

By Nick Heinle
1st Edition September 1997
256 pages, Includes CD-ROM
ISBN 1-56592-300-6

Written by the author of the "JavaScript Tip of the Week" web site, this new Web Review Studio book focuses on the most useful and applicable scripts for making truly interactive, engaging web sites. You'll not only have quick access to the scripts you need, you'll finally understand why the scripts work, how to alter the scripts to get the effects you want, and, ultimately, how to write your own groundbreaking scripts from scratch.

GIF Animation Studio

By Richard Koman
1st Edition October 1996
184 pages, Includes CD-ROM
ISBN 1-56592-230-1

GIF animation is bringing the Web to life—without plug-ins, Java programming, or expensive authoring tools. This book details the major GIF animation programs, profiles work by leading designers (including John Hersey, Razorfish, Henrik Drescher, and Erik Josowitz), and documents advanced animation techniques. A CD-ROM includes freeware and shareware authoring programs, demo versions of commercial software, and the actual animation files described in the book. *GIF Animation Studio* is the first release in the new Web Review Studio series.

Shockwave Studio

By Bob Schmitt
1st Edition March 1997
200 pages, Includes CD-ROM
ISBN 1-56592-231-X

This book, the second title in the new Web Review Studio series, shows how to create compelling and functional Shockwave movies for web sites. The author focuses on actual Shockwave movies, showing how the movies were created. The book takes users from creating simple time-based Shockwave animations through writing complex logical operations that take full advantage of Director's power. The CD-ROM includes a demo version of Director and other software sample files.

Designing Web Graphics with Photoshop

By Mikkel Aaland
1st Edition March 1998 (est.)
264 pages (est.), ISBN 1-56592-350-2

While Adobe Photoshop is the graphics tool of choice for web graphics, its full potential is rarely tapped. Mikkel Aaland has interviewed dozens of top web designers and distilled their best tips and techniques for creating great effects in Photoshop in the smallest possible GIF and JPEG files. Besides sharing techniques from designers at clnet, HotWired, Discovery Channel, Second Story, and others, Aaland explains how to set up Photoshop for web work, how to process photographs, how to work with vector graphics in Photoshop, and how to use Photoshop as a layout tool. While this book is accessible to beginning users, it provides advanced explanations and techniques.

Web Navigation: Designing the User Experience

By Jennifer Fleming
1st Edition April 1998 (est.)
250 pages (est.), Includes CD-ROM
ISBN 1-56592-351-0

Web Navigation: Designing the User Experience offers the first in-depth look at designing web site navigation. Through case studies and designer interviews, a variety of approaches to navigation issues are explored. The book focuses on designing by purpose, with chapters on entertainment, shopping, identity, learning, information, and community sites. The accompanying CD-ROM includes a tour of selected sites, a "netography," and trial versions of popular software tools.

O'REILLY™

TO ORDER: **800-998-9938** • *order@oreilly.com* • *http://www.oreilly.com/*
OUR PRODUCTS ARE AVAILABLE AT A BOOKSTORE OR SOFTWARE STORE NEAR YOU.
FOR INFORMATION: **800-998-9938** • **707-829-0515** • *info@oreilly.com*

How to stay in touch with O'Reilly

1. Visit Our Award-Winning Web Site

http://www.oreilly.com/

★ "Top 100 Sites on the Web" —*PC Magazine*
★ "Top 5% Web sites" —*Point Communications*
★ "3-Star site" —*The McKinley Group*

Our web site contains a library of comprehensiveproduct information (including book excerpts and tables of contents), downloadable software, background articles, interviews with technology leaders, links to relevant sites, book cover art, and more. File us in your Bookmarks or Hotlist!

2. Join Our Email Mailing Lists

New Product Releases
To receive automatic email with brief descriptions of all new O'Reilly products as they are released, send email to:
listproc@online.oreilly.com
Put the following information in the first line of your message (*not* in the Subject field):
subscribe oreilly-news

O'Reilly Events
If you'd also like us to send information about trade show events, special promotions, and other O'Reilly events, send email to:
listproc@online.oreilly.com
Put the following information in the first line of your message (*not* in the Subject field):
subscribe oreilly-events

3. Get Examples from Our Books via FTP

There are two ways to access an archive of example files from our books:

Regular FTP
- ftp to:
  ftp.oreilly.com
  (login: anonymous
  password: your email address)
- Point your web browser to:
  ftp://ftp.oreilly.com/

FTPMAIL
- Send an email message to:
  ftpmail@online.oreilly.com
  (Write "help" in the message body)

4. Contact Us via Email

order@oreilly.com
To place a book or software order online. Good for North American and international customers.

subscriptions@oreilly.com
To place an order for any of our newsletters or periodicals.

books@oreilly.com
General questions about any of our books.

software@oreilly.com
For general questions and product information about our software. Check out O'Reilly Software Online at **http://software.oreilly.com/** for software and technical support information. Registered O'Reilly software users send your questions to: **website-support@oreilly.com**

cs@oreilly.com
For answers to problems regarding your order or our products.

booktech@oreilly.com
For book content technical questions or corrections.

proposals@oreilly.com
To submit new book or software proposals to our editors and product managers.

international@oreilly.com
For information about our international distributors or translation queries. For a list of our distributors outside of North America check out:
http://www.oreilly.com/www/order/country.html

O'Reilly & Associates, Inc.
101 Morris Street, Sebastopol, CA 95472 USA
TEL   707-829-0515 or 800-998-9938
        (6am to 5pm PST)
FAX   707-829-0104

Titles from O'Reilly

Please note that upcoming titles are displayed in italic.

WEBPROGRAMMING

Apache: The Definitive Guide
Building Your Own Web Conferences
Building Your Own Website
CGI Programming for the World Wide Web
Designing for the Web
HTML: The Definitive Guide, 2nd Ed.
JavaScript: The Definitive Guide, 2nd Ed.
Learning Perl
Programming Perl, 2nd Ed.
Mastering Regular Expressions
WebMaster in a Nutshell
Web Security & Commerce
Web Client Programming with Perl
World Wide Web Journal

USING THE INTERNET

Smileys
The Future Does Not Compute
The Whole Internet User's Guide & Catalog
The Whole Internet for Win 95
Using Email Effectively
Bandits on the Information Superhighway

JAVA SERIES

Exploring Java
Java AWT Reference
Java Fundamental Classes Reference
Java in a Nutshell
Java Language Reference, 2nd Edition
Java Network Programming
Java Threads
Java Virtual Machine

SOFTWARE

WebSite™ 1.1
WebSite Professional™
Building Your Own Web Conferences
WebBoard™
PolyForm™
Statisphere™

SONGLINE GUIDES

NetActivism   NetResearch
Net Law   NetSuccess
NetLearning NetTravel
Net Lessons

SYSTEM ADMINISTRATION

Building Internet Firewalls
Computer Crime: A Crimefighter's Handbook
Computer Security Basics
DNS and BIND, 2nd Ed.
Essential System Administration, 2nd Ed.
Getting Connected: The Internet at 56K and Up
Linux Network Administrator's Guide
Managing Internet Information Services
Managing NFS and NIS
Networking Personal Computers with TCP/IP
Practical UNIX & Internet Security, 2nd Ed.
PGP: Pretty Good Privacy
sendmail, 2nd Ed.
sendmail Desktop Reference
System Performance Tuning
TCP/IP Network Administration
termcap & terminfo
Using & Managing UUCP
Volume 8: X Window System Administrator's Guide
Web Security & Commerce

UNIX

Exploring Expect
Learning VBScript
Learning GNU Emacs, 2nd Ed.
Learning the bash Shell
Learning the Korn Shell
Learning the UNIX Operating System
Learning the vi Editor
Linux in a Nutshell
Making TeX Work
Linux Multimedia Guide
Running Linux, 2nd Ed.
SCO UNIX in a Nutshell
sed & awk, 2nd Edition
Tcl/Tk Tools
UNIX in a Nutshell: System V Edition
UNIX Power Tools
Using csh & tsch
When You Can't Find Your UNIX System Administrator
Writing GNU Emacs Extensions

WEB REVIEW STUDIO SERIES

Gif Animation Studio
Shockwave Studio

WINDOWS

Dictionary of PC Hardware and Data Communications Terms
Inside the Windows 95 Registry
Inside the Windows 95 File System
Windows Annoyances
Windows NT File System Internals
Windows NT in a Nutshell

PROGRAMMING

Advanced Oracle PL/SQL Programming
Applying RCS and SCCS
C++: The Core Language
Checking C Programs with lint
DCE Security Programming
Distributing Applications Across DCE & Windows NT
Encyclopedia of Graphics File Formats, 2nd Ed.
Guide to Writing DCE Applications
lex & yacc
Managing Projects with make
Mastering Oracle Power Objects
Oracle Design: The Definitive Guide
Oracle Performance Tuning, 2nd Ed.
Oracle PL/SQL Programming
Porting UNIX Software
POSIX Programmer's Guide
POSIX.4: Programming for the Real World
Power Programming with RPC
Practical C Programming
Practical C++ Programming
Programming Python
Programming with curses
Programming with GNU Software
Pthreads Programming
Software Portability with imake, 2nd Ed.
Understanding DCE
Understanding Japanese Information Processing
UNIX Systems Programming for SVR4

BERKELEY 4.4 SOFTWARE DISTRIBUTION

4.4BSD System Manager's Manual
4.4BSD User's Reference Manual
4.4BSD User's Supplementary Documents
4.4BSD Programmer's Reference Manual
4.4BSD Programmer's Supplementary Documents
X Programming
Vol. 0: X Protocol Reference Manual
Vol. 1: Xlib Programming Manual
Vol. 2: Xlib Reference Manual
Vol. 3M: X Window System User's Guide, Motif Edition
Vol. 4M: X Toolkit Intrinsics Programming Manual, Motif Edition
Vol. 5: X Toolkit Intrinsics Reference Manual
Vol. 6A: Motif Programming Manual
Vol. 6B: Motif Reference Manual
Vol. 6C: Motif Tools
Vol. 8 : X Window System Administrator's Guide
Programmer's Supplement for Release 6
X User Tools
The X Window System in a Nutshell

CAREER & BUSINESS

Building a Successful Software Business
The Computer User's Survival Guide
Love Your Job!
Electronic Publishing on CD-ROM

TRAVEL

Travelers' Tales: Brazil
Travelers' Tales: Food
Travelers' Tales: France
Travelers' Tales: Gutsy Women
Travelers' Tales: India
Travelers' Tales: Mexico
Travelers' Tales: Paris
Travelers' Tales: San Francisco
Travelers' Tales: Spain
Travelers' Tales: Thailand
Travelers' Tales: A Woman's World

O'REILLY™

TO ORDER: **800-998-9938** • **order@oreilly.com** • **http://www.oreilly.com/**
OUR PRODUCTS ARE AVAILABLE AT A BOOKSTORE OR SOFTWARE STORE NEAR YOU.
FOR INFORMATION: **800-998-9938** • **707-829-0515** • **info@oreilly.com**

International Distributors

UK, EUROPE, MIDDLE EAST AND NORTHERN AFRICA (EXCEPT FRANCE, GERMANY, SWITZERLAND, & AUSTRIA)

INQUIRIES

International Thomson Publishing Europe
Berkshire House
168-173 High Holborn
London WC1V 7AA
United Kingdom
Telephone: 44-171-497-1422
Fax: 44-171-497-1426
Email: itpint@itps.co.uk

ORDERS

International Thomson Publishing Services, Ltd.
Cheriton House, North Way
Andover, Hampshire SP10 5BE
United Kingdom
Telephone: 44-264-342-832 (UK)
Telephone: 44-264-342-806 (outside UK)
Fax: 44-264-364418 (UK)
Fax: 44-264-342761 (outside UK)
UK & Eire orders: itpuk@itps.co.uk
International orders: itpint@itps.co.uk

FRANCE

Editions Eyrolles
61 bd Saint-Germain
75240 Paris Cedex 05
France
Fax: 33-01-44-41-11-44

FRENCH LANGUAGE BOOKS

All countries except Canada
Telephone: 33-01-44-41-46-16
Email: geodif@eyrolles.com
English language books
Telephone: 33-01-44-41-11-87
Email: distribution@eyrolles.com

GERMANY, SWITZERLAND, AND AUSTRIA

INQUIRIES

O'Reilly Verlag
Balthasarstr. 81
D-50670 Köln
Germany
Telephone: 49-221-97-31-60-0
Fax: 49-221-97-31-60-8
Email: anfragen@oreilly.de

ORDERS

International Thomson Publishing
Königswinterer Straße 418
53227 Bonn, Germany
Telephone: 49-228-97024 0
Fax: 49-228-441342
Email: order@oreilly.de

JAPAN

O'Reilly Japan, Inc.
Kiyoshige Building 2F
12-Banchi, Sanei-cho
Shinjuku-ku
Tokyo 160-0008 Japan
Telephone: 81-3-3356-5227
Fax: 81-3-3356-5261
Email: kenji@oreilly.com

INDIA

Computer Bookshop (India) PVT. Ltd.
190 Dr. D.N. Road, Fort
Bombay 400 001 India
Telephone: 91-22-207-0989
Fax: 91-22-262-3551
Email: cbsbom@giasbm01.vsnl.net.in

HONG KONG

City Discount Subscription Service Ltd.
Unit D, 3rd Floor, Yan's Tower
27 Wong Chuk Hang Road
Aberdeen, Hong Kong
Telephone: 852-2580-3539
Fax: 852-2580-6463
Email: citydis@ppn.com.hk

KOREA

Hanbit Media, Inc.
Sonyoung Bldg. 202
Yeksam-dong 736-36
Kangnam-ku
Seoul, Korea
Telephone: 822-554-9610
Fax: 822-556-0363
Email: hant93@chollian.dacom.co.kr

SINGAPORE, MALAYSIA, AND THAILAND

Addison Wesley Longman Singapore PTE Ltd.
25 First Lok Yang Road
Singapore 629734
Telephone: 65-268-2666
Fax: 65-268-7023
Email: daniel@longman.com.sg

PHILIPPINES

Mutual Books, Inc.
429-D Shaw Boulevard
Mandaluyong City, Metro
Manila, Philippines
Telephone: 632-725-7538
Fax: 632-721-3056
Email: mbikikog@mnl.sequel.net

CHINA

Ron's DataCom Co., Ltd.
79 Dongwu Avenue
Dongxihu District
Wuhan 430040
China
Telephone: 86-27-3892568
Fax: 86-27-3222108
Email: hongfeng@public.wh.hb.cn

ALL OTHER ASIAN COUNTRIES

O'Reilly & Associates, Inc.
101 Morris Street
Sebastopol, CA 95472 USA
Telephone: 707-829-0515
Fax: 707-829-0104
Email: order@oreilly.com

AUSTRALIA

WoodsLane Pty. Ltd.
7/5 Vuko Place, Warriewood NSW 2102
P.O. Box 935
Mona Vale NSW 2103
Australia
Telephone: 61-2-9970-5111
Fax: 61-2-9970-5002
Email: info@woodslane.com.au

NEW ZEALAND

Woodslane New Zealand Ltd.
21 Cooks Street (P.O. Box 575)
Waganui, New Zealand
Telephone: 64-6-347-6543
Fax: 64-6-345-4840
Email: info@woodslane.com.au

THE AMERICAS

McGraw-Hill Interamericana Editores, S.A. de C.V.
Cedro No. 512
Col. Atlampa 06450
Mexico, D.F.
Telephone: 52-5-541-3155
Fax: 52-5-541-4913
Email: mcgraw-hill@infosel.net.mx

SOUTH AFRICA

International Thomson Publishing South Africa
Building 18, Constantia Park
138 Sixteenth Road
P.O. Box 2459
Halfway House, 1685 South Africa
Telephone: 27-11-805-4819
Fax: 27-11-805-3648

O'REILLY™

O'Reilly & Associates, Inc.
101 Morris Street
Sebastopol, CA 95472-9902
1-800-998-9938

Visit us online at:
http://www.ora.com/
orders@ora.com

O'REILLY WOULD LIKE TO HEAR FROM YOU

Which book did this card come from?

_____

Where did you buy this book?
- ❏ Bookstore   ❏ Computer Store
- ❏ Direct from O'Reilly   ❏ Class/seminar
- ❏ Bundled with hardware/software
- ❏ Other _____

What operating system do you use?
- ❏ UNIX   ❏ Macintosh
- ❏ Windows NT   ❏ PC(Windows/DOS)
- ❏ Other _____

What is your job description?
- ❏ System Administrator   ❏ Programmer
- ❏ Network Administrator   ❏ Educator/Teacher
- ❏ Web Developer
- ❏ Other _____

❏ Please send me O'Reilly's catalog, containing
a complete listing of O'Reilly books and
software.

Name _____   Company/Organization _____

Address _____

City _____   State _____   Zip/Postal Code _____   Country _____

Telephone _____   Internet or other email address (specify network) _____

Nineteenth century wood engraving
of a bear from the O'Reilly &
Associates Nutshell Handbook®
Using & Managing UUCP.

PLACE
STAMP
HERE

NO POSTAGE
NECESSARY IF
MAILED IN THE
UNITED STATES

BUSINESS REPLY MAIL

FIRST CLASS MAIL   PERMIT NO. 80   SEBASTOPOL, CA

Postage will be paid by addressee

O'Reilly & Associates, Inc.
101 Morris Street
Sebastopol, CA 95472-9902